ADVANCES IN WEB-BASED GIS, MAPPING SERVICES AND APPLICATIONS

International Society for Photogrammetry and Remote Sensing (ISPRS) Book Series

Book Series Editor

Paul Aplin
School of Geography
The University of Nottingham
Nottingham, UK

information from imagery

Advances in Web-based GIS, Mapping Services and Applications

Editors

Songnian Li
Department of Civil Engineering, Ryerson University,
Toronto, Ontario, Canada

Suzana Dragićević
Department of Geography, Simon Fraser University, Burnaby,
British Columbia, Canada

Bert Veenendaal
Department of Spatial Sciences, Curtin University,
Perth WA, Australia

CRC Press
Taylor & Francis Group
Boca Raton London New York

CRC Press is an imprint of the
Taylor & Francis Group, an **informa** business

CRC Press
Taylor & Francis Group
6000 Broken Sound Parkway NW, Suite 300
Boca Raton, FL 33487-2742

First issued in paperback 2017

CRC Press/Balkema is an imprint of the Taylor & Francis Group, an informa business

© 2011 Taylor & Francis Group, London, UK

Typeset by Vikatan Publishing Solutions (P) Ltd., Chennai, India

No claim to original U.S. Government works

ISBN-13: 978-0-415-80483-7 (hbk)
ISBN-13: 978-1-138-11781-5 (pbk)

Published by: CRC Press/Balkema
P.O. Box 447, 2300 AK Leiden, The Netherlands
e-mail: Pub.NL@taylorandfrancis.com
www.crcpress.com – www.taylorandfrancis.co.uk – www.balkema.nl

British Library Cataloguing in Publication Data

Library of Congress Cataloging-in-Publication Data

Advances in web-based GIS, mapping services and applications / editors Songnian Li, Suzana Dragicevic, Bert Veenendaal.
         p. cm.

Includes bibliographical references and index.

   ISBN 978-0-415-80483-7 (hardback)
   1. Geographic information systems. 2. Geography--Computer network resources. 3. World Wide Web. I. Li, Songnian. II. Dragicevic, Suzana, III. Veenendaal, Bert.

G70.212.A426 2011

910.285--dc22

                                             2011012049

Visit the Taylor & Francis Web site at
http://www.taylorandfrancis.com

and the CRC Press Web site at
http://www.crcpress.com

Table of contents

Preface   vii

Acknowledgements   ix

Contributors   xi

Introduction

Advances, challenges and future directions in web-based GIS, mapping services
and applications   3
Songnian Li, Bert Veenendaal & Suzana Dragićević

Analytical and geospatial services

Geography 2.0—A mash-up perspective   15
T. Edwin Chow

GeoClustering: A web service for geospatial clustering   37
Jing Wang, Xin Wang & Steve H.L. Liang

Creating GIS simulation models on a TeraGrid-enabled geospatial web portal:
A demonstration of geospatial cyberinfrastructure   55
Ming Hsiang Tsou & Ick-Hoi Kim

An OGC Web Processing Service for automated interpolation   71
Jan Dürrfeld, Jochen Bisier & Edzer Pebesma

GeoGlobe: A Virtual Globe for multi-source geospatial information
integration and service   85
Jianya Gong, Longgang Xiang, Jin Chen, Peng Yue & Yi Liu

Building web services for public sector information and the geospatial web   109
David Pullar, David Torpie & Tim Barker

Performance

WebGIS performance issues and solutions   121
*Chaowei Yang, Huayi Wu, Qunying Huang, Zhenlong Li, Jing Li, Wenwen Li,
Lizhi Miao & Min Sun*

Data reduction techniques for web and mobile GIS   139
Michela Bertolotto & Gavin McArdle

A load balancing method to support spatial analysis in XML/GML/SVG-based WebGIS   153
Haosheng Huang, Yan Li & Georg Gartner

Augmentation and location-based services

Geolocating for web based geospatial applications   171
Bert Veenendaal, Jacob Delfos & Tele Tan

The mobile web: Lessons from mobile augmented reality   185
Sylvie Daniel & Robin M. Harrap

A survey on augmented maps and environments: Approaches, interactions
and applications   207
Gerhard Schall, Johannes Schöning, Volker Paelke & Georg Gartner

Collaboration and decision making

Map-chatting within the geospatial web   229
G. Brent Hall & Michael G. Leahy

Jump-starting the next level of online geospatial collaboration: Lessons
from AfricaMap   255
Benjamin Lewis & Weihe Guan

A geospatial Web application to map observations and opinions
in environmental planning   277
Claus Rinner, Jyothi Kumari & Sepehr Mavedati

Web-based collaboration and decision making in GIS-built virtual environments   293
Christian Stock, Ian D. Bishop, Haohui Chen, Marcos Nino-Ruiz & Peter Wang

Development and challenges of using web-based GIS for health applications   311
Sheng Gao, Darka Mioc, Xiaolun Yi, Harold Boley & François Anton

Open standards for geospatial services

OGC standards: Enabling the geospatial web   327
Carl Reed

Vector data formats in internet based geoservices   349
Franz-Josef Behr, Kai Holschuh, Detlev Wagner & Rita Zlotnikova

Geospatial catalogue web/grid service   369
Aijun Chen & Liping Di

Author index   381

Keyword index   383

ISPRS Book Series   385

Preface

Web mapping/GIS is the process of designing, implementing, generating and delivering maps, geospatial data and Geographic Information Systems (GIS) functionality or services on the Web. Primarily focusing on technological issues, this field increasingly includes theoretic aspects such as cartographic design, theory and principles, social and organizational issues and applications. Given the recent advances led by mainstream Information Technology (IT) developers, the need to examine these issues becomes increasingly critical.

This book volume, *Advances in Web-based GIS, Mapping Services and Applications*, aims at examining both theoretical/technological advancements and social/organizational issues in the field of web-based GIS and mapping services and applications. It presents an overall view of current progress and achievements with considerable technical details and examples. The contents address: 1) constant updating of related web and geospatial technologies as well as the revolution of web mapping caused by mainstream IT vendors such as Google, Yahoo and Microsoft; 2) increased interest in geospatial information technologies from the industry; and 3) increasing demand from the general public for prompt and effective online access to geospatial information. All contributing chapters were advised to consider: 1) inclusion of recent technological advancements, especially new developments under Web 2.0, map mashups, neogeography, and the like; 2) balanced theoretical discussions and technical implementations; 3) commentary on the current stages of development; and 4) prediction of future developments over the next decade.

The original recommended topics and themes, as listed in the call for chapter proposals, include:

- Web 2.0, neogeography, and map mashups
- Technologies providing new service-oriented, distributed architectures, e.g., web services, SOA, P2P, grid computing, etc.
- Technologies enhancing web interaction with maps and spatial representations, e.g., Ajax, SVG, GeoRSS, etc.
- Advances in virtual earth technologies
- Open source and open standards as related to web GIS/mapping
- Web-based spatial decision support
- Applications in public participation
- Geospatially-enabled workflow processes for automating web-based geospatial services
- Content and knowledge mapping
- Social mapping and networking
- New service and application models such as SaaS
- Data quality and integration, date policies, privacy and ownership
- Quality of web-based geospatial services and processes
- Impact of Web 2.0 on enterprise-wide web GIS/mapping and location services

While many of these topics have been addressed in this book volume, we feel that what needs to be further studied is related to social and organizational issues in the field of web-based GIS and mapping services and applications, as well as an assessment of the impact of Web 2.0 and more recently emerged web 3.0 on enterprise-wide web GIS/mapping and location services. We would like to see more research into, for example, data policies, privacy and ownership, quality assurance and acceptable use policies, especially for crowd-sourcing and community-generated geographic content technologies and applications.

In addition to the introductory chapter, the book includes 20 accepted chapters after double-blind peer review processes, which are organized into the following six sections:

SECTION 1: INTRODUCTION

An introductory chapter is included to present an overview of recent advances in web-based GIS, mapping services and applications, and identified some of issues and challenges faced by researchers and professionals in the field.

SECTION 2: ANALYTICAL AND GEOSPATIAL WEB SERVICES

This section includes six chapters focusing on various state-of-the-art geographic information web and processing services, ranging from analytical, simulation and virtual visualization uses to building web services and mashups.

SECTION 3: PERFORMANCE

Three chapters included in this section present some recent studies on techniques and solutions to enhance the performance of web mapping and services.

SECTION 4: AUGMENTATION AND MOBILE MAPPING

Mobile applications are increasingly using better positioning techniques and augmented reality. The three chapters in this section describe recent developments and advances in geolocating using a range of positioning systems, and augmented systems and environments for use in mobile mapping.

SECTION 5: COLLABORATION AND DECISION MAKING

This section covers the developments in both traditional application areas such as decision support and public participation, and new emerging areas such as social mapping and collaboration. Application cases exemplify the developments in some emerging areas. Five chapters are included in this section.

SECTION 6: OPEN STANDARDS FOR GEOSPATIAL SERVICES

The three chapters in this section retrieve some aspects of open standards and their use in developing web-based geospatial services.

The volume is aimed at researchers, application specialists and developers, practitioners and students who work or have an interest in the area of web GIS/mapping and its application for business, government services, communities, enterprise computing and social networking.

Acknowledgements

The editors of this **ISPRS** book volume would like to acknowledge many colleagues who have contributed to the publication of the book in preparing, authoring and reviewing chapters as well as providing administrative assistance in the final compilation. Without their invaluable help and support, the book would not have been published.

All chapters included in the book have gone through a rigorous double-blind peer review process. We are grateful to all the reviewers including chapter authors and many external researchers in the field who kindly agreed to review the chapters. Our special thanks go to the following reviewers who are not a chapter author: Shivanand Balram, Mustafa Berber, Soheil Borousshaki, Guoray Cai, Rosaline Canessa, Rob Corner, Matt Duckham, Mark Gahegan, Nick Hedley, Shunfu Hu, Bin Jiang, Anthony Jjumba, Ari Jolma, Carsten Kessler, Menno-Jan Kraak, Jane Law, Lingkui Meng, Martin Meyers, Ahmed Mohamed, Mir Abolfazl Mostafavi, Zigiang Ou, Theresa Rhyne, Xianfeng Song, Tele Tan, Jianguo Wang, Geoff West, Stephan Winter, and Paul Zandbergen.

We wish to acknowledge the editorial and professional guidance given by the **ISPRS** book series editor, Mr. Paul Aplin, and the editing and management staff at CRC Press / Balkema, Taylor & Francis Group, Mr. Léon Bijnsdorp, Mrs. Désirée de Blok, and Mr. Richard Gundel. Mr. Paul Aplin provided much editorial advice and an overall review of all accepted chapters to ensure the quality of the book. Mr. Léon Bijnsdorp, Mrs. Désirée de Blok, and Mr. Richard Gundel assisted us by kindly answering questions during the book editing and publication processes.

All the chapter authors deserve our special thanks for their time and effort spent on chapter proposals, full chapter preparations, and chapter revisions addressing reviewers' feedback. We thank them for their patience in this long and sometimes arduous editing process. It was our great pleasure to work with 60 authors on different aspects of web mapping, GIS and services, to collectively produce this book and make it available on the desks of many readers. Finally, we would like to acknowledge the National Science and Engineering Research Council (NSERC) and GEOmatics for Informed DEcisions (GEOIDE) for providing support to editors' research programs that are related to this book project.

Advances in Web-based GIS, Mapping Services
and Applications – Li, Dragićević & Veenendaal (eds)
© 2011 Taylor & Francis Group, London, ISBN 978-0-415-80483-7

Contributors

Anton, François
Department of Informatics and Mathematical Modelling, Technical University of Denmark,
Denmark, Email: fa@imm.dtu.dk

Barker, Tim
Queensland Treasury, Brisbane, Australia, 4000, Email: tim.barker@derm.qld.gov.au

Behr, Franz-Josef
Faculty of Geomatics, Computer Science and Mathematics, Stuttgart University of
Applied Sciences Schellingstr 24, 70174 Stuttgart, Germany, Email: franz-josef.behr@hft-
stuttgart.de

Bertolotto, Michela
School of Computer Science and Informatics, University College Dublin, Belfield,
Dublin 4, Ireland, Email: michela.bertolotto@ucd.ie

Bishop, Ian D.
Department of Geomatics, The University of Melbourne, Parkville, VIC 3010, Australia,
Email: i.bishop@unimelb.edu.au

Bisier, Jochen
Institute for Geoinformatics, University of Muenster, Weseler Strasse 253, 48151 Mnnster,
Germany, Email: j.m.b@uni-muenster.de

Boley, Harold
Institute for Information Technology, NRC, Fredericton, NB, Canada, Email: Harold.
Boley@nrc-cnrc.gc.ca

Chen, Aijun
Center for Spatial Information Science and Systems, George Mason University, 6301 Ivy
Lane, Ste. 620, Greenbelt, MD 20770, USA, Email: achen6@gmu.edu

Chen, Haohui
Department of Geomatics, The University of Melbourne, Parkville, VIC 3010, Australia,
Email: h.chen22@pgrad.unimelb.edu.au

Chen, Jin
State Key Laboratory of Information Engineering in Surveying, Mapping and Remote
Sensing, Wuhan University, 129 Luoyu Road, Wuhan, Hubei, 430072, China, Email:
jchen@whu.edu.cn

Chow, T. Edwin
Department of Geography, Texas State University—San Marcos, ELA 374, 601 University
Dr., San Marcos, TX 78666, USA, Email: chow@txstate.edu

Daniel, Sylvie
Department of Geomatics Science, Laval University, Quebec City, Quebec, Canada
G1V 0A6, Email: sylvie.daniel@scg.ulaval.ca

Delfos, Jacob

Department of Spatial Sciences, Curtin University, Box U1987, Perth, Western Australia, Australia 6845, Email: Jacob.delfos@postgrad.curtin.edu.au

Di, Liping

Center for Spatial Information Science and Systems, George Mason University, 6301 Ivy Lane, Ste. 620, Greenbelt, MD 20770, USA, Email: ldi@gmu.edu

Dragićević, Suzana

Department of Geography, Simon Fraser University, 8888 University Drive, Burnaby, BC, V5A 1S6 Canada, Email: suzanad@sfu.ca

Dürrfeld, Jan

Institute for Geoinformatics, University of Muenster, Weseler Strasse 253, 48151 Muenster, Germany, Email: jan.duerrfeld@uni-muenster.de

Gao, Sheng

Department of Geodesy and Geomatics Engineering, UNB, Fredericton, New Brunswick, Canada, Email: sheng.gao@unb.ca

Gartner, Georg

Institute of Geoinformation and Cartography, Vienna University of Technology, Gusshausstrasse 30/E127-2, Vienna, A-1040, Austria, Email: georg.gartner@tuwien.ac.at

Gong, Jianya

State Key Laboratory of Information Engineering in Surveying, Mapping and Remote Sensing, Wuhan University, 129 Luoyu Road, Wuhan, Hubei, 430072, China, Email: geogjy@163.net

Guan, Weihe (Wendy)

Center for Geographic Analysis, Harvard University, 1737 Cambridge Street, Suite 350, Cambridge, Massachusetts, 02138, USA, Email: wguan@cga.harvard.edu

Hall, G. Brent

School of Surveying, University of Otago, P.O. Box 56 Dunedin, New Zealand, Email: brent.hall@otago.ac.nz

Harrap, Robin M.

Department of Geological Sciences, Queen's University, Kingston, Ontario, Canada K7L 3N6, Email: harrap@geol.queensu.ca

Holschuh, Kai

Marketing and Intercultural Management, Karlshochschule International University, Karlstr 36, 76133 Karlsruhe, Germany, Email: kholschuh@karlshochschule.de

Huang, Haosheng

Institute of Geoinformation and Cartography, Vienna University of Technology, Gusshausstrasse 30/E127-2, Vienna, A-1040, Austria, Email: huanghaosheng@gmail.com

Huang, Qunying

Joint Center for Intelligent Spatial Computing, George Mason University, Fairfax, VA, 22030, USA, Email: qhuang1@gmu.edu

Kim, Ick-Hoi

Department of Geography, San Diego State University, 5500 Campanile Drive, San Diego, California, USA, Email: ikim@rohan.sdsu.edu

Kumari, Jyothi

Department of Geography, Ryerson University, 350 Victoria Street, Toronto, Ontario, M5B 2K3, Canada, Email: jyothi.kumari@bordeaux.inra.fr

Leahy, Michael G.
Department of Geography and Environmental Studies, Wilfrid Laurier University, Waterloo, Ontario N2L 3C5, Canada, Email: mgleahy@alumni.uwaterloo.ca

Lewis, Benjamin
Center for Geographic Analysis, Harvard University, 1737 Cambridge Street, Suite 350, Cambridge, Massachusetts, 02138, USA, Email: blewis@cga.harvard.edu

Li, Jing
Joint Center for Intelligent Spatial Computing, George Mason University, Fairfax, VA, 22030, USA, Email: jlih@gmu.edu

Li, Songnian
Department of Civil Engineering, Ryerson University, 350 Victoria Street, Toronto, Ontario, M5B 2K3, Canada, Email: snli@ryerson.ca

Li, Wenwen
Joint Center for Intelligent Spatial Computing, George Mason University, Fairfax, VA, 22030, USA, Email: wli6@gmu.edu

Li, Yan
Spatial Information Research Center, South China Normal University, Shipai, Tianhe, Guangzhou, 510631, China, Email: yanli@scnu.edu.cn

Li, Zhenlong
Joint Center for Intelligent Spatial Computing, George Mason University, Fairfax, VA, 22030, USA, Email: zli1@gmu.edu

Liang, Steve H.L.
Department of Geomatics Engineering, University of Calgary, 2500 University Drive NW, Calgary, Alberta, Canada, Email: steve.liang@ucalgary.ca

Liu, Yi
School of Geodesy and Geomatics, Wuhan University, 129 Luoyu Road, Wuhan, Hubei, 430072, China, Email: liuyiwhu@gmail.com

Mavedati, Sepehr
Department of Geography, University of Toronto, 100 St. George Street, Toronto, Ontario, M5S 3G3, Canada, Email:

McArdle, Gavin
School of Computer Science and Informatics, University College Dublin, Belfield, Dublin 4, Ireland, Email: gavin.mcardle@ucd.ie

Miao, Lizhi
Joint Center for Intelligent Spatial Computing, George Mason University, Fairfax, VA, 22030, USA, Email: lmiao@gmu.edu

Mioc, Darka
National Space Institute, Technical University of Denmark, Denmark, Email: mioc@space.dtu.dk

Nino-Ruiz, Marcos
Department of Geomatics, The University of Melbourne, Parkville, VIC 3010, Australia, Email: m.ninoruiz@pgrad.unimelb.edu.au

Paelke, Volker
Institut de Geomàtica, Barcelona, Av. Carl Friedrich Gauss, 11 – Parc Mediterrani de la Tecnologia – E-08860 Castelldefels, Spain, Email: Volker.paelke@ideg.es

Pebesma, Edzer

Institute for Geoinformatics, University of Muenster, Weseler Strasse 253, 48151 Münster, Germany, Email: edzer.pebesma@uni-muenster.de

Pullar, David

Geography, Planning and Environmental Management, The University of Queensland, Brisbane, Australia 4072, Email: d.pullar@uq.edu.au

Reed, Carl

Open Geospatial Consortium, Inc., Wayland, MA 01778-5037, USA, Email: creed@opengeospatial.org

Rinner, Claus

Department of Geography, Ryerson University, 350 Victoria Street, Toronto, Ontario, M5B 2K3, Canada, Email: crinner@ryerson.ca

Schall, Gerhard

Graz University of Technology, Inffeldgasse 16, 8010 Graz, Austria, Email: schall@icg.tugraz.at

Schöning, Johannes

DFKI GmbH, Campus D3_2, Stuhlsatzenhausweg 3, D-66123 Saarbruecken, Germany, Email: johannes.schoening@dfki.de

Stock, Christian

Department of Geomatics, The University of Melbourne, Parkville, VIC 3010, Australia, Email: cstock@skm.com.au

Sun, Min

Joint Center for Intelligent Spatial Computing, George Mason University, Fairfax, VA, 22030, USA, Email: msun@gmu.edu

Tan, Tele

Department of Computing, Curtin University, Box U1987, Perth, Western Australia, AUSTRALIA 6845, Email: t.tan@curtin.edu.au

Torpie, David

Queensland Treasury, Brisbane, Australia, 4000.

Tsou, Ming Hsiang (Ming)

Department of Geography, San Diego State University, 5500 Campanile Drive, San Diego, California, USA, Email: mtsou@mail.sdsu.edu

Veenendaal, Bert

Department of Spatial Sciences, Curtin University, Perth Western, Australia 6845, Email: B.Veenendaal@curtin.edu.au

Wagner, Detlev

Faculty of Geomatics, Computer Science and Mathematics, Stuttgart University of Applied Sciences, Schellingstr 24, 70174 Stuttgart, Germany, Email: detlev.wagner@hft-stuttgart.de

Wang, Jing

Department of Geomatics Engineering, University of Calgary, 2500 University Drive NW, Calgary, Alberta, Canada, Email: wangjing@ucalgary.ca

Wang, Peter

Department of Geomatics, The University of Melbourne, Parkville, VIC 3010, Australia, Email: p.wang2@pgrad.unimelb.edu.au

Wang, Xin
Department of Geomatics Engineering, University of Calgary, 2500 University Drive NW, Calgary, Alberta, Canada, Email: xcwang@ucalgary.ca

Wu, Huayi
Joint Center for Intelligent Spatial Computing, George Mason University, Fairfax, VA, 22030, USA, Email: hwu8@gmu.edu

Xiang, Longgang
State Key Laboratory of Information Engineering in Surveying, Mapping and Remote Sensing, Wuhan University, 129 Luoyu Road, Wuhan, Hubei, 430072, China, Email: lgxiang@lmars.whu.edu.cn

Yang, Chaowei
Joint Center for Intelligent Spatial Computing, George Mason University, Fairfax, VA, 22030, USA, Email: cyang3@gmu.edu

Yi, Xiaolun
Service New Brunswick, Fredericton, New Brunswick, Canada, Email: xiaolun.yi@snb.ca

Yue, Peng
State Key Laboratory of Information Engineering in Surveying, Mapping and Remote Sensing, Wuhan University, 129 Luoyu Road, Wuhan, Hubei, 430072, China, Email: geopyue@gmail.com

Zlotnikova, Rita
Blom Romania, 130010—Targoviste Str. I.H. Radulescu, Nr. 3-5, Jud Dambovita, Romania, Email: rita.zlotnikova@blominfo.ro

Introduction

Advances in Web-based GIS, Mapping Services
and Applications – Li, Dragićević & Veenendaal (eds)
© *2011 Taylor & Francis Group, London, ISBN 978-0-415-80483-7*

Advances, challenges and future directions in web-based GIS, mapping services and applications

Songnian Li
Department of Civil Engineering, Ryerson University, Toronto, Ontario, Canada

Bert Veenendaal
Department of Spatial Sciences, Curtin University, Perth, Western Australia

Suzana Dragićević
Department of Geography, Simon Fraser University, Burnaby, Canada

ABSTRACT: This introductory chapter presents an overview of recent advances in web-based Geographic Information Systems (GIS), mapping services and applications, and identifies some of the issues and challenges faced by researchers and professionals in the field. Primarily driven by current advances in web technologies and the expressed needs from geospatial communities, the field has evolved rapidly over the last two decades. The uncertainty of future technology developments makes it difficult to predict what the future holds. Nevertheless, we identify some trends and potential directions for web-based mapping and services.

Keywords: Web-based, mapping, GIS, advances, issues, future

1 INTRODUCTION

Web mapping is the process of designing, implementing, generating and delivering maps, geospatial data, and web map services on the World Wide Web, hereafter referred to as *the Web* (Wikipedia contributors 2011). Web GIS, often used as an interchangeable term for web mapping, brings in additional functionality for spatial analysis and exploration supported by geoprocessing functions. Over the past two decades, web mapping and GIS have evolved dramatically from the simple presentation of static map data to more dynamic and semi-dynamic data visualization and data analysis, and more recently, data and map services. Driven by the constantly-available new web technologies and the expressed needs from geospatial communities, a wide array of web mapping and GIS technologies have been developed by government agencies, companies, research institutions and open user communities. These technologies have been employed to implement applications for: (1) spatial data access and dissemination, (2) spatial data exploration and visualization, (3) spatial data processing, analysis and modeling, (4) collaborative spatial decision support using public participatory GIS, and (5) integration of web-based geo-spatial services in mainstream and enterprise computing processes and environments (Li 2008). Efforts to structure these developments have been summarized in constructs such as the collaborative GIS cube and the notion of Geo-Collectives where participation levels, map and GIS usage, and internet delivery platforms intersect to reinforce and further expand existing GIS services (Balram et al. 2009).

Web mapping and GIS appear, at first glance, to be technical in nature. However, many of their technical issues and challenges require studies into corresponding new theories, algorithms, and methods. For example, the theories and principles of cartographic design in web mapping are among these new studies (Kraak 2004). Other examples include simulation modeling (see Chapter 4), load balancing and data reduction methods (see Chapters 9 and 10), and geospatial semantics and ontologies (Egenhofer 2002, Mata & Claramunt 2011). An increasing number of studies have also looked into the social factors affecting the design of web mapping/GIS applications and the impact of these applications on the organizations and their business processes. Given the recent advances as described in Section 2, the need to examine these issues becomes more significant than ever before.

This introductory chapter presents brief reviews of some recent advances in web-based GIS, mapping services and applications, and identifies some of the issues and challenges faced by researchers and professionals in the field. Finally, some potential directions for future web mapping and GIS developments are presented.

2 RECENT ADVANCES

Commencing with static web map publishing, web mapping has evolved through a number of important development stages from static web mapping through interactive web mapping to distributed web mapping services (Peng & Tsou 2005, Moseley 2007). Web mapping development, to a large extent, has mostly been driven by innovations and developments of web technologies, architecture models, specifications, and standards. The client-server technology has long dominated the development of web mapping since it started, and will continue to play an important role. Over the last decade, some recent web technologies have enabled new types of web mapping and GIS developments towards open access and software solutions, content and knowledge mapping, and virtual earth (globes), in addition to more interactive web mapping services and more functional rich web-based GIS services. These include more general web technologies such as web services, Service-Oriented Architecture (SOA), grid computing, peer-to-peer (P2P), as well as some emerging technologies especially suitable for web mapping, *e.g.*, Asynchronous JavaScript and XML (Ajax), Services Oriented Access Protocol (SOAP), and Really Simple Syndication (RSS). Notably, grid computing helps manage massive distributed computing and storage resources, provides high-performance, and facilitates "collaborative" virtual computers and organizations. Ajax enables more interactive mapping interfaces for users to obtain better experiences and to be able to access more intuitive functionality.

The continuing advances in web-based mapping, GIS and geospatial services have generated ever increasing access, diffusion, usage and processing of geographically-referenced data, with corresponding standards to promote interoperability between different technologies and easy sharing of vast amounts of data. Inspired by the success of Yahoo, Google, MapQuest and Microsoft's efforts of providing free access to basic map data online, governments at different levels have increasingly supplied their data holdings online using new web technologies such as, for example, Ordnance Survey OpenData (Ordnance Survey 2011). On the other hand, user-generated content, under the umbrella of Web 2.0 and Volunteered Geographic Information (VGI) (Goodchild 2007), have created massive amounts of geographic information linked to geospatial locations. Much geospatial data has been captured by people who are themselves from a non-traditional geospatial data collection domain. In responding to this ever-increased accessibility of geospatial data over the Web, the traditional internet map server technologies from commercial GIS software vendors have been redesigned. New and cost-effective solutions to web-based geospatial information and GIS services have been developed by open software communities. Open specifications and standards have been continuously released and enhanced with the aim of supporting more efficient and interoperable data fusion and service integration processes.

On the development side, necessary open Application Programming Interfaces (API), such as Google API, Bing Maps API, Yahoo Maps API and OS OpenSpace API, are being provided to help users and developers to access data and "mash up" web-based applications using such data. The rich collection of APIs allows for easy customized web GIS services, inclusion of maps into various online services, and "mash-ups" of web mapping applications that combine contents from one or more online sources (Li & Gong 2008, Chapter 2 in this volume). Further to this "API" phenomenon, a Rich Internet Application (RIA) approach has been explored to enhance web mapping and GIS interfaces to catch up with the rich geospatial data content. An RIA is a web application that has an appearance and characteristics of desktop applications. RIA technologies support the idea of a "rich client", a user interface that is more robust, responsive and visually interesting than an HTML-based interface (O'Rourke 2004). A number of studies have developed applications of RIA techniques in web-based mapping and GIS (Wang & Hu 2009, Li et al. 2011). One of the leading RIA technologies, the Adobe Integrated Runtime (AIR) platform, has been used by ArcGIS Server from ESRI to provide two RIA APIs: ArcGIS API for Flex and API for Microsoft Silverlight (ESRI 2011).

Virtual globe technologies, such as Google Earth, ESRI ArcGIS Explorer, Microsoft Virtual Earth, and NASA World Wind, have been developed to provide users with the ability to visualize data and information on 3D earth models over the Web or Internet. While most of these technologies focus on visualization, SkylineGlobe allows for real-time, synchronous sharing of views within a group of people joining a conference session (Deiana 2009). The science benefits of using virtual globes due to their data presentation and visualization power are also asserted by many applications. For example, Google Earth was used to visualize and combine live data about the density and drift of Arctic ice (*e.g.*, ice maps) and animal tracking data to study the effects of changes in ice drift on the movement and behavior of animals (Butler 2006). Virtual globes are continuing to emerge. One such example is GeoGlobe, which forms part of the recently launched Chinese Map World (天地图), an equivalent to the combination of Google Maps and Google Earth (MapWorld 2011, see also Chapter 6 in this volume).

The *semantic web*, as one of the key components of Web 3.0, represents a group of methods and technologies to allow machines to understand the meaning of information and to talk to other machines over the Web (Berners-Lee et al. 2001). The last few years have witnessed growing efforts in studying and developing geospatial ontologies, together with some formal or standard semantics (Egenhofer 2002, Arpinar et al. 2006, W3C 2007). A particular focus relevant to web mapping and GIS is the semantic geospatial web which not only allows users to more precisely retrieve the data they need based on the semantics associated with these data, but also facilitates interaction of data among software agents or machines.

The importance of web mapping and web geospatial services have been well recognized by various professional and learning societies. The latest International Society for Photogrammetry and Remote Sensing (ISPRS) Congress Resolutions (ISPRS 2008) note "the fast emergence of Web 2.0" and "the paramount role of the Internet and location-based services, and 'virtual global' in society". It further recognizes "the increasing potential of the Web for dissemination of spatial information" and "that 'virtual globes' offer a more intuitive view of spatial phenomena for a wider audience than conventional maps". With these understandings, the Resolutions make recommendations on the development of "geospatial data processing techniques using distributed services and grid computing" and "web search engines for spatio-temporal data". The International Cartographic Association (ICA) Commission on Maps and the Internet has a focus on the effects of the Internet and the Web on mapping processes and map uses, and the importance of cartographic quality of web-based mapping (ICA 2011). At the International Federation of Surveyors (FIG), working groups 3.1 and 3.2 in the Commission on Spatial Information Management are catering for web mapping and the required geospatial data infrastructures (FIG 2011).

Over the past two decades, web mapping and GIS have received an important focus in numerous academic conferences, symposiums, workshops and trade events, including annual conferences and meetings of most geomatics professional and learned societies. A number of these events are dedicated to this topic, for example, workshops organized by ICA Commission on Maps and the Internet, a series of bi-annual International Workshops on Web and Wireless Geographical Information Systems (W2GIS), a series of GeoWeb conferences and the International Workshops on Pervasive Web Mapping, Geoprocessing and Services (WebMGS).

3 ISSUES AND CHALLENGES

Despite the rapid development of web mapping, web-based GIS and more recently web geospatial services, there remains a variety of issues that challenge researchers, developers, professionals and public users in the field. While some of these issues have been well discussed and addressed, others are still emerging. A thorough treatment of detailed issues is beyond the scope of this introductory chapter. Rather the following discussion examines dominant issues with particular reference to data, technology and social/organizational aspects.

3.1 *Data-related issues*

One of the oldest and still continuing issues in geospatial systems and applications is access to good quality geospatial data. The web environment has both helped to address some of the issues and compounded and created others. By the provision of geospatial data via the Web, it has become more accessible to users, but there are also greater demands on its quality, scale, dimensionality, integration, timely provision and ease of discovery for users and applications (Cartwright 2011).

The development of open standards and services has enabled the interoperability and access of geospatial data to users and applications in the form of web services, mashups and spatial data infrastructures. In turn, applications have developed the need for access and integration (or data reduction) of a broader range of data acquired from a wide array of sources (see Chapter 9 in this book). For example, a disaster management application may require, not only physical environment data involving topography, wind, temperature, etc., but also information about the water levels, built environment, infrastructure, activities, etc., which are sourced from (aerial, terrestrial and mobile) mapping, imagery, sensor networks, location based and GPS enabled devices, and others. Such information is dynamic and continually managed and updated over periods of time, and visualized in multiple dimensions, to reflect current conditions and support appropriate responses and decision-making. Standards and procedures, such as for sensor networks, mobile internet and indoor positioning, need to be further developed to enable information interoperability and integration.

Data on its own is of limited usefulness and the real value lies in piecing together multiple data to produce information, knowledge and intelligence as services. The challenge with the geospatial web is to intelligently utilize existing and new data from a range of sources, extract relevant and critical information, and build meaning and context that can be utilized in geospatial querying and processing to improve our understanding of the world. A vision for what is known as the semantic web was outlined over a decade ago by Berners-Lee (1998), the founder of the Web. We are currently in the midst of the interactive web (Web 2.0) and the challenge is to move forward towards the semantic web (Web 3.0) (Scharl & Tochtermann 2007).

Access to data and services must also be considered in the broader business context of markets, ownership, pricing, copyright, etc. The concept of a *spatial marketplace* was raised and outlined over a decade ago by Abel (1997). Since then, similar models have been adopted for non-spatial applications such as Amazon.com. Currently the Cooperative Research Center for Spatial Information in Australia is putting together a plan for the design and implementation of a *Spatial Marketplace* which will provide a front end for access by geospatial users and applications (CRCSI 2011).

3.2 Issues pertaining to technology

The accessibility and usability of geospatial data, information, knowledge and intelligence as services are underpinned and influenced by technological developments. As technology continues to advance and develop, the opportunities for geospatial systems and applications continue to expand. Web-based geospatial developments are at the confluence of technologies including the interactive and semantic web (Web 2.0 and Web 3.0), data and processing web services, cloud computing, location-based systems, mobile and GPS-enabled devices, 3D and multidimensional visualization, grid computing environments, data infrastructures and Service Oriented Architectures (SOA). Issues pertaining to technology lie with the infrastructure needed to support the geospatial web, as well as the processing and utilization that build on this infrastructure by users and applications.

With an increasing number of concurrent users/applications in an online environment providing and retrieving an increasing amount geospatial data from a vast array of sources, there are increasing pressures on managing and delivering such information in a timely manner. Grid computing and cloud computing resources are being explored and developed in order to provide transparent, scalable and reliable access. Appropriate and efficient distributed components and data architectures need to be designed for both fixed and mobile web environments. Performance management and load balancing must be achieved through efficient workflows (see Chapters 9–10 in this book, for example).

As mobile technologies including cellular phones and GPS-enabled devices are becoming ubiquitous, the technology infrastructures need to be expandable, scalable and reliable (Delfos et al. 2010). Given that access to "current location" is a central focus to mobile devices and location based systems, there needs to be more emphasis on obtaining "location" by enhancing positioning technologies and geolocating techniques. Issues include accessing high quality positions both indoor and outdoor, embedding such locational information into applications, and geolocating which includes attaching location coordinates to objects, events and services in the Web and mobile environments (see Chapter 11 in this book).

Web services technology is still relatively young although it has been rapidly developing within the last decade or so. In particular, Web Mapping Services (WMS) and Web Feature Services (WFS) following Open Geospatial Consortium standards are prevalent across the industry, whereas the Web Processing Services (WPS) implementations are more recent. The current challenge is web services *orchestration* which involves the selection of appropriate web services, the determination of how they will relate and interact, and placing them in workflows that provide solutions for users, organizations and applications.

Access and processing of geospatial information is not complete without a means to present and visualize it. The range of virtual globe clients and geobrowsers available over the past half decade has given the general public easy and intuitive access to geospatial information. The virtual globe viewers such as Google Earth and Microsoft Virtual Earth are able to integrate a range of image, raster and vector geospatial data types a into a seamless environment that allows users to integrate their own data and events. In addition to desktop environments, these are moving rapidly into mobile environments with location based data that allows the blending of virtual and physical environments through augmented reality and the integration of geolocated and geotagged information. Issues involve the integration of multisource, multiscale and multidimensional information, including 3D data, pictometry, Building Information Management (BIM) models, etc. in such environments. Another important issue that requires attention from the research community is how cartographic theories and principles may be re-examined in dealing with new web mapping technologies and development of new web maps.

3.3 Social and organizational issues

Privacy issues have drawn much attention over the last decade with the new developments in putting online high resolution satellite images, street views along with map services, user-generated data content, as well as mobile location-based services that require users' real-time location information. These issues are further complicated by associated ownership and copyright challenges. Some of the major issues include: the balance between privacy/copyright protection and the free flow of information, ownership of user-added, collective intelligence, and potential legal implications or problems with remixing geospatial content. Two more recent special journal issues, edited by Elwood (2008) and Feick & Roche (2010), published a number of papers specifically addressing the limitations and implications of user-generated data or volunteered geographic information, from different perspectives of social and political opportunities.

An area that requires further investigation is related to the acceptable use policies that govern free or licensed web map services, especially in the context of orchestrating web mapping services using map and service APIs (Li & Yan 2010). Acceptable use policies, defining code of conduct, consequences of violations, unacceptable uses, copyright rules, disclaimers, etc., are normally scattered in general Terms of Use, specific Terms of Use or Terms of Service and legal notices. The major issues encountered by the web mapping community with respect to many acceptable use policies lie in their clauses on restrictive use of free data and map services, no guarantee on quality and service stability, and the interpretation of policies.

Data quality issues, especially with the current free online data and user-generated data content, are well documented in the literature. As we continue to move into a service-oriented web mapping paradigm, the quality of online mapping and geoprocessing services has emerged as an equally important issue. With the technology advances described in Section 2, technical issues that affecting the quality of services, such as network performance, throughput and response time, are likely to be lessened. However, more empirical studies are still needed to assess the quality of web map and geospatial processing services within actual business, use and application contexts.

4 FUTURE DIRECTIONS

The directions of web mapping have progressed from more data-rich, interactive client/server systems that overcame technical limitations with the increased industry interest and user communities (Plewe 1997, Hardie 1998), to the expectations of delivering and integrating distributed GIS and services with component-based applications (Peng & Tsou 2005). Technologies have played an indispensable role in shaping the development of web mapping and GIS, but because technological developments are highly unpredictable, it is difficult to predict the future directions of web mapping and GIS. Instead, we would like to identify and/or revisit trends and potential directions for web-based mapping and services (Dragicevic 2004, Li 2008).

4.1 *Service-oriented, workflow-enabled services on the cloud*

As the service-driven paradigm continues to dominate web mapping/GIS development, the orchestration of web mapping and geoprocessing services from different sources will become increasingly important. Old workflow principles and technologies have been explored as one of the solutions to tackle this problem. Workflows can help to achieve a better understanding of business and service processes and automate the service execution process. It is expected that workflow principles and engines will be seamlessly integrated into the web mapping service-oriented architecture.

More and more geospatial data and applications will reside on the *cloud*, moving towards the Web as a GeoWeb platform and providing traditional GIS capabilities in a non-traditional way (blog.fortiusone.com). Grid computing will continue to be an active research subject in Web mapping/GIS because it links disparate computers to form one large infrastructure, harnessing unused resources and forming a part of cloud computing (Myerson 2009). Emerging research in the area of CyberGIS and Geospatial Cyberinfrastructure is addressing some of these challenges (Wang 2010). While performance issues are important, given the nature of geospatial data processing (see Chapter 8 in this volume), other issues related to security, privacy, currency and interoperability are also to be actively studied. The emerging of web-centric operating systems, such as Google's Chrome OS, is likely to stimulate many new interesting developments.

4.2 *More analytical, functionally-rich, intelligent and collaborative*

Analysis other than simple access and visualization will be the trend for the entire web community (Berners-Lee 2007). We will see more traditional GIS functions being integrated into web mapping/GIS applications in a way transparent to users, using distributed analytical and geoprocessing services coordinated by well-defined workflows. Especially with the increasing use of Rich Internet Applications (RIA), both the interface and functionality of web-based mapping/GIS will be greatly enhanced, allowing users to perform more spatial analysis on the Web without accessing traditional desktop GIS. However, "the traditional approach to publishing maps on the Internet 'by GIS experts, for GIS experts' is outdated, ineffective and unusable as Web mapping becomes more and more mainstream" (van der Vlugt & Stanley 2005). The audience and their requirements must be carefully studied and the functionally-rich interfaces must be adaptive to different user groups.

Artificial intelligence will play an important role in making future web-based mapping/GIS more analytical. Intelligent analytical capability will be enhanced with further developments in Geographic Knowledge Discovery (GKD), spatial data mining techniques and tools, web-based modeling and simulation, augmented reality applications, agent-based systems and paradigms (Hutchinson & Veenendaal 2010), etc. The ability of performing intelligent spatial analysis will assist location-based intelligent search and personalization of web mapping or geoprocessing according to user and application preferences.

The Web was born with collaboration in mind. We have already seen the successful developments in collaborative web mapping, *e.g.*, OpenStreetMap (www.openstreetmap.org), Wikimapia (wikimapia.org) and Google's My Maps, and asynchronous/synchronous sharing of maps and map-based commentary. As the research and development in the area of collaborative GIS continues to generate solid outcomes in theories, designs, and insights on related social issues (Balram & Dragicevic 2006), collaborative web mapping/GIS tools with rich analytical functions to support various levels of decision support that build on the functionality of current collaborative systems will become more pervasive.

4.3 *Move to more personalized services*

Driven by the "attention economy" where users agree to receive services in exchange for their attention, the future web will be more personalized (Iskold 2007). The key ingredients of the attention economy are "relevance" and "personalization" which both depend on how much information can be obtained from users and how this information is used to personalize web services. Personalization based on preferences not only helps ease the challenging problem of "information overload", but also improves performance of web mapping services by reducing the amount of geospatial data needed to be transmitted on the Web (see Chapter 9 in this volume for some discussions). The future personalized web mapping/GIS applications should allow users to control what data and services they want to see and how they want to see them, and customize their web interface/navigation, geospatial data and information content, formats, etc. Personalization is especially important for future mobile geospatial applications, as built-in GPS becomes more prevalent in mobile devices.

4.4 *More open source based*

Although the debate between open source and proprietary solutions will continue, the choice of implementing open source in GIS in general and web mapping in particular has been broader. Open source GIS has become well established, over the past few years. While the next generation web mapping/GIS will be more interoperable, open source solutions are likely to provide complete implementation of open standards because they are intimately connected with open standards and embody open standards in their development (Simon 2005). As the web develops towards being more collaborative and collective, open source software will be one of the future directions because its success has been created through constantly growing *communities of shared interest* (Ramsey 2007).

4.5 *Social mapping and networking*

The development of social mapping and geo-referencing user-added and user-controlled data content will continue to be an important area for web mapping researchers to focus on. Harmonization, ubiquity and context awareness will be some of the key words in the evolution of geospatially-enabled social networks. Accordingly, data policies, privacy and ownership issues, technical challenges as well as searching for new paradigms will be important research areas.

The internet will continue to be a more useable and enjoyable environment with rich social and realistic environments that utilize authentic 3D environments, augmented reality and mobile technologies (see Chapters 12, 14 and 15 in this volume). Rather than simply being a data repository or data sharing environment, it creates an environment that involves and engages the user communities (*e.g.*, whether discipline-based, geographically-located, research-focused, etc.) in shared application environments in both space and time. Further study will focus on what these future collaborative and social environments will look like, how they interact in virtual and real space, and what tools, resources and services need to be provided.

5 CONCLUSION

In this chapter, we have established broad connections regarding the advances being made and challenges being confronted in web-based GIS, mapping services and applications. These connections form the background and context to the remaining chapters of this book, which addresses many of these issues and explores solutions in unique and innovative ways. The rapid improvements in technologies related to the geospatial web world means that a snapshot in time is overtaken almost as soon as these technologies are released. However, the sharing and exchange of ideas and solutions is an important step on the fast moving path that web GIS is taking us. This book volume is a step along that road as we progress to understand what the geospatial web will hold for us in the immediate future.

REFERENCES

Abel, D.J. (1997) Spatial internet marketplaces: a grand challenge? In: Scholl, M. & Voisard, A. (eds.) *Advances in Spatial Databases*, Lecture Notes in Computer Science, Springer-Verlag, Heidelberg, 1262/1997, pp. 3–10.

Arpinar, B., Sheth, A., Ramakrishnan, C., Usery, E.L., Azami, M. & Kwan, M. (2006) Geospatial Ontology Development and Semantic Analytics. *Transactions in GIS*, 10 (4), 551–575.

Balram, S. & Dragicevic, S. (2006) *Collaborative Geographic Information Systems*. London: Idea Group Publishing.

Balram, S., Dragicevic, S. & Feick, R. (2009) Collaborative GIS for spatial decision support and visualization. *Journal of Environmental Management*, 90, 1963–1965.

Berners-Lee, T. (1998) *Semantic Web Road Map*. [Online] Available from: http://www.w3.org/DesignIssues/Semantic.html, [Accessed 11th February 2011].

Berners-Lee, T. (2007) Future of the World Wide Web. Testimony to the United States House of Representatives, Internet and Telecommunications Subcommittee, 1st March 2007.

Berners-Lee, T., Hendler, J. & Lassila, O. (2001) The Semantic Web. *Scientific American Magazine*. 17th May 2001.

Butler, D. (2006) The web-wide world. *Nature: International weekly journal of science*, 439 (7078), 776–778.

Cartwright, W. (2011) Possibilities and issues in contemporary mapping. *Geospatial World* January: 44–48.

CRCSI. (2011) *Cooperative Research Centre for Spatial Information.* [Online] Available from: http://www.crcsi.com.au

Deiana, A. (2009) Skylineglobe: 3D Web Gis Solutions For Environmental Security and Crisis Management. *GeoSpatial Visual Analytics*. Springer, Netherlands. pp. 363–373.

Delfos, J, Tan, T. & Veenendaal, B. (2010) Design of a Web-Based LBS Framework Addressing Usability, Cost, and Implementation Constraints, *World Wide Web*, 13 (4), 391–418.

Dragicevic, S. (2004) The potential of web-based GIS. *Journal of Geographical Systems*, 6, 79–81.

Egenhofer, M.J. (2002) Toward the semantic geospatial web. *Proceedings of the 10th ACM international symposium on Advances in geographic information systems*. 04–09 November 2002, McLean, Virginia, USA. pp. 1–4.

Elwood, S. (2008) Volunteered Geographic Information: Future Research Directions Motivated by Critical, Participatory, and Feminist GIS. *Geo Journal*, 72 (3&4), 173–183.

ESRI. (2011) *Creating ArcGIS Server Solutions: Roadmap for Developers.* [Online] Available from: http://resources.esri.com/help/9.3/arcgisserver/adf/dotnet/roadmap_developers.htm, [Accessed 22nd Feb 2011].

FIG. (2011) *FIG Commission 3 Spatial Information Management, International Federation of Surveyors*, [Online] Available from: http://www.fig.net/commission3, [Accessed 22nd February 2011].

Fu, P. & Sun, J. (2011) *Web GIS Principles and Applications*. Redlands: ESRI Press.

Goodchild, M.F. (2007) Citizens as sensors: the world of volunteered geography. *Geo Journal*, 69 (4), 211–221.

Hardie, A. (1998) The development and present state of web-GIS. *Cartography*, 27 (2), 11–26.

Hutchinson, M. & Veenendaal, B. (2011). An agent based framework to enable intelligent geocoding services. Applied Geomatics, in press.

ICA. (2011) Commission on Maps and the Internet, International Cartographic Association, maps. unomaha.edu/ica, [Accessed 22nd February 2011].

Iskold, A. (2007) *The Attention Economy: An Overview. Read Write Web Blog*, [Online] 1st March 2007, Available from: http://www.readwriteweb.com/archives/attention_economy_overview.php.

ISPRS Congress Resolutions. (2008) *Resolutions of the XXIth Congress of ISPRS in Beijing*, [Online] 2008. Available from: http://www.isprs.org/documents/resolutions.aspx, [Accessed 14th Feb 2011].

Kraak, M.J. (2001) Settings and needs for web cartography. In: Kraak, M.J. & Brown, A. (eds.), *Web Cartography*. New York, Francis and Taylor. pp. 3–4.

Li, S. (2008) Web Mapping/GIS services and applications. In: Li, Z. Chen, J. & Baltsavias, M. (Eds.), *Advances in photogrammetry remote sensing and spatial information science: 2008 ISPRS congress book*. Balkema, CRC Press. pp. 335–354.

Li, S. & Gong, J. (2008) Mashup: A new way of providing web mapping and GIS services. *Proceedings of XXI ISPRS Congress, Commission IV*, XXXV-B4:639–648, 3–11th July 2008, Beijing, China.

Li, S., Xiong, C. & Ou, Z. (2011) A Web GIS for Sea Ice Information and Ice Service Archive. *Transactions in GIS,* in press.

Li, S. & Yan, W. (2010) Mashing up geospatial data services: implications of acceptable use policies. *Geomatica*, 64 (1), 111–122.

MapWorld. (2011) *Map World*, [Online] Available from: http://www.tianditu.com, [Accessed 10th Feb 2011].

Mata, F. & Claramunt, C. (2011) GeoST: Geographic, thematic and temporal information retrieval from heterogeneous Web data sources. In: Tanaka, K., Frohlich, P. & Kim, K-S (Eds.), *Web and Wireless Geographical Information Systems*: LNC Science 6574. Berlin, Springer-Verlag. pp. 5–20.

Moseley, R. (2007) *Developing Web Applications*. West Sussex, John Wiley & Sons.

Myerson, J. (2009) Cloud computing versus grid computing. *IBM developer Works*. [Online] 03rd Mar 2009. Available from: http://www.ibm.com/developerworks/web/library/wa-cloudgrid/

Ordnance Survey. (2011) *OS Open Data: Mapping data and geographic information from Ordnance Survey*. [Online] Available from: http://www.ordnancesurvey.co.uk/oswebsite/opendata/index.html, [Accessed 22nd Feb 2011].

O'Rourke, C. (2004) A Look at Rich Internet Applications. *Oracle Magazine.* July/August 2004.

11

Peng, Z.R. & Zhang, C. (2005) Services-oriented architecture in Internet GIS. GIS Development (online) 9 (10).

Plewe, B. (1997) *GIS Online: information retrieval, mapping, and the Internet*. Albany: OnWord Press.

Ramsey, P. (2007) The State of Open Source GIS. Refractions Research Inc., Suite 300–1207 Douglas Street, Victoria, BC, V8W-2E7, email: pramsey@refractions.net

Scharl, A. & Tochtermann, K. (eds.) (2007) *The geospatial web: How geobrowsers, social software and the web 2.0 are shaping the network society (Advanced Information and Knowledge Processing Series)*. Springer-Verlag, Heidelberg.

Simon, K. (2005) The value of open standards and open-source software in government environments. *IBM System Journal*, 44 (2), 227–238.

van der Vlugt, M. & Stanley, I. (2005) Trends in Web Mapping: It's all about usability. *Directions Magazine*. [Online] Tuesday, 18th October 2005, Available from: http://www.directionsmag.com/articles/trends-in-web-mapping-its-all-about-usability/123318, [Accessed 22nd Feb 2011].

W3C. (2007) *W3C Geospatial Ontologies. W3C Incubator Group Report* [Online] 23rd October 2007, Available from: http://www.w3.org/2005/Incubator/geo/XGR-geo-ont-20071023/, [Accessed 22 Feb 2011].

Wang, L. & Hu, D. (2009) Research and Realization of RIA WebGIS Based on Flex. *Proceedings of the International Workshop on Intelligent Systems and Applications*. 23–24 May 2009, Wuhan, China. pp. 1–4.

Wang, S. (2010) A CyberGIS Framework for the Synthesis of Cyberinfrastructure, GIS, and Spatial Analysis. *Annals of the Association of American Geographers*, 100 (3), 535–557.

Wikipedia contributors. (2011) *Web mapping. Wikipedia, The Free Encyclopedia*. [Online] Available from: http://en.wikipedia.org/w/index.php?title=Web_mapping&oldid=412909484, [Accessed 22 Feb 2011].

Analytical and geospatial services

*Advances in Web-based GIS, Mapping Services
and Applications – Li, Dragićević & Veenendaal (eds)*
© 2011 Taylor & Francis Group, London, ISBN 978-0-415-80483-7

Geography 2.0—A mash-up perspective

T. Edwin Chow

Department of Geography, Texas State University—San Marcos, San Marcos, USA

ABSTRACT: Web 2.0 represents the emergence of diverse web technologies that proliferates the interconnectivity and interactivity of dynamic content over the Internet. Various Web 2.0 technologies are often integrated to create mash-up—a hybrid web application combining services and/or data from multiple sources. In fact, Web 2.0 accelerates the paradigm shift of Geographic Information Systems (GIS) from desktop solutions to online Geographic Information Services (GIServices). In this paper, Geography 2.0 is defined as the deployment of mash-ups that integrate multiple third-party GIServices, including geographic data (e.g. map) and spatial operations (e.g. routing) for Internet GIS applications. The objectives of this chapter are to 1) highlight the evolution of mash-ups in the revolutionary trend of Web 2.0, 2) review recent advancement of representative mash-up technologies related to Geography 2.0 by examining the data, programming interfaces and integration tools, and 3) comment on the role of mash-ups in developing Internet GIS applications and shed some insight into the frontiers of research in future web development. For illustration purposes, a mash-up prototype has been created to visualize the cases of H1N1 flu in United States onto a map.

Keywords: Geography 2.0, GIServices, Internet GIS, Mash-up, Web 2.0

1   INTRODUCTION

1.1   *Web 2.0, Geography 2.0 and mash-ups*

Web 2.0 represents the emergence of diverse web technologies that proliferates the interconnectivity and interactivity of dynamic content over the Internet. O'Reilly (2007) outlined seven key principles of Web 2.0 technologies that enable users to contribute and co-develop Rich Internet Applications (RIA) by utilizing lightweight and scalable online services. Despite some dispute about the meaning of Web 2.0 (Berners-Lee 2006, Schmidt 2007), the general concept implies a revolutionary web culture that attempts to enrich the user experience from merely accessing to exercising control over the web content by information sharing and public collaboration.

The social nature of Web 2.0 is a critical aspect of its success. Many thriving Web 2.0 applications, such as flickr (http://www.flickr.com) and facebook (http://www.facebook.com), are founded upon a network of effective and active user contribution in addition to social tagging (i.e. folksonomy). Moreover, these web sites often provide an easy platform and the infrastructure needed to create, search, and share user-generated content. Hence, Web 2.0 represents a paradigm shift from "users" to "producers" within a vast social network among people. In fact, the power of the Internet is nested within the linkage created not only among the web documents but also the social network and the interactions arising from these synergistic activities (Hendler & Golbeck 2008).

In light of Geographic Information Systems (GIS), Web 2.0 accelerates the paradigm shift of GIS "from an isolated architecture to an interoperable framework, from a standalone solution to a distributed approach, from individual proprietary data formats to open specification exchange of data, from a desktop platform to an Internet environment" (Chow 2008). For example, an amateur cartographer could easily geocode the latitude/longitude coordinates from Angelina Jolie's tattoo, which bears the coordinates of her children's place of birth, visualize the geocoded markers onto a 3-dimensional (3D) virtual globe (e.g. Google Earth), and distribute the geographic information on the Internet through Keyhole Markup Language (KML). Sui (2008) described this trend of "massive and voluntary collaboration among both amateurs and experts" as the wikification of GIS. The word "wikification" has its roots in the innovative movement of user-contributed dynamic content that is created and maintained by the interested public, such as the well-known wikipedia project. Sui (2008) envisioned that the wikification of GIS will be empowered by continual development of open-source GIS software, availability of conventional and "volunteered geographic information" (Goodchild 2007a, Goodchild 2007b), and increasing involvement of citizens as sensors and voluntary GIS corps. Given the realization and expectation of Web 2.0 technologies to continue as the core platform for the wikification of GIS, Geography 2.0 represents a mature stage that emphasizes the deployment of Web 2.0 technologies to respond to new frontiers in geography.

The notion of Geography 2.0 originates from O'Reilly's Where 2.0 2006 conference and was loosely associated with virtual globe applications. Despite its powerful visualization capabilities, the 3D representation of the virtual globe does not reflect the full spectrum of Web 2.0 technologies, which also include RIA, web syndication, Application Programming Interface (API), web services, Asynchronous JavaScript and XML (AJAX), etc. In fact, various Web 2.0 technologies are often integrated to create so-called mash-up—a hybrid web application combining services (Rinner et al. 2008) and/or data (Cheung et al. 2008, Sui 2008) from multiple sources. In this chapter, Geography 2.0 is defined as the deployment of mash-ups that integrate multiple third-party Geographic Information Services (GIServices), including geographic data (e.g. map) and spatial operations (e.g. routing) for Internet GIS applications. Therefore, mash-up is an important manifestation of Geography 2.0 in the Web 2.0 era.

The goal of this chapter is to examine emerging web technologies that have been, or can be, developed for Internet GIS applications and their roles in supporting potential research agenda in GIScience. As such, the objectives of this chapter and thus how it is organized are three-fold. The first is to introduce Geography 2.0 in the context of Web 2.0 and to highlight the evolution of mash-ups in this revolutionary trend. The second is to review recent advancement of representative mash-up technologies related to Geography 2.0 by examining the data, programming interfaces and integration tools. The third objective is to comment on the role of mash-ups in developing Internet GIS applications and shed some insight into the frontiers of research in future web development.

1.2 *The paradigms of mash-up*

Mash-ups have been an emerging Internet phenomenon since 2005. Many web developers applauded the freely adaptable tools and spawned numerous Internet applications. One of the famous mash-up websites was the former chicagocrime.org which plots the Chicago crime data onto a Google map and allows attribute and spatial queries near a user-defined location (or along a digitized path).

There has not been an official record of the total number of mash-ups on the Internet; but it is evident that the number of new sites has been increasing every day (Fig. 1). By mingling multiple data sources and visualizing the extracted information over the Internet, mash-ups are application-oriented. Among most of the mash-up applications that have been explored, more than one-third is tied to mapping applications (Fig. 2). Common mash-ups serving multimedia are often seen as photos, video, music, or a hybrid of multiple data types (e.g. photos tagged in a map).

In order to better understand the trend of mash-ups, it is perhaps more insightful to gain a perspective on the adopted paradigms. The concept of mash-ups originates from the notion of web service, which is intended to provide an interoperable framework to connect

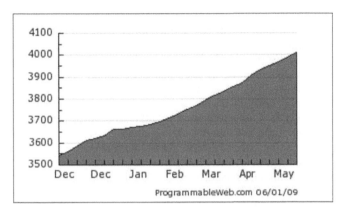

Figure 1.   Number of mash-ups over time (courtesy of ProgrammableWeb 2009).

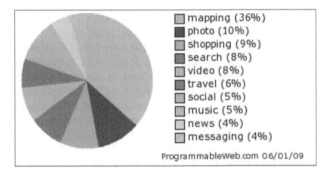

Figure 2.   Mash-up applications (courtesy of ProgrammableWeb 2009).

existing software and reusable application components through Internet protocols. In this research, web service does not convey the broader sense of providing a service over the web. From a technical perspective, however, web service is based on the Service-Oriented Architecture (SOA) to discover, communicate, and utilize external services distributed over the network. Users' requests and interaction could be handled by using Simple Object Access Protocol (SOAP), a platform- and language-independent web protocol that exchanges information over the Hyper-Text Transfer Protocol (HTTP) based on eXtensible Markup Language (XML) communication. In SOAP applications, a server hosts a service with XML specifications while the client discovers and requests the needed service (or information) by exchanging XML-encoded messages. For example, the Web Feature Services (WFS), an Open Geospatial Consortium (OGC) standard for accessing (e.g. create, query, edit, delete) geographic features, can support both SOAP- and conventional HTTP get/post requests. Web Map Service (WMS) and Web Coverage Service (WCS) are similar OGC standards that will allow access to non-editable map images and raster data. In addition to merely granting access to geographic data, web services also present a framework of distributed and reusable computational services to build complex models (Fensel & Bussler 2002). Smiatek (2005) described the implementation of web services as a neutral platform for the climate model in accessing the GIS database.

   Evolving from the XML-oriented paradigm of web service, mash-ups gain momentum from the launching of Application Programming Interface (API), another form of emerging GIService. In conventional software development, API is a set of programming libraries that allows an application to communicate with another application or the underlying operation system to perform a service, like accessing a file. An example is to grant access

17

to some routines/methods, data structure, protocols or class/objects used in a programming library, which builds the foundation of any application. For web applications, API is a source code interface that grants web developers access to the programming library hosted by a web server and to request services over the Internet. Thus, API offers the most flexibility to the web developers by documenting detailed instructions and operations. In 2005, Google released the Google Maps API and allowed web developers to tap into their powerful web map servers and mapping services. Due to "its no cost policy, the availability of global data coverage, dynamic navigation, query capability, and ease of implementation" (Chow 2008), web developers globally took advantage of this valuable asset to combine their data with the map interface in creating new mash-ups. Since then, there have been thousands of APIs offered for web development in various capacities (ProgrammableWeb 2009). Mapping, however, remains the most dominant type of mash-up applications (Fig. 2).

An interesting trend of Web 2.0 is that technologies are moving towards simpler approaches. Unlike the SOAP-based web service, Representational State Transfer (REST) does not adopt the message-oriented paradigm but view representation of specific information on the Internet (e.g. a file, a query that returns an item, an abstract object) as resources. REST is an architecture style that promotes lightweight web services by using a global identifier to address each resource (e.g. Universal Resource Identifier in HTTP) to make calls between machines (Fielding 2000). Web developers can deploy a RESTful web service or API, one that follows REST principles, to do the four CRUD operations: Create/Read/Update/Delete. For example, the famous Twitter REST API provides a method to show the most recent status of a user by sending a request in the form of URI, such as: http://twitter.com/statuses/show/id.format (Twitter 2009). One may use any scripting language, such as PHP: Hypertext Preprocessor (PHP) to send the HTTP request and interpret the returned representation in a specific format, commonly in XML, JavaScript Object Notation (JSON), Real Simple Syndication (RSS), etc. The websites twittervision and flickrvision (Troy 2009a, Troy 2009b) demonstrates the binding of multiple APIs, including Google Maps API, Twitter REST API, Flickr REST API, and Poly9 Globe API, in creating powerful mashups—focus of the next section.

2 MASH-UPS

2.1 *Data*

Data is an important component of mash-ups, with one of its' primary functions being to blend into multiple data sources. While many GIService providers offer valuable geographic data (e.g. traffic overlay) through API, it is often desirable to integrate custom data (e.g. census data) into the mash-up applications. As illustrated from the paradigm evolution of mashups, there is an extensive list of web data expressed in various formats. In the context of Geography 2.0, this paper will limit the discussion to the spatial/mapping component directly relevant to GIServices.

2.1.1 *Desktop GIS data*
GIS evolved from a long tradition of thematic cartography as a desktop application. The initial effort of the U.S. Census Bureau in creating the Dual Independent Map Encoding (DIME) file format and Fisher's (1966) pioneer mapping program Synegraphic Mapping System (SYMAP) laid down the framework for early GIS data and software. On the one hand, DIME is a data structure for large-scale digital geographic databases, the precursor to the Topological Integrated Geographic Encoding and Reference System (TIGER), which serves free base maps for many value-added map products. On the other hand, the inspirational SYMAP sparked off decades of continual development in GIS software. In general, the GIS data utilized by commercial computer systems are proprietary in nature and hence is software-specific. Common vector data (i.e. point, line or polygon) include, but are not limited to, TIGER line files, ESRI shapefiles, Arc/Info coverage, ESRI geodatabases, AutoCAD

DXF, MapInfo TAB, etc. Raster (i.e. pixel-based imageries) examples are ESRI grid, ERDAS Imagine, USGS DEM, Digital Raster Graphics, Idrisi RST, etc.

Ideally, a GIService provider that empowers mash-up applications can parse the binary GIS file directly from a web server. This assumption may be valid given that the GIService provider is one of the desktop GIS software vendors. For example, the ESRI ArcGIS JavaScript API can overlay custom map tiles based on shapefiles and geodatabases into their mash-ups. Unfortunately, most proprietary GIS data are not supported by external mash-up providers for direct parsing. Some GIS data are compliant to Object Linking (e.g. geodatabase), and Embedding Database (OLEDB) or ActiveX Data Object (ADO), however, the requirement for deployment licenses may present resistance to many web developers. Therefore, it is often necessary to convert the original dataset into the file format(s) acceptable to the GIService provider. A strategy is to convert shapefile, an open-format dataset (ESRI 1998), to text files, conventional databases, or XML-formatted data *a priori* by using a scripting language. In the Web 2.0 era that promotes sharing and open standards, web developers warmly embrace geographic data that have open specification and are ready for mash-up applications. Some emerging data-interchange formats are included below.

2.1.2 *GML*

Geography Markup Language (GML) is an XML extension to encode geographic features with spatial and attribute information (Peng and Tsou, 2003). The GML 3.1 encoding specification supports geometric primitives (e.g. Point, LineString and Polygon), geometric complexes (i.e. closed collection of geometric primitives), geometric aggregate (e.g. MultiPoint, MultiLineString, MultiPolygon, MultiGeometry) and raster imageries (grid). The CityGML demonstrates a GML showcase to encode 3-dimensional (3D) urban objects in a virtual city. Thus, GML embodies sophisticated data models and geometries to represent real world objects. Some websites offer programming interface to develop mash-ups by utilizing GML for modeling spatial objects. For example, hostip.info (http://www.hostip.info) offers a mash-up interface to look up an Internet Protocol (IP) address and return spatial information, such as latitude, longitude, country encoded in GML. Chow (2008) demonstrated a prototype to visualize GML onto Google Maps that illustrated great potential for developing Internet GIS solutions around open specification.

Similar to XML, GML must also be well-formed and validated by an external reference of a GML application schema (Fig. 3). Fortunately, there are existing freeware (e.g. GeoCon) and commercial software (e.g. TatukGIS Editor) that readily converts most vector and raster database formats into GML. An eXtensible Stylesheet Language Transformation (XSLT) can be used to parse the XML-based data and convert GML into other open specification formats, such as Scalable Vector Graphics (SVG).

2.1.3 *KML*

Keyhole Markup Language (KML), similar to GML, adopts a custom XML application schema to describe geographic data. One major difference between the two markup languages is that KML also contains XML tags which define visualization and navigation parameters (e.g. `<Camera>`, `<LookAt>`, `<gx:FlyTo>`). By using a 3D virtual globe application, such as Google Earth, to interpret these custom tags, KML provides the data and visualization parameters to display geographic phenomenon and to explore their spatial context in 3D form (Fig. 4). However, KML can only support simple geometries, such as marker (i.e. point), polyline, polygon and raster tile. Hence, KML is comparatively limited, in representing complex real-world objects and their spatial relationship, than GML. Nevertheless, other earth browsers, including Google Maps, NASA World Wind, Microsoft Bing Maps 3D (i.e. successor of Virtual Earth), support KML encoding (fully or partially) and allow KML overlay. The Open Geospatial Consortium (OGC) endorsed KML version 2.2 as an implementation standard in 2008. An OGC workgroup is considering to harmonize future versions of KML with relevant GML specification and OGC standards, such as sharing the same geometry representation. (OGC 2009).

```
<?xml version="1.0" encoding="UTF-8"?>
<Cities xmlns:gml="http://www.opengis.net/gml">
   <gml:featureMember>
      <City fid="City0">
         <ID>1</ID>
         <Name>Point1</Name>
         <TheGeometry>
            <gml:Point>
               <gml:pos>-83.688, 43.019</gml:pos>
            </gml:Point>
         </TheGeometry>
      </City>
   </gml:featureMember>
</Cities>
```

Figure 3.   A GML representation of a point feature.

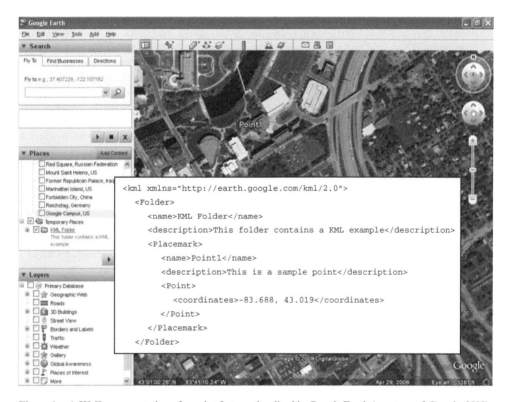

Figure 4.   A KML representation of a point feature visualized in Google Earth (courtesy of Google, 2009).

Due to the availability of global data coverage (e.g. aerial/satellite imageries, vector base map, terrain, street view photos), ease of KML creation and distribution, KML have been used for data conglomeration in hazardous events—when time is crucial in soliciting and releasing geographic information to the community. Shortly after the Sichuan earthquake (Luo 2008), California wildfire (Los Angeles Times 2008) and the recent H1N1 flu outbreak (Niman 2009), KML-based mash-ups were created to track and update the latest disper-sion of the geographic phenomenon. KML files containing satellite imageries, time-series animation of polygon overlay, point cloud of reported incidents, aggregated charts, etc were distributed over the Internet within days. The content of Volunteered Geographic Information (VGI) is also added to by grassroots movements and is being updated regularly. The sharing of knowledge through KML enables politicians, emergency response managers, scientists, and the community to analyze the phenomenon and make informed decisions.

2.1.4 *GeoRSS*

GeoRSS originates from Really Simple Syndication (RSS), a very common XML-based web feed and syndication format that automatically updates the web content of an Internet application by an external provider. A web feed is a lightweight version of the original text that includes a headline, an abstract, a hyperlink to the original content and ancillary metadata such as publishing date and time. The prefix "geo" in GeoRSS refers to Geographically Encoded Objects for RSS feeds, implying additional spatial information embedded within the web feed content as in a RSS file. The spatial information typically contains the location content related to the full content. Ordinary RSS/GeoRSS applications are created for news update and channel broadcast. A user can subscribe to a GeoRSS feed by storing the URI into a RSS aggregator—a desktop or web client application that checks the web feeds for update and renders the retrieved content. For example, the U.S. Geological Survey (USGS 2009) provides GeoRSS feeds (as well as KML files) to update global earthquake events in real-time.

There are currently two encodings of GeoRSS—the GeoRSS Simple and GeoRSS GML. The Simple serialization of GeoRSS supports simple geometries, including point, line, polygon and box (i.e. rectangular bounding box). Additional properties like feature type, feature relationship, elevation and radius can be used to represent geographic features in 3D space and reflect their inter-relationships, such as containment and buffering. The GeoRSS GML supports the same geometries as in the Simple counterpart, but it adopts the GML application schema and their tags in defining the geometric coordinates. Figure 5 describes the two GeoRSS encodings in reference to a point feature. The GeoRSS Simple accepts latitude/longitude point pair in decimal degree based on World Geodetic System (WGS) 1984, whereas the GML encoding also supports other coordinate systems, such as Universal Transverse Mercator (UTM) or State Plane coordinate system. The compliance of GeoRSS with GML implies better representation of geometries (e.g. the inner ring of a complex polygon) and high mapping accuracy by using a local coordinate system in feet/meter. In conjunction with its real-time update properties, GeoRSS is an emerging input data feed for developing mapping mash-ups and holds great potential for future development. It is currently supported by major GIService providers, such as Google Maps, Yahoo! Maps and Microsoft Bing Maps 3D (i.e. former Virtual Earth). Both encodings of GeoRSS can be extended from either RSS 1.0, RSS 2.0 or Atom, an alternative format of RSS.

2.1.5 *GeoJSON*

GeoJSON is a lightweight data-interchange format for encoding geographic data based on JavaScript Object Notation (JSON), akin to the relationship between GeoRSS and RSS. Unlike XML-based data description languages (e.g. GML and KML) that use a well-formed tag structure, JSON (and hence GeoJSON) is easily read/parsed by both human and machine. The syntax of a GeoJSON object follows a very simple structure of name/value pairs (Fig. 6). Despite being a subset of JavaScript in formatting the data structure, GeoJSON is language-independent that can be used in conjunction with other web technologies, such as ActionScript, Active Server Page (ASP), PHP, etc in requesting the GeoJSON object. A GeoJSON object may denote a geometry, a feature (i.e. geometry with additional properties) or a feature collection (i.e. an array of features). GeoJSON supports various geometry types, including geometric primitives (e.g. Point, LineString and Polygon) and geometric aggregate (e.g. MultiPoint, MultiLineString, MultiPolygon, GeometryCollection). Similar to GeoRSS GML, GeoJSON enables any coordinate reference system other than the default WGS 84 datum.

```
GeoRSS Simple:
<?xml version="1.0"?>
<rss version="2.0" xmlns:georss="http://www.georss.org/georss/">
    <channel>
        <title>GeoRSS Example</title>
        <description>This file contains a GeoRSS item</description>
        <link>http://www.georss.org</link>
        <item>
            <title>Point1</title>
            <description>This is a sample point</description>
            <georss:point>43.019 -83.688</georss:point>
        </item>
    </channel>
</rss>

GeoRSS GML:
... ... ...
<rss version="2.0" xmlns:georss="http://www.georss.org/georss/"
xmlns:gml="http://www.opengis.net/gml">
... ... ...
        <item>
            ... ... ...
            <georss:where>
                <gml:Point><gml:pos>43.019 -83.688</gml:pos></gml:Point>
            </georss:where>
        </item>
```

Figure 5.   GeoRSS examples encoded for RSS 2.0 to represent a point feature.

```
{
    "type": "Feature",
    "id": "Point1",
    "properties": {       },
    "geometry": {
        "type": "Point",
        "coordinates": [-83.688, 43.019]
    },
    "crs": {
        "type": "OGC",
        "properties": {
            "urn": "urn:ogc:def:crs:OGC:1.3:CRS84"
        }
    }
}
```

Figure 6.   A GeoJSON representation of a point feature.

When the GIService provider returns a GeoJSON object (by using the HTTP GET method, AJAX XMLHttpRequest object, or other methods), the client application will need to parse the data. Using JavaScript as the client scripting language as an example, one may use the eval() method to parse the GeoJSON data as an object to get access to its properties and values. Since JSON text adopts JavaScript syntax and may contain JavaScript code other

22

than data serialization, parsing JSON data from an unknown or malicious source may cause severe browser security issues. Crockford (2006) described some safer alternatives to validate and access the JSON data. Since GeoJSON is a subset of an object-oriented programming language, there is no need to restructure the JSON-formatted data into an object or an array after parsing. Thus, the simple format of GeoJSON is self-describing and does not require an ancillary "user manual" as in XML schema for XML data. GeoJSON is still in its infancy, with only a few mash-up providers (e.g. FireEagle, OpenLayer) currently supporting GeoJSON. Given its simple and efficient format, this new data format may gain momentum and further spotlight from GIService providers to incorporate it as part of the data alternatives.

2.2   *Application Programming Interface*

Application Programming Interface (API), as explained from the previous section, is an emerging technology that is prevailing among web developers, including geographers and non-geographers. It enables web developers to access the existing programming libraries and add new functionalities to create mash-up applications. In some instances, like Flickr, it is possible to recreate the functionalities of the hosting web site. This section introduces the Maps APIs and other APIs (i.e. non-map) that are common in creating mash-ups related to Geography 2.0. Due to the changing trend of web technologies, it is important to note that some specific examples may be subjected to frequent updates.

2.2.1   *Maps APIs*

Maps APIs are the source code interfaces that grant web developers access to powerful web map servers for geographic data and spatial operations. The geographic data include both the satellite/aerial imageries and map data (e.g. city points, road network, and political boundaries) rendered as raster data in most cases. The spatial data is global coverage in general but the finest details, including street-level geographic features and high resolution imagery, are available only around major metropolitan areas. It is noted that the requested geographic data is often returned by the API as raster tiles, similar to WMS, in most cases. However, the web developer may also integrate custom geographic data as overlays in any compatible data formats described in the previous section. Thus, a Maps API enables the web developer to request spatial data of a select geographic region and embed the resulting map as an object in any external web site. The Maps API also supports adding map controls, such as a navigation slide bar for zooming in/out or a toggle button to switch between map/aerial and hybrid view, for dynamic navigation by the map users.

Maps API can be implemented by using JavaScript, ActionScript, or other scripting languages. To enhance user experience, Maps API can be combined with web development technique like Asynchronous JavaScript and XML (AJAX) and REST architecture to parse, exchange, and query XML-formatted data (i.e. GML, KML, GeoRSS) as well as other data types (e.g. GIS data, GeoJSON). Common examples of Maps API include Google Maps API, Yahoo! Map Developer API, Mapquest OpenAPI, Microsoft Bing Maps Control and ESRI ArcGIS JavaScript API. Among the maps APIs, ArcGIS JavaScript API offers functionalities commonly available in convention GIS, such as buffering, viewshed analysis, thematic mapping, etc. The OpenLayers JavaScript API is a free API "developed for and by the Open Source software community" (OpenLayers 2009). A unique characteristic of this new API is the separation of map tool from the map data, which remains very flexible about the data formats and sources of base map. Another Maps API worthy of note, and is complementary to the OpenLayers JavaScript API is the OpenStreetMap API, a RESTful web service for viewing, editing and using the "free wiki world map" (OpenStreetMap 2009). The complete features of OpenLayers JavaScript API and their usefulness have yet to be unfolded and tested, interested readers can refer to Table 1 to look up the API documentation and road map. Table 2 compares the common built-in Maps APIs available to web developers. A detailed comparison of Maps API functionalities can be found elsewhere (Mapstraction 2009).

Table 1. A list of maps API and link to their documentation.

Maps API	Documentation URL
Google Maps API	http://code.google.com/apis/maps
Yahoo! Maps API	http://developer.yahoo.com/maps
MapQuest OpenAPI	http://www.mapquest.com/openapi
Bing Maps Control	http://msdn.microsoft.com/en-us/library/bb429619.aspx
ArcGIS JavaScript API	http://resources.esri.com/arcgisserver/apis/javascript/arcgis
OpenLayers JavaScript API	http://openlayers.org

There is limited research on the Maps API in the context of Geography 2.0. One obvious contributions of Maps API is that the wealth of geospatial data opens new possibilities for scientists to explore and visualize spatial data in new dimensions. Among other advantages such as simplicity and maintenance-free to the users, Mielke and Burger (2007) also highlighted the limitations of Maps API from a thematic cartography perspective. Nong et al. (2007) illustrated the use of GeoRSS as an easy and effective way for sharing geographic information, such as house renting and selling information. Given the advantages of automatic web syndication and its compliance with GML (i.e. the GeoRSS GML encoding) in representing more complex geographic features, GeoRSS has the potential to be one of the data formats to support practical Internet GIS applications. Moreover, the open specification of GeoRSS and aggregator programs can assist GIS/web professionals and amateurs to create and share georeferenced content through mash-up technologies. Thus, this emerging data format can help in narrowing the technological gap between GIS and web communities.

In exploring GML as an alternative representation of geographic information, Chow (2008) evaluated the potential of Maps API for developing Internet GIS applications. Although existing Maps APIs are effective in parsing GML and displaying vector as well as raster tiles, parsing raster GIS data at the pixel level is not efficient leading to serious performance issues over the Internet traffic. As table 1 reveals, most Maps APIs are also lacking in spatial operations common in GIS solutions, such as spatial interpolation, geoprocessing, map algebra, etc. Chow's (2008) study proposed some workaround solutions and recommended mash-up development around GML open specification because XML-technology can be useful to expand the spectrum of GIS operations of the Maps APIs, such as feature identification, joining and attribute query.

Rinner et al. (2008) developed a mash-up consisting of a map-based forum to evaluate its usefulness for a university's future master plan collaboration. While the stakeholders in campus planning issues can explore the geography of surrounding vicinity and their socio-economic and cultural context, they are also encouraged to participate and deliberate their inputs in a discussion forum by exchanging dialogue throughout the process of spatial decision making. It was reported that the multiple views of geographic data (i.e. map or satellite view) as well as the ease of geographical referenced comments were constructive in clarifying the participants' arguments and foster better understanding and collaboration.

Cheung et al. (2008) applied mash-up creation/editing tools to integrate multiple Web 2.0 technologies for Health Care and Life Science (HCLS) data visualization. The authors gave a brief overview of Web 2.0 technologies and discussed their strengths and weaknesses in creating mash-ups for HCLS applications. Their observations in mash-up tools and semantic web will be discussed in the later sections.

2.2.2 *Non-map APIs*

To create a mash-up that dynamically binds multiple data sources together, Geography 2.0 applications typically employ other non-map APIs that offer georeferenced content and user-enriched functionalities. These additional data/functionalities are just as important as the mapping component. Creativity is arguably the most important factor in developing a successful mash-up; the list of possible APIs is infinite and hence cannot be referenced

Table 2. Comparison of the common built-in Maps APIs. (Adopted and modified from Chow (2008)).

	Google maps API	Yahoo! maps API	MapQuest OpenAPI	Bing maps control	ArcGIS JavaScript API**	OpenLayers JavaScript API**
Remotely Sensed Imagery						
Vertical angle	Yes	Yes	–	Yes	Yes	Use other sources
Other angle	Street view	–	–	Oblique	–	–
Map data						
State/City	International					Use other sources
Highway/Street	U.S., Canada and selected countries					Use other sources
Overlay						
Geometry	Point, polyline, polygon	Point, polyline	Point	Point, polyline, polygon	Point, Multipoint, polyline, polygon	Point, polyline, polygon
Raster tile	Yes	Yes		Yes	Yes	Yes
Geographic data	KML, GeoRSS	KML, GeoRSS, JSON	–	KML, GeoRSS	JSON	GML, KML, GeoRSS, GPX, GeoJSON, WFS, WMS, etc.
Other layer(s)****	Terrain, traffic, wikipedia	Traffic	–	–	Terrain, census, weather, etc.	–
Address matching	Yes	Yes	Yes	Yes	Yes	–
Routing	Yes	–	Yes	Yes	Yes	–
Spatial Query of GIS data	–	–	–	–	Yes	Yes?
Buffering	–	–	–	–	Yes	–
Thematic mapping	–	–	–	–	Yes	Yes?
3D Visualization	Yes	–	–	Yes	–	–
Scripting language	JavaScript	JavaScript ActionScript	JavaScript	JavaScript	JavaScript	JavaScript
AJAX	Yes	Yes	Yes	Yes	Yes	Yes
SOAP/REST Option	–	REST	REST	SOAP	SOAP/REST	–
Plug-in required	3D viewer (optional)	Flash player (optional)	–	3D viewer (optional)	Flash player	–
Registration required	Yes*	Yes	Yes	–	Yes	–

*   Registration for a map API key is required in version 2 but not in the latest version 3.
**   Only the free version is provided here. The ArcGIS web mapping APIs also have options for Flex and Silverlight.
***   The OpenLayers API is still in development, and hence some features are speculated from the documentation but not finalized. Please refer to the text for details.
****   The other layers are only available in U.S. and selected countries.

completely. Table 3 illustrates a selected API example from each of the popular category (Fig. 2) and is intended to inspire creative minds for further exploration.

The content desirable from the non-map APIs are often not georeferenced in a format that is readily integrated into a mapping engine (e.g. lat/long, GML, KML, etc). However, the content at least should have some location references such as place names, street addresses etc. It is possible to address match (often regarded as geocoding), a process in assigning map coordinates to any location reference, by using an online address matching service. For example, GeoNames offers a web service to address match most forms of location reference into a data format ready for most Maps APIs. GeoNames web service also offers a method to convert any RSS location content to the corresponding latitude and longitude information in the output GeoRSS. Despite the challenge of resolving unique place remains, (e.g. London, Canada and London, United Kingdom), such web services are helpful for mash-up creation especially in the context of semantic web which can potentially mitigate confusion in similar place names.

2.3 Tools

A skillful web developer can take full advantage of the permissible access provided by APIs to create powerful mash-ups. A unique characteristic of Web 2.0, however, is the prominent promotion of grassroots participation and contribution for the mash-up content and applications themselves. Prompted by such enthusiasm and aspiration, there are tools that provide not only a graphic user interface (GUI) to identify the available APIs for mash-up, but also integrate them in a seamless environment as well. Below describes some tools that are useful to users of varying experience in the GIS and/or web domain.

2.3.1 WSDL

Web Service Description Language (WSDL) is a XML-based World Wide Web Consortium (W3C) recommendation to describe and locate an existing SOAP-based web service (Christensen et al. 2001). WSDL is not a tool *per se* but the metadata that describes how to access a web service. The relationship of WSDL to web service (i.e. SOAP) is akin to that of XML schema to XML. To utilize an existing web service, one may refer to the <port-Type>, <message>, <types> and <binding> elements in WSDL to learn the name of operation(s) (i.e. function as in traditional programming), input parameters, output format and communication port. Moreover, WSDL comprises the crucial information needed for compiling a directory of available web services and their specifications (e.g. functionalities, data types) in XML format. Thus, WSDL enables the Universal Description, Discovery and Integration (UDDI), a registry service to enlist appropriate web services based on keyword query described in WSDL. WSDL does not provide the tool for direct integration of web services, however, it is an important component to search, identify and details the specification of messages and procedures in conjunction with SOAP.

Table 3. A list of API examples and illustrated mash-ups.

Category	API example	Mash-up URL
Messaging	Twitter	http://beta.twittervision.com
Music	Last.fm	http://lastmusicmap.com
News	Digg	http://flutracker.rhizalabs.com
Photos	Flickr	http://www.earthalbum.com
Search	Google	http://www.2realestateauctions.com
Shopping	Amazon	http://www.buyitnearby.com
Social	Facebook	http://apps.facebook.com/mapmate
Travel	Freebase	http://city-guide.freehostia.com
Video	Youtube	http://www.virtualvideomap.com
Others	Zillow	http://www.diversesolutions.com/zillow-neighborhood-demo

For example, the GIS Department in the city of Orem, Utah, USA hosts two web services and their corresponding WSDLs (City of Orem 2009). An abstract of the web service to address match a location is demonstrated in Figure 7. The conforming request must include a street address as text and an offset distance in feet (i.e. numerical value). In return, the web service will perform address matching by using the in-house GIS data and return a custom data type consisted of a point expressed in *x*- and *y*-coordinate defined in the corresponding coordinate system. It is also possible to create a web service to identify the polygon (e.g. parcel, school district) in which an address-matched location lies within or other GIS operations.

2.3.2 *Aggregators*

With the wide range of web content increasingly available, it is becoming an extremely trivial task to look for updates from various content providers, sort out relevant subjects, convert them into a compatible data format and centralize the information into a single location for displaying. An aggregator is a web-based or desktop program that performs such tasks and collects syndicated web content through subscribed RSS feeds. It is often used to syndicate news headlines, blog updates, podcasts and multimedia distribution. Some aggregator programs apply sophisticated computing algorithm(s) to filter and rank the content based on customized user behaviors and analyze usage trend, such as Google Reader.

Instead of receiving updates through passive subscription of existing RSS feeds, there are aggregator programs that will search for web content and actively extract useful information. Such data mining procedure on web content is known as "screen-scraping". Typically, the output of such operation(s) can be converted into a specified web data format (e.g. XML-formatted data, JSON, tab-delimited values, etc), which provides full or partial content

```
… … …
  <types>
    <xs:schema targetNamespace="urn:OremGeocoderIntf">
    <xs:complexType name="TNAD27Point">
      <xs:sequence>
        <xs:element name="X" type="xs:double"/>
        <xs:element name="Y" type="xs:double"/>
      </xs:sequence>
    </xs:complexType>
  </xs:schema>
  </types>
  <message name="GeocodeOremAddress0Request">
    <part name="sAddress" type="xs:string"/>
    <part name="FeetOffset" type="xs:double"/>
  </message>
  <message name="GeocodeOremAddress0Response">
    <part name="return" type="ns1:TNAD27Point"/>
  </message>
  <portType name="IOremGeocoder">
    <operation name="GeocodeOremAddress">
      <input message="tns:GeocodeOremAddress0Request"/>
      <output message="tns:GeocodeOremAddress0Response"/>
    </operation>
  </portType>
… … …
```

Figure 7. A WSDL example to request a SOAP web service (Courtesy of City of Orem 2009).

for many APIs. Dapper (http://www.dapper.net/open) is a useful web-based application to search, create, and customize syndicated web content. Through online sharing and massive collaboration, a web developer can reuse and link multiple screen-scraping services to blend in multiple data sources and create powerful mash-ups by mixing them in a mash-up integration tool.

2.3.3 *Mash-up integration tools*

As discussed from the previous section, API opens up the gateway of programming libraries and provides valuable data and/or functionalities over the Internet. Despite the fact that most heavy-duty programming burden has been shared by the APIs, conventional mash-up development still requires good understanding of the server-client communication, communication protocols and some programming skills. Mash-up integration tools further lessen the burden on web developers and provide GUI in a friendly environment to mash up web content. Moreover, users can develop, test, debug, preview, and deploy mash-ups over a web-based platform, such as Yahoo! Pipes and JackBe.

In general, the mash-up integration tools contain a panel of operations (e.g. pipes in Yahoo! Pipes) that can perform discrete simple tasks, such as fetching a RSS from a given URL, filtering the results by a keyword, extracting the data by tag names, etc. These tasks are equivalent to a method or function in a conventional programming language or API. Users can drag and drop these operations from the list onto a flat panel to define the workflow (Fig. 8). By specifying the input parameters, passing the outputs and linking them into the next operations, the users can construct a data flow diagram similar to the one used in a Unified Modeling Language (UML). Thus, the mash-up integration tools enable the users with minimal or no programming background to create dynamic mash-up applications. The following section illustrates a case study to create a mash-up by utilizing some mash-up tools.

2.4 *Case study: H1N1 flu map*

In early 2009, there has been an outbreak of Influenza A, a new strain of virus also known as "H1N1 flu" or "swine flu". Despite the unknown origin in place and species, the early cases were first detected in Mexico and United States. Through close contacts with infected

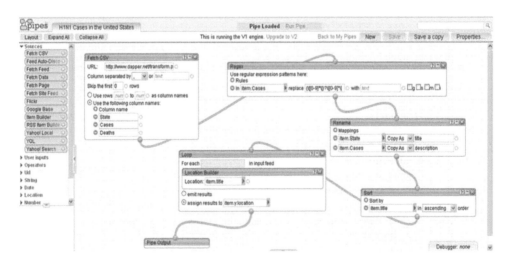

Figure 8. A screen shot of the Yahoo! Pipes interface and an illustration of the workflow (courtesy of Yahoo! 2009).

people such as coughing, sneezing, and touching, the outbreak became an epidemic with thousands of cases reported worldwide. One month after the first confirmed case, the World Health Organization (WHO) announced a phase 5 alert level, indicating that the spread has become human-to-human in multiple countries and a pandemic is imminent. Government officials, scientists, and public media were racing against time to track down the spread of this virus to better understand its epidemiology for mitigation planning. Averaging at least hundreds of confirmed cases per day around the globe, it has been a challenging task to update the latest epidemic development and communicate the information to the general public.

To illustrate the effectiveness of mash-up technologies, a mash-up has been created to update the status of H1N1 flu in United States onto a map. As the H1N1 epidemic became widespread in the U.S., the Centers for Disease Control and Prevention (CDC) had stopped reporting the number of confirmed H1N1 cases found in each state. Thus, the unofficial tabulated summary of reported cases by state was used (Wikipedia 2009a). The technique of screen-scraping was adopted to create a customized web feed by using the Dapp Factory. By analyzing the HyperText Markup Language (HTML) code and other web content on the H1N1 summary page, the dapper engine categorizes similar HTML elements (e.g. rows of table cells) and allow user to define the content to be "scraped off" from. Each set of selected web content would be grouped into a separate item and can be transformed into heterogeneous data formats, such as XML, RSS, etc. The output web feed can be accessed by a URI and the web-based Yahoo! Pipes was used to bind the web feed onto an Internet map. The parsed web feed underwent some text formatting to extract the state names, confirmed cases and deaths. The output strings will then be fed into the address matching web service to return the latitude and longitude of each state. Based on the georeferenced location, pushpins were added onto the map by combining the H1N1 flu information (Fig. 9). Figure 10 demonstrate the data flow diagram of the automated process in creating this mash-up application.

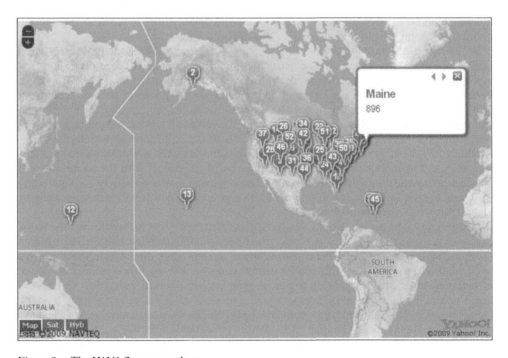

Figure 9.   The H1N1 flu map mash-up.

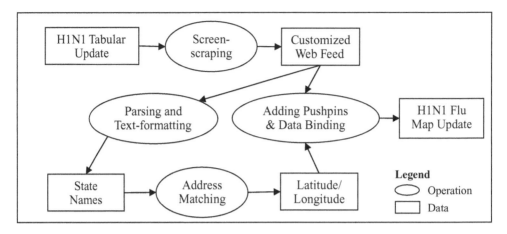

Figure 10.   The data flow diagram of the H1N1 flu map mash-up.

3   FRONTIERS IN GEOGRAPHY 2.0

3.1   *PPGIS, neogeography and GIScience*

The extravaganza of Geography 2.0 and the promotion of geospatial technologies (e.g. Google Earth, Maps API) proliferate the principles of Public Participation GIS (PPGIS)—a formal term coined at a special meeting on Empowerment, Marginalization and PPGIS hosted by National Center for Geographic Information and Analysis (NCGIA) in 1998. Originated from the research initiative for "GIS and Society" by the University Consortium for Geographic Information Science (UCGIS), PPGIS was concerned with the societal impacts of GIS on the general public. PPGIS covers a wide range of topics to investigate how GIS technologies empower and/or marginalize individuals and communities in terms of their social, political, economical, and technical conditions (Sieber 2006).

In the report submitted at the NCGIA meeting, Kingston (1998) argued that web-based GIS applications can engage public participation in environmental decision making by a progressive public participation ladder through their rights to 1) know, 2) be informed, 3) object, 4) define interests, actors, and determine agenda, 5) assess risk and recommend solutions, and 6) make final decision. In fact, numerous online projects, like the master plan collaboration mash-up (Rinner et al. 2008)—which involved all tasks in the public participation ladder but the final decision—are a realization of the bottom-up approach of PPGIS. Facilitated by the improved web infrastructure, the general public was more aware of, attentive, able, and assiduous in using innovative mash-ups to interact with the social web and physical environment more than ever before. The enhanced interconnectivity and interactivity extend the research agenda enlisted by PPGIS and open up new frontiers in neogeography and GIScience as discussed below.

Neogeography, or new geography, is a loosely-defined expression that represents the rediscovery of geospatial tools to aid in "conveying understanding through knowledge of place" (Turner 2006). Whether or not the true meaning of such practice differs from the objectives of conventional geography is still open for debate; neogeography implies extending the power of mapping and spatial exploration to non-geographers. In political geography, Shin (2009) urged rethinking the concept of democracy as a result of electoral neogeography. In 2008, the proposition 8 in California, a state ballot that prohibits same-sex marriage, spawned the creation of mapping mash-ups that reveal donors information (e.g. name, address, occupation, employer). Shin (2009) argued that the information distribution and abuse possibly "facilitated vandalism, harassment, in the form of death threats and hate mail, and spurred the boycott of local businesses". Spiro (2009) solicited the spatial distribution of fast food restaurants, which can be used to study the relationship these restaurants have with obesity rate among other socioeconomic factors and health status of individuals. Boulos et al. (2008) demonstrated a playlist of 2D and 3D showcases to inspire the use of mash-up technologies in conjunction with virtual reality software and semantic web for public health neogeographers. As online cartography has been made easy and quick in the Geography 2.0 era, neogeography fosters enormous demand for map production and consumption, as well as rethinking the ontology of cartography in general (Kitchin and Dodge 2007).

Regardless of the disciplinary nature, one of the natural consequences of neogeographic application is the flooding of Volunteered Geographic Information (VGI, Goodchild 2007a). For example, shortly after the California wildfire and H1N1 flu outbreak, mash-ups were created that utilized Maps API and VGI to compile grassroots efforts into map databases (e.g. Nimen, 2009). The spatial information acquired and compiled by these websites present a unique resource for scientists to study the diffusion of large-scale phenomenon. Similarly, the VGI of gas price (http://www.gasbuddy.com) contains a wealth of spatio-temporal data to examine the hypothesis of gas price disparity (e.g. decreasing gas price with increasing distance from highway). While one may share the skeptical view of Jack Dangermond regarding the accuracy and reliability of VGI (Hall, 2007), collaborative projects like OpenStreetMap (OSM), which intend to provide free editable world map, would embrace free data and valuable insights. In referring to this wave of "Maps 2.0", Crampton (2009) discussed other dimensions of neogeography in crowdsourcing, net neutrality, digital divide and their implications on cartography as a profession.

Echoing with the OSM slogan of upholding "free wiki-world map", Sui (2008) described the bottom-up approach of VGI as the "wikification of GIScience". While the analogy sufficiently describes the massive contribution of amateur-controlled web content through interactive platforms, the strengths and weaknesses of wiki-projects are not directly parallel to the Geography 2.0 phenomenon. In fact, it is interesting to reflect on the potential pitfalls and recommended remedies for Geography 2.0 from wiki-projects. One important component that is lacking in Geography 2.0 is the enforcement of quality control and quality assurance (QC/QA). In most wiki-projects, in addition to contributors, there is also a "peer review" process that ensures web content is well-written, comprehensive, well-researched, neutral, and stable on the subject (Wikipedia 2009b). Notwithstanding the imperfection of Wikipedia's reviewing process, such QC/QA standards can be applied on mash-up applications in terms of depth, scope, precision, accuracy, neutrality, stability, and security, etc.

By utilizing the programming library of a third party web service/API, however, the hosting web server in essence has less sovereignty over its published content or functionalities by "outsourcing" the tasks. Thus, the quality of derived information may be compromised through error propagation after series of data serialization, transformation, processing, and intermingling in the procedure of mashing up several web services/APIs from multiple sources. For example, the infamous "GoogleBot" algorithm, the news searching and ranking automation, mistakenly posted the bankruptcy of United Airline from 6 years ago as the headline in September 2008. The rumor caused panic reaction among the stockholders and the stock of troubled airline dropped 75% and did not fully recover (McCarthy 2008). Similarly, a security loophole on the hosting web server may exist especially if the mash-up requires interchanging or parsing JSON-formatted data from an untrusted source. Moreover, APIs do not have self-explanatory metadata structure like WSDL for SOAP-based web services. Few API providers supply detailed documentation, code samples, and other technical resources needed for a diverse background of mash-up developers. There is neither W3C official standard for mash-up specification nor central depository to keep track of all web services/APIs available for mash-up development. Nevertheless, the emergence of mash-up integration tools is a good start to compile, search, evaluate and examine the performance of web services/APIs in a structured Integrated Development Environment (IDE).

On the other hand, web data can benefit from semantic enhancements. In general, the XML-formatted data use XML application schema to validate the legitimate structure on top of the verification in a well-formed XML syntax. However, these measures fall short of capturing the full connotation of web content and its relevance to a specific application in a given context. Due to its simplicity, JSON-formatted data face the same scrutiny and do not have an "application schema" to check for invalid data structure. Due to this inefficiency, mash-up cannot understand the meaning of web content and hence the user requests. Semantic web is the vision to create machine-understandable data (instead of merely interpreting HTML code) so that computers can understand the semantics of web data and their interrelationship in a human context. Many regard the semantic web as the future version of Web 3.0—a topic to be discussed in the next section.

3.2 *Semantic web and the future web development*

The first generation of World Wide Web (WWW), as well as the improvised version of what many consider Web 2.0 nowadays, is a web of documents linking to each other with separate references of data (e.g. image, GML, etc) residing in each document. To the web browser, a hyperlink to a curriculum vitae is no different than a hyperlink to a recipe. The current web infrastructure does not enable the understanding that the curriculum vitae should be uniquely associated with a specific person, nor the linkage between a recipe and a dish. The semantic web, a vision shared by Tim Berners-Lee and others as a major component of the future WWW, is intended to be a web of data "in which information is given well-defined meaning, better enabling computers and people to work in cooperation" (Berners-Lee et al. 2001). Quoting from a popular example that illustrates the limitation of existing web documents, keyword searches for "Paris Hilton" would return all web documents containing information related to a celebrity the same as those referencing to a hotel brand at a location. By precisely defining the vocabulary, the syntax rules in the appropriate context, semantic web is about adding meaning (i.e. semantics) to the web content and linking them together. In essence, semantic web has a noble goal to transform all web content, visible or invisible for human when browsing a web site, into a machine-readable database. It is important to establish a mechanism to define the relationship about the data and the set of inference rules to automate meaningful query (from human perspective).

W3C, which shares Berners-Lee's view, perceives the semantic web as an interoperable framework to share and reuse web data across applications. As part of the W3C's semantic web activity, Resource Description Framework (RDF) is a model designed for use by computers to describe how information on the Internet is related. An entity in a relationship (e.g. a person) is regarded as a resource and is uniquely identified by using a URI. A RDF statement adopts the encoding of a triple set to link semantics by using a resource, its property and the corresponding property value to represent the subject, predicate, and the object in a statement. A simple statement can be "the capital of United States is Washington D.C.", in which "United States" is the subject; "capital" is the predicate, and "Washington D.C." is the object. RDF/XML is a XML-based specification, among other formats of RDF, which implements the semantics of web data. Web Ontology Language (OWL) is another building block of semantic web that enriches the "vocabulary" in describing the relationship of resources and their properties. To enhance the machine interpretability of web content, OWL can be used to construct taxonomy, including classes of objects, their properties, as well as subclasses. The defined taxonomical relationship supplies the inference rules needed for processing web content. Similar to the development of a specific XML schema for geographic features (i.e. GML), it is necessary to construct an ontological structure (i.e. schema) for geographic information, such as the GeoNames Ontology (GeoNames 2009). The GeoNames ontology is still in its infancy and hence it is very lightweight in the context of semantic web development. It is also possible to use OWL to represent more sophisticated ontology for GIS-based environmental modeling (Fallahi et al. 2008). Given the explicit meaning and relationship of web content, it is possible to construct effective queries in the semantic web by using SPARQL, a recursive acronym for SPARQL Protocol and RDF Query Language.

Since the birth of WWW, the web community has been slow to adopt the semantic web initiatives in spite of the early introduction of the semantic web concept in the 90's.

The W3C bottom-up approach of re-annotating the web document into a structured format is a massive undertaking to many web masters. There are some existing tools, such as Calais (http://opencalais.com), that can generate automatic semantic metadata to any published web content. On the other hand, some semantic web proponents suggest an alternative top-down approach to devise smarter agents that can better understand and utilize the existing web structure with social networking tools, such as Glue from Adaptive-Blue (http://www.adaptiveblue.com). Although the browser add-on is far from the perfect semantic web assistant, it can be viewed as an application of collective knowledge system (Gruber 2007). Skeptics of semantic web suggest that the resistance may be attributed to the incompetence of many web masters to adopt the semantic web technology, egocentrism from commercial providers to comply with standards, and the deceptive nature of human in manipulating the semantic web system (Lombardi 2006). Nevertheless, many search engines have adopted the semantic technologies to improve their search engines (e.g. Hakia) to better organize structured data into useful information for more effective decision making.

The semantic web can be useful in developing "smarter" mash-ups. Despite the fact that mash-up application has already been realized using the existing web platform, the burden of identifying appropriate content and its update largely falls onto the mash-up developers. Semantic technologies laid down the infrastructure needed to allow automatic retrieval of related data across application boundaries. In solving environmental problems, such as "What (and where) are the sources of contamination found in a river?", it is essential to understand the context of the problems and access relevant data and appropriate operations needed for the task. Semantic technologies hold the promise of enabling semantic plug and play—an integrated framework in which data are processed in an interoperable environment based on a defined conceptual model. Examples of possible implementation of semantic technologies include the development of semantic data model (Mennis 2003) and ontology-driven data integration (Buccella et al. 2009). As such, semantic technologies are essential for the development of a Spatial Decision Support System (SDSS) that can provide valuable information for decision making.

While there is little doubt that the semantic web is important in shaping the evolving web technology, the future web is not necessarily the synonym of the semantic web. The direction for future web development is nascent and remains diverse. The list below describes some alternative visions (Metz 2007):

- 3D Web—3D visualization of the web content in distributed applications like Google Earth or NASA World Wind. Combining virtual reality with the collaborative roleplaying element in gaming, one may explore the 3D digital landscape with first-person experience or a digital representation of oneself as an "avatar" (Buolos et al. 2008).
- Media-centric Web—A data-rich web protocol that includes non-textual data as well, including images, video, audio, etc. Conventional query will be complemented by pattern recognition algorithms (e.g. image classification, voice segmentation) to search media data by using media data.
- Pervasive Web (or Augmented Web)—a web infrastructure that can be accessed by numerous computing devices so that users can always stay connected. On top of ordinary devises like computers, cell phone, and Personal Digital Assistant (PDA), the pervasive web is enabled by intelligent software agent through many unconventional devices such as window, mirror, sticky notes etc (Mistry & Maes 2008).

Given the long tradition of the GIScience community in geographic exploration, scientific visualization, spatial analysis, and digital cartography, 3D Web seems to be a natural arena for future development of Geography 2.0. Many researchers embrace the valuable visualization tools to model geographic phenomenon in 3D space and develop up-to-date Internet GIS applications (Nourbakhsh et al. 2006, Pearce et al. 2006). Buolos et al. (2008) demonstrated a mash-up prototype in Second Life to illustrate a seamless integration of 3D virtual reality known as the "metaverse". However, it is also possible that future development of Geography 2.0 is not confined to any particular version, but a hybrid. For example,

Cheung et al. (2008) proposed the extension of Web 2.0 technologies, including mash-up and 3D visualization tools, with the semantic web in developing the future version in the Health Care and Life Science (HCLS) domain. Lee et al. (2009) envisioned an augmented-reality system interoperable to a variety of media data to support location-based decision making.

Schmidt (2007) suggested that the future generation of web applications will be running not only on an interoperable platform and have access to a remote data-cloud, but will also be small, fast, and highly customizable to be distributed virally through social networks or emails. While there may be infinite possibilities of the prospect of future web development, the mash-up technologies seem to be a viable approach to mingle the aforementioned web technologies and remain flexible and scalable. In the context of Geography 2.0, the mash-up approach will be the fundamental building block to build up a true Spatial Decision Support System (SDSS) based on VGI distributed on the web. No matter how the future development of Geography 2.0 unfolds, Internet GIS are heading towards the direction where spatial models will utilize VGI in an open-source, open-standard framework and will communicate the result in real-time with the audience through distributed GIServices in an interoperable environment.

ACKNOWLEDGEMENT

The author would like to thank Dhruba J. Baishya for his inspiring work on a collaborative mash-up project. The author is in debt to Richard Sadler and Dr. Niem Huynh for proof-reading this paper. The constructive comments from Derek Chan and the two anonymous reviewers improved the draft version of the manuscript.

REFERENCES

Berners-Lee, T. (2006) Interview *Developer Works Interviews: Tim Berners-Lee*. By Scott Lanningham, IBM developerWorks. [Online] 22nd August 2006. Available from: <http://www.ibm.com/developerworks/podcast/dwi/cm-int082206txt.html> [Accessed 30th January 2009].

Berners-Lee, T., Hendler, J. & Lassila, O. (2001) The Semantic Web. *Scientific American Magazine* [Online] Available from: < http://www.scientificamerican.com/article.cfm?id=the-semantic-web > [Accessed 26th May 2009].

Buccella, A., Cechich, A. & Fillottrani, P. (2009) Ontology-driven geographic information integration: A survey of current approaches, *Computers & Geosciences*, 35 (4), 710–723.

Buolos, M.N.K., Scotch, M., Cheung, K.-H. & Burden, D. (2008) Web GIS in Practice VI: a demo playlist of geo-mash-ups for public health neogeographers. *International Journal of Health Geographics*, [Online] 7 (38) Available from: doi:10.1186/1476–072X-7-38.

Cheung, K.-H., Yip, K.Y., Townsend, J.P. & Scotch, M. (2008) HCLS 2.0/3.0: Health care and life sciences data mash-up using Web 2.0/3.0. *Journal of Biomedical Informatics*, 41, 694–705.

Chow, T.E. (2008) The potential of Maps APIs for Internet GIS. *Transactions in GIS*, 12 (2), 179–191.

Christensen, E., Curbera, F., Meredith, G. & Weerawarana, S. (2001) W3C: Web Services Description Language (WSDL) 1.1 [Online] Available from: < http://www.w3.org/TR/wsdl > [Accessed 11th May 2009].

City of Orem, Utah, (2009) GIS Web Services [Online] Available from: < http://gis.orem.org/webservices.html > [Accessed 6th May 2009].

Clarke, K.C. (2003) Geocomputation's future at the extremes: High performance computing and na-clients. *Parallel Computing*, 29 (10), 1281–1295.

Crampton, J. (2009) Cartography: maps 2.0. *Progress in Human Geography*, 33 (1), 91–100.

Crockford, D. (2006) JSON: The fat-free alternative to XML. [Online] Available from: < http://www.json.org/fatfree.html > [Accessed 14th May 2009].

ESRI, (1998) ESRI Shapefile Technical Description: An ESRI White Paper – July 1998. [Online] Available from: < http://www.esri.com/library/whitepapers/pdfs/shapefile.pdf > [Accessed 14th May 2009].

Fallahi, G.R., Frank, A.U., Mesgari, M.S. & Rajabifard, A. (2008) An ontological structure for semantic interoperability of GIS and environmental modeling. *International Journal of Applied Earth Observation and Geoinformation*, 10, 342–357.

Fensel, D. & Bussler, C. (2002) The web service modeling framework WSMF. *Electronic Commerce Research and Applications*, 1, 113–137.

Fielding, R. (2000) Architectural styles and the design of network-based software architectures, Ph.D. dissertation, University of California, Irvine.

Fisher, H.T. (1966) SYMAP. In: *Selected Projects: 1966–1970*, Laboratory for Computer Graphics and Spatial Analysis, Harvard Graduate School of Design, Cambridge, MA.

GeoNames, (2009) GeoNames Ontology [Online] Available from: < http://www.geonames.org/ontology > [Accessed 28th May 2009].

Goodchild, M.F. (2007a). Citizens as voluntary sensors: Spatial data infrastructure in the world of Web 2.0. *International Journal of Spatial Data Infrastructures Research*, 2, 24–32.

Goodchild, M.F. (2007b) Citizens as sensors: The world of volunteered geography. *Geo. Journal*, 69 (4), 211–221.

Gruber, T. (2007) Collective knowledge system: where the Social Web meets the Semantic Web. *Journal of Web Semantics*, 6, 4–13.

Hall, M. (2007) On the mark: will democracy vote the experts off the GIS island? ComputerWorld [Online] Available from: < http://www.computerworld.com/action/article.do?command=viewArticleBasic&articleId=299936 > [Accessed 25th May 2009].

Hendler, J. & Golbeck, (2008) Metcalfe's Law, Web 2.0, and the Semantic Web, *Web Semantics: Science, Services and Agents on the World Wide Web*, 6 (1), 14–20.

Kingston, R. (1998) *Web based GIS for public participation decision making in UK*. Papers submitted for the NCGIA Specialist Meeting on Empowerment, Marginalization, and Public Participation GIS, [Online] Santa Barbara, California. 14–17th October 1998. < Available from: http://www.ncgia.ucsb.edu/varenius/ppgis/papers/kingston/kingston.html > [Accessed 23rd May 2009].

Kitchin, R. & Dodge, M. (2007) Rethinking maps. *Progress in Human Geography*, 31 (3), 331–344.

Lee, R., Kitayama, D., Kwon, Y.-J. & Sumiya, K. (2009) Interoperable augmented web browsing for exploring virtual media in real space, In: Wilde, E., Boll, S., Cheverst, K., Fröhlich, P., Purves, R. & Schöning J. (Ed.), *Proceedings of the Second International Workshop on Location and the Web* pp. 20–23.

Lombardi, C. (2006) Google exec challenges Berners-Lee, *CNET News Digital Media* [Online] Available from: < http://news.cnet.com/Google-exec-challenges-Berners-Lee/2100–1025_3–6095705.html > [Accessed 28th May 2009].

Los Angeles Times, (2009) Southern California Wildfires [Online] Available from: < http://www.latimes.com/news/local/la-me-regionfires-map,0,2173230.htmlstory > [Accessed 13th May 2009].

Luo, W. (2009) Imagery of Sichuan, China earthquake [Online] Available from: < http://google-latlong.blogspot.com/2008/05/imagery-for-sichuan-china-earthquake.html > [Accessed 13th May 2009].

Mapstraction, (2009) Mapstraction Provider Feature Matrix: JavaScript mapping abstraction library. [Online] Available from: < http://www.mapstraction.com/features.php > [Accessed 25th May 2009].

McCarthy, C. (2008) September 8. Google News snafu leads to airline stock plunge, *CNET News Digital Media*. [Online] Available from: < http://news.cnet.com/8301–1023_3–10036131–93.html > [Accessed 25th May 2009].

Metz, C. (2007) Web 3.0, *PC Magazine* [Online] Available from: < http://www.pcmag.com/article2/0,2817,2102852,00.asp > [Accessed 23rd May 2009].

Mennis, J. (2003) Derivation and implementation of a semantic GIS data model informed by principles of cognition, *Computers, Environment and Urban Systems*, 27 (5), 455–479.

Mielke, L. & Burger H. (2007) Thematic web-mapping potential and limitations of Google Maps API. *Proceedings of 12th Conference of International Association for Mathematical Geology*, Beijing, China. pp. 715–718.

Mistry, P. & Maes. P. (2008) Intelligent Sticky Notes that can be Searched, Located and can Send Reminders and Messages. *In the Proceedings of the ACM International Conference on Intelligent User Interfaces (IUI2008)*. Canary Islands, Spain.

Nimen, H. (2009) *FluTracker: Tracking the progress of H1 N1 swine flu* [Online] Available from: < http://flutracker.rhizalabs.com > [Accessed 13th May 2009].

Nong, Y., Wang, K., Miao, L. & Chen F. (2007) Using GeoRSS feeds to distribute house renting and selling information based on Google map. Geoinformatics 2007: Geospatial Information Technology and Applications. *Proceedings of SPIE*, 6754, p. 675422.

35

Nourbakhsh, I., Sargent, R., Wright, A., Cramer, K., McClendon, B. & Jones M. (2006) Mapping disaster zones, *Nature*, 439, 787–788.

OpenLayers, (2009). OpenLayers: Free Maps for the Web [Online] Available from: < http://openlayers.org > [Accessed 21st May 2009].

Open Geospatial Consortium, (2009) KML [Online] Available from: < http://www.opengeospatial.org/ standards/kml > [Accessed 12th May 2009].

OpenStreetMap, (2009). *The free wiki world map.* [Online] Available from: http://www.openstreetmap. org> [Accessed 9th November 2009].

O'Reilly, T. (2007) *What is Web 2.0: Design Patterns and Business Models for the Next Generation of Software. Communications and Strategies* 65(1st Q), pp. 17–37.

Pearce, J.M., Johnson, S.J. & Grant G.B. (2007) 3D-mapping optimization of embodied energy of transportation. *Resources. Conservation and Recycling*, 51 (2), pp. 435–453.

ProgrammableWeb, (2009) ProgrammableWeb: Mash-up Timeline [Online] Available from: <http:// www.programmableweb.com> [Accessed 1st June 2009].

Rinner, C., Keßler, C. & Andrulis, S. (2008) The Use of Web 2.0 Concepts to Support Deliberation in Spatial Decision-Making. *Computers, Environment and Urban System,* 32 (5), 386–395.

Schmidt, E. (2007) Speech at Seoul Digital Forum, Seoul, South Korea, 31st May 2007. [Online] Available from: < http://www.youtube.com/watch?v=T0QJmmdw3b0 > [Accessed 28th May 2009].

Sieber, R. 2006. Public participation and Geographic Information Systems: a literature review and framework. *Annals of the American Association of Geographers*, 96 (3), pp. 491–507.

Shin, M.E. (2009) Democratizing electoral geography: Visualizing votes and political neogeography. *Political Geography*, Available from: doi:10.1016/j.polgeo.2009.03.001.

Spiro, I. (2009) Personal communication. 13th November 2009.

Sui, D.Z. (2008) The wikification of GIS and its consequences: Or Angelina Jolie's new tattoo and the future of GIS. *Computers, Environment and Urban System*, 32, pp. 1–5.

Troy, D. (2009a) Twittervision. [Online] Available from: < http://twittervision.com >, [Accessed 5th May 2009].

Troy, D. (2009b) Flickrvision. [Online] Available from: < http://flickrvision.com >, [Accessed 5th May 2009].

Turner, A. (2006). *Introduction to Neogeography*. O'Reilly Media, Inc., p. 54.

Twitter, (2009) Twitter API Documentation [Online] Available from: < http://apiwiki.twitter.com/ Twitter-API-Documentation > [Accessed 5th May 2009].

USGS, (2009). Latest Earthquakes: Feed & Data [Online] Available from: < http://earthquake.usgs.gov/ eqcenter/catalogs > [Accessed 14th May 2009].

Wikipedia, (2009a) Wikipedia: 2009 flu pandemic in the United States [Online] Available from: < http:// en.wikipedia.org/wiki/2009_flu_pandemic_in_the_United_States > [Accessed 10th November 2009].

Wikipedia, (2009b) Wikipedia: Featured article criteria [Online] Available from: < http://en.wikipedia. org/wiki/Wikipedia: What_is_a_featured_article%3F > [Accessed 18th April 2009].

*Advances in Web-based GIS, Mapping Services
and Applications – Li, Dragićević & Veenendaal (eds)*
© *2011 Taylor & Francis Group, London, ISBN 978-0-415-80483-7*

GeoClustering: A web service for geospatial clustering

Jing Wang, Xin Wang & Steve H.L. Liang
Department of Geomatics Engineering, University of Calgary, Calgary, Alberta, Canada

ABSTRACT: Geospatial clustering is a very important research topic in the field of "Geospatial Knowledge Discovery." Its main purpose is to group similar objects, from large geospatial datasets, into clusters, based on the objects' spatial and non-spatial attributes. This capability is very useful for better understanding the patterns and distributions of geographical phenomena. With the emerging availability of large amounts of geospatial data on the Internet, the demand for powerful methods of extracting meaningful information from large amounts of web-based geospatial data is thus also rapidly increasing. In this chapter, we present a geospatial clustering web service called GeoClustering. With this service, users are able to cluster online data sources easily and to then conveniently visualize the clustering results through a friendly map-centric user interface. Since it is web-based, GeoClustering offers a platform that can be accessed by anyone from anywhere at any time. To achieve better interoperability, GeoClustering offers an API and follows the OGC standard data format for spatial data exchange.

Keywords: Geospatial clustering, web service, web GIS, spatial data mining, clustering

1 INTRODUCTION

Geographic information science has entered an age of "data-rich and information-poor." Unprecedented amounts of geospatial data are gathered quickly and easily through various methods, such as sensors, GPS receivers and cell phones (Miller 2008). Especially on the World Wide Web, the volume and scope of geospatial data is growing rapidly. People need to access the information contained in the data as opposed to the data itself. Consequently they are usually overwhelmed by and disinterested in large collections of raw data since these are of not much direct use to them. For example, online stores are interested in identifying the geographical distribution of their customers or residents may be interested in the crime patterns in their communities.

The strong need for actual knowledge, rather than for just the raw data from which the knowledge was derived, explains the development and emergence of Geographic Knowledge Discovery (GKD). GKD is the process of extracting information and knowledge from massive amounts of geo-referenced data. Due to the complex nature of the underlying data, such geographic information is often novel and useful in that it reveals unexpected patterns and trends (Miller 2008). Among major GKD techniques, geospatial cluster analysis is an important and very useful method. It is the process of grouping similar objects based on their distance, connectivity, or relative density in space (Han et al. 2001). Clusters create an abstract representation of the original data, which merge similar points within the same groups, based on location as well as on other non-spatial attributes. Some clustering methods can also identify noise, which are typically unimportant to the study, within the data. In addition, clustering methods can discern interesting spatial patterns and features; capture intrinsic relationships between spatial and non-spatial data; and present data regularity concisely and at higher conceptual levels (Ng & Han 2002). Geospatial clustering has been commonly used in disease surveillance, spatial epidemiology, population genetics, landscape ecology, crime analysis and many other fields (Jacquez 2008).

With the emergence of vast amounts of diverse and distributed geospatial data on the web, the demand for effective geospatial clustering methods is increasing significantly. Although, due to the importance of clustering, various clustering tools have been proposed and developed, a key limitation is that most of these have not been developed for use specifically in online geospatial clustering. Limitations of existing clustering tools are summarized as follows.

First, most clustering tools are not designed for web-based geospatial clustering tasks and also lack visualization capability. Most existing clustering tools are desktop-based software or packages, such as SaTScan (Kulldorff 2009) or ClusterSeer (ClusterSeer 2009). These are effective for dealing with data available at local machines but cannot handle online data sources. For example, Flicker (www.flickr.com) is a website for photo sharing. If users need to find spatial clusters of geo-tagged photos published by other Flickr users, these traditional software cannot handle the request directly. Rather, users need to download and transform the data before performing the clustering. In addition, the clustering results provided by these tools are mainly in text form. Users cannot visualize the clustering results.

Second, most of the tools are not designed with open and interoperable principles in mind. Openness means that the system can be widely accessed through open interfaces. Current clustering applications can only be accessed by a limited number of local users. However, web users need a service that can be accessed at anytime from anywhere.

Interoperability is the ability of a system, or components of a system, to provide information portability and inter-application cooperative process control (Bishr 1998). Current clustering functions are usually packaged as part of a specific and proprietary system. It is difficult to be utilized by other applications without knowing the system API (Application Programming Interface) specifications. The input and output files are usually proprietary and defined by the system. Users cannot choose to use common spatial data standards or to use them as third-party clustering service components. Consequently, exchanging data between different systems or reusing the service is difficult, if not impossible.

Third, existing online clustering APIs are not designed for general users, i.e. they are not easy to use. Some clustering function APIs are available on the Internet. For example, Microsoft Virtual Earth v6.2 Map Control API offers built-in clustering functions (Microsoft 2009). Google Maps Clustering API Project (Google Maps API Projects 2009) and ACME cluster JavaScript library (ACME Labs 2009) add clustering functions to Google Maps. However, these are designed for developers, not for general users. Thus, in order to implement their own clustering assignment, general users need to learn specific programming languages and API usages.

Fourth, in terms of clustering functions, most web-based map applications such as ClustrMap (ClustrMap 2009) provide embedded clustering functions only for cartographic generalization and do not focus on clustering patterns. Grid-based clustering is widely used to combine neighbor points into one single cluster to reduce the total number of symbols being displayed in the current map view. These tools are not concerned with patterns or with showing patterns on the map.

In order to solve the problems mentioned above, we propose and develop a new web service called GeoClustering, which offers a web-based spatial data clustering service for general users. With this service, users can cluster their data online and visualize the clustering result on the map.

As an online service, GeoClustering can handle distributed data resources on the Internet. GeoClustering is an independent and reusable web resource that allows Internet users to invoke geospatial knowledge discovery from anywhere at any time. It can be considered as one of the basic geographic processing function modules available. It is centered on the concepts of "loose coupling and reuse," and is designed as a component that can be readily embedded into complex web-based geographic information systems.

In order to enable interoperability between heterogeneous data sources on the Internet, International organizations such as the World Wide Web Consortium (W3C), the International Organization for Standardization (ISO) and the Open Geospatial Consortium (OGC), have been making significant efforts to define data exchange standards and protocols. With the proliferation and implementation of various ISO and OGC standards, spatial data in the form of standard XML (extensible markup language) format has become more and more popular. GeoClustering supports XML files drawn directly from different websites and can also save clustering results in the same general format. With the API, the GeoClustering platform can offer clustering services to any website or online application.

GeoClustering has the function showing clustering patterns on the map by using online map services API. The current GeoClustering prototype offers a user friendly GUI (Graphic User Interface), through which users can easily perform online clustering, and have an intuitive result, even without any knowledge of programming.

2 RELATED WORK

2.1 Spatial clustering

In this section we briefly introduce the distance function, which defines the similarity between spatial objects. Then we present current research in the area of spatial clustering methods.

2.1.1 Distance functions in spatial clustering

The "distance function" is the key component of any clustering method in that it measures the similarity among spatial objects. Distance is a numerical description of how similar two objects are in space. According to Tobler's First Law (Tobler 1970), geometric distance is usually used as the scale of similarity in the "ideal model." However, sometimes the non-spatial attributes (alphanumeric attributes) of objects are also incorporated into the distance function.

The character of geometric distance is that it is defined by exact mathematical formulas that reflect the physical length between two objects in defined coordinate systems, such as Euclidean Distance, Manhattan Distance, Great Circle Distance, etc. Spatial clustering methods do not always use geometric distance; for example, if the distance can be defined as the shortest traveling time between two different addresses in a city. In this case, the distance function should take into account road networks, speed limits, volume of traffic, number of traffic lights, and stop signs. In fact, the distance function is always tailored to different clustering purposes.

Spatial objects may have significantly different non-spatial attributes that distinguish them from each other and influence the clustering result. Consequently geometric distance will sometimes be extended to include not only x y coordinates but also non-spatial attributes. Non-spatial attributes can be classified into two categories: numerical and non-numerical. For numerical non-spatial attributes, the numerical values can usually be transformed into some standardized values, and calculated by using geometric distance functions as additional dimensions. For the non-numerical attributes, new functions are defined to transform non-numerical values to numerical, such as "weight" in GDBSCAN (Sander et al. 1998) or "purity" in DBRS (Wang & Hamilton 2003).

2.1.2 Spatial clustering methods

Spatial clustering algorithms exploit spatial relationships among data objects to discern groupings inherent within the input data. The spatial clustering methods can be classified into five categories, based on the underlying clustering technique used (Han et al. 2001, Han & Kamber 2006).

1. Partitioning methods: Partitional clustering methods partition points into clusters, such that the points in a cluster are more similar to each other than to points in different clusters. They start with some arbitrary initial clusters and iteratively reallocate points to clusters until a stopping criterion is met. These methods tend to find clusters with hyperspherical shapes. Examples of partitional clustering algorithms include k-means (Lloyd 1982), PAM (Kaufman & Rousseeuw 1990), CLARA (Kaufman & Rousseeuw 1990), CLARANS (Ng & Han 2002) and EM (Kaufman & Rousseeuw 1990).

2. Hierarchical methods: Hierarchical clustering methods can be either agglomerative or divisive. An agglomerative method starts with each point as a separate cluster, and successively performs merging until a stopping criterion is met. A divisive method begins with all points in a single cluster and performs splitting until a stopping criterion is met. The result of a hierarchical clustering method is a tree of clusters called a dendogram. Examples of hierarchical clustering methods are CURE (Guha et al. 1998), BIRCH (Zhang et al. 1996) and CHAMELEON (Karypis et al. 1999).

3. Density based methods: Density-based clustering methods try to find clusters based on the density of points in regions. Dense regions that are reachable from each other are merged to form clusters. Density-based clustering methods excel at finding clusters of arbitrary shapes. Examples of density-based clustering methods include DBSCAN (Ester et al. 1996), OPTICS (Ankerst et al. 1999) and DBRS (Wang & Hamilton 2003).

4. Grid-based methods: Grid-based clustering methods quantize the clustering space into a finite number of cells and then perform the required operations on the quantized space. Cells containing more than a certain number of points are considered to be dense. Contiguous dense cells are connected to form clusters. Examples of grid-based clustering methods include CLIQUE (Agrawl et al. 1998), STING (Wang et al. 1997) and WaveCluster (Sheikholeslami et al. 1998).

5. Model-based methods: Model-based methods hypothesize a model for each of the clusters and attempt to optimize the fit between the data and some mathematical models such as Autoclass (Cheeseman et al. 1993) and COBWEB (Fisher 1987).

2.1.3 *DBSCAN algorithm*

The current GeoClustering prototype uses DBSCAN as the clustering method. Hence we provide a detailed description of this algorithm as follows.

DBSCAN (Density Based Spatial Clustering of Applications with Noise) (Ester et al. 1996) was the first density-based spatial clustering method proposed. The key idea is to define a new cluster, or extend an existing cluster, based on a neighborhood. The neighborhood around a point of a given radius (*Eps*) must contain at least a minimum number of points (*MinPts*).

Given a dataset D, a distance function *dist*, and parameters *Eps* and *MinPts*, the following definitions are used to define DBSCAN. An arbitrary point p, $p \in D$, the neighborhood of p is defined as $N_{Eps}(p) = \{q \in D \mid dist\,(p, q) \le Eps\}$. If $|N_{Eps}(p)| \ge MinPts$, then p is a core point of a cluster. If p is a core point and q is p's neighbor, q belongs to this cluster and each of q's neighbors is examined to see if it can be added to the cluster. Otherwise, point q is labeled as noise.

The expansion process is repeated for every point in the neighborhood. If a cluster cannot be expanded further, DBSCAN chooses another arbitrary unlabelled point and repeats the process. This procedure is iterated until all points in the dataset have been placed in clusters or labeled as noise. The pseudocode of DBSCAN can be found in Appendix 1.

2.2 *Current online clustering systems*

Clustering techniques have already been used in some geospatial online applications. Scale reduction from source maps to target maps inevitably leads to conflict and congestion of map symbols. To make the maps legible, appropriate operations (e.g. selection, simplification, aggregation, etc.) must be employed to simplify map features (Yan & Weibel 2008). The most common example is cartographic generalization, which is meant to generalize map symbols for online map applications. Here clustering is more like a technique with which several points of interest can be represented by a single icon when they are close to one another. For example, as shown in Figure 1, when all photos are given for a geographic region such as the downtown San Francisco area, users may want to find a set of "representatives" to improve the display when there are huge numbers of photos on the map.

Client-Server architecture is the most popular among online clustering systems. It is a distributed application architecture that partitions tasks or workloads between service requesters (clients) and service providers (servers).

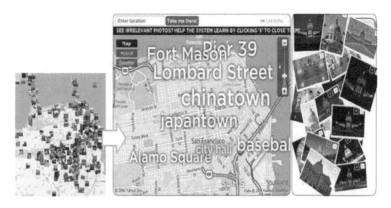

Figure 1.   A clustering example—Tag Map (Jaffe et al. 2006).

Basically, the client is responsible for interacting with users. A server is usually a high-performance host that offers functions and/or resources. When users submit a request to the client, the client requests the server's content or service function. The server will respond to the user's request. In terms of implementation, the underlying techniques can be divided into client-side and server-side methods.

2.2.1   *Clustering on the client side*

Client-side clustering methods are usually attached to specific online map services APIs. The following discussion will use Microsoft Virtual Earth and Google Maps, the most popular mapping platforms, as examples to introduce client-side methods.

Microsoft Virtual Earth V6.2 Map Control API offers a built-in method, `VEShape-Layer.SetClusteringConfiguration(type, options)`, which can set the method to determine which symbols are clustered, as well as how the clustering result is displayed (Microsoft 2009). The first parameter `type` has two values: `None` and `Grid`. In case of `None`, this method will return the original symbols. In case of `Grid`, a simple grid-based clustering algorithm will be used. In addition, users can override this method with the name of other clustering method functions in the form of `VEShapeLayer.SetClustering Configuration(algorithm, options)`, where `algorithm` is the name of the clustering method functions developed by users.

Although the Google Maps API does not offer built-in functions, some Google Maps supporters published several similar projects to help users manage the symbols on the map, such as Google Maps Clustering API Project (Google Maps API Projects 2009) and ACME cluster JavaScript library (ACME Labs 2009). Similar to the clustering function of the Virtual Earth API, users can call these APIs from the client side to cluster symbols on the map.

The clustering techniques at the client side are relatively simple. Grid-based clustering algorithms are the most common method. Most of these are implemented in JavaScript and the clustering work is usually performed by the browser. Subsequently the browser and online map services APIs display the clustering result as a layer superimposed on the map. This mechanism restricts the size of clustering data to be relatively small, usually under few thousand points.

2.2.2   *Clustering on server side*

Server-side clustering methods are relatively independent from online map services. The implementation technique can be divided into either real-time methods or pre-processing methods. For real-time methods, the server responds to the user's query and performs clustering on the fly. This method cannot be used for applications with huge data since the response time is much too long. To improve the clustering response time for huge datasets, pre-processing methods are applied in which pre-processing methods pre-cluster the dataset at different zoom levels. Usually the server first converts each symbol's position to a pair of pixel coordinates on the screen at each zooming level. It then calculates the pixel distance between symbols and combines the symbols closest to one another into one cluster symbol. Finally, these cluster symbols are saved on the server and are displayed at their pre-determined optimal zoom levels, depending on users' queries. Generally, there are two ways to save the pre-processing result: static raster images and vector point files.

ClustrMap (ClustrMap 2009), shown in Figure 2, illustrates the use of a static raster image. It is an archived clustering map based on the visitors to the website clustrmap.com

ClustrMaps archive for http://clustrmaps.com   35833 visits from 1 May 2009 to 1 Jun 2009

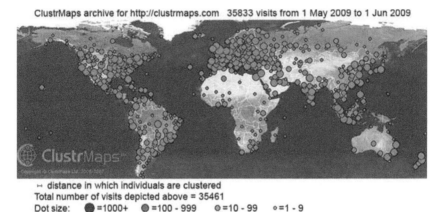

↦ distance in which individuals are clustered
Total number of visits depicted above = 35461
Dot size: ● =1000+   ● =100 - 999   ◎ =10 - 99   ○ =1 - 9

(a) http://www2.clustrmaps.com/counter/maps.php?url=http://clustrmaps.com&clusters=yes&hist=2009-05-01_to_2009-06-01&type=small&category=plus&map=world

ClustrMaps archive for http://clustrmaps.com   35833 visits from 1 May 2009 to 1 Jun 2009

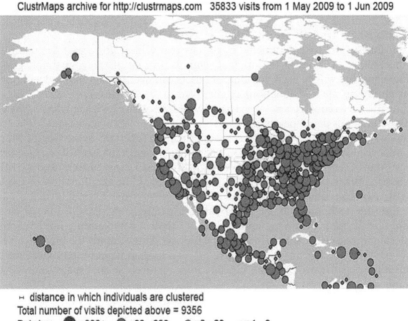

↦ distance in which individuals are clustered
Total number of visits depicted above = 9356
Dot size: ● =300+   ● =30 - 299   ◎ =3 - 29   ○ =1 - 2

(b) http://www2.clustrmaps.com/counter/maps.php?url=http://clustrmaps.com&clusters=yes&hist=2009-05-01_to_2009-06-01&type=small&category=plus&map=North%20America

Figure 2.   Clustering results, in the form of static raster images, for two zooming levels: (a) Global, (b) Continental (ClustrMap 2009).

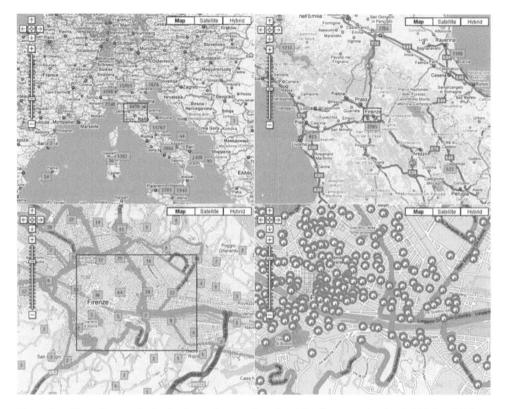

Figure 3. Clustering results in the form of vector "markers" for four zooming levels (Maiom 2009).

from May 1st to June 1st 2009. This clustering map has two zooming levels: global and continental. Different images are returned according to the user's request. Figure 2a is the clustering map at the global scale while Figure 2b is the clustering map at the continental scale. The static images are usually updated after a fixed time period, for example every 24 hours.

Figure 3 shows an example of using vector point files to display zoom-scale-dependent point-location information (Maiom 2009). This sequence of maps shows the location of real estate for sale or rent in Italy. Each real estate can be considered as a point. To avoid overlap and to make the map more legible, the system only shows the clustering result of points instead of the real locations at the small zoom level. During pre-processing, the system saves the clustering result at different zooming levels as "markers" into different xml files. Each "marker" has a new coordinate to represent the points in this cluster. When users view the map at different zooming levels, the server returns the corresponding xml files. Figure 3 shows "markers" for the city "Firenze" at different zoom levels.

2.2.3 Limitations of current systems

Current systems have some important limitations for general users: First, most current systems are designed for specific applications. These are closed platforms that can be used by only a limited number of individuals and that do not offer any interoperability. Second, most of these systems work simply by combining nearby symbols into one single cluster symbol. Their clustering algorithms are used to generalize the points on the map instead of for supporting the discovery of underlying cluster patterns. Third, some systems provide APIs for clustering functions, such as Microsoft Virtual Earth v6.2 Map Control API. However, users need to do extra programming to be able to use them.

3 GEOCLUSTERING

In this section, we will briefly introduce the framework of the online clustering service called GeoClustering.

We draw on the Web Service concept to design the architecture of GeoClustering. Web Service represents a convergence of the service oriented architecture (SOA) and the web (W3C 2004). Beyond this, the data exchange between GeoClustering and the outside uses XML-based language files. The Web Service and the XML schema satisfy the requirement to build an online geospatial clustering platform with openness and interoperability to deal with distributed data.

Users can interact with GeoClustering via two types of interfaces. One is the GeoClustering API, and the other is the user-friendly GUI. The GeoClustering API is the lowest level of the interface for using the clustering service and is developed around the "reuse" design paradigm. With this API, other web applications or services can call the geospatial clustering service directly. The GUI is a higher interface for the clustering service and works like a wrapper of the GeoClustering API. It provides general users with an easy and friendly user interface. It includes the basic interactive elements that users need to input necessary clustering information, to turn this input into parameters used by the GeoClustering API, and to also call the GeoClustering API. In order to enhance the user experience and to ensure better visualization, the GUI also offers a map viewer that draws on third party Web Map APIs. Using client-side script, users can add and view the original spatial data on the map. The map viewer can also help people to understand the clustering result. Each cluster will be shown on the map with its region covered by a convex hull. Users can control the visibility of each layer and each cluster on the map.

A GeoClustering prototype has already been implemented using Client-Server architecture. The current version can perform density-based clustering and visualize the results. The prototype is available at http://geoclustering.jingking.net. A user guide is provided at http://geoclustering.jingking.net/help.php. In the GeoClustering prototype we only consider the clustering objects as points. The distance function between two objects is defined as the shortest geometric distance on the spherical Earth surface. DBSCAN has been chosen as the first algorithm to be implemented. This section presents the implementation of the prototype shown in Figure 4. The data formats for user input and system output will be discussed in Section 3.1. Users could submit clustering requests via browser or other applications as well. The client component will be discussed in detail in Section 3.2. The server will perform the clustering procedure after receiving the GeoClustering API request and parameters and then give a returned result. This will be discussed in Section 3.3.

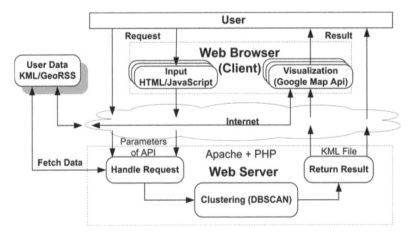

Figure 4.   Prototype of GeoClustering architecture.

3.1 *Data formats*

To meet the requirement for interoperability, the standard XML file format is used in the data transaction in the GeoClustering platform. The system accepts two kinds of standard xml files from users: KML (Keyhole Markup Language) and GeoRSS. The clustering results are returned in the form of KML.

KML is a tag-based structure with nested elements and attributes and is based on the XML standard. It is used for expressing geographic annotation and visualization on existing or future Web-based, two-dimensional maps and three-dimensional Earth browsers (OGC 2008). KML v2.2 was adopted as an official OGC implementation standard on April 14, 2008.

Appendix 2 is part of a KML input file for GeoClustering. It includes a point with the name of "Point001", which is contained by a "Placemark." The time property of the point is under the `<TimeStamp>` node. The geometry property of the point is described in the `<coordinates>` element under the `<Point>` node. The geographic location is defined by longitude, latitude, and altitude (optional).

GeoRSS means geographically encoded objects for RSS (really simple syndication) feeds. It is based on RSS 2.0 and adding location information to data items (OGC 2006). An example of a GeoRSS input file, which includes the same point in the example KML file, is shown as Appendix 3. The name of that point is contained by `<title>`. Time property is described in `<description>`. The geographic location is defined by `<geo:lat>` and `<geo:long>` element.

3.2 *Client*

3.2.1 *GeoClustering API*

In the GeoClustering prototype, GeoClustering API will be offered in the form of an endpoint URL address. Via this users can access the clustering service with the assigned clustering parameter values. Simple HTTP GET or POST actions are the only supported request formats. The API endpoint URL is as follows:

```
http://localhost/cluster.php?<Data     URL>&[File     Type]&<Algorithm
Name>&<Clustering Parameters>
```

To request the clustering service, invoke as follows:

```
http://localhost/cluster.php?url=http://earthquake.usgs.gov/
eqcenter/catalogs/shakerss.xml&filetype=georss&algorithm=DBSCAN&
param1=3&param2=300
```

```
Data URL :: = "url =" <the url of remote point data file>
File Type :: = "filetype =" <kml | georss>
Algorithm Name :: = "algorithm =" <algorithm name: DBSCAN |
K-mean | ....>
Clustering Parameters :: = {"param"<number>"=" <the value of
parameter for the chosen algorithm>}
```

In the example above, when `algorithm=DBSCAN`, then param1 is the minimum number of points (*MinPts*) and param2 is the radius (*Eps*) in km.

To respond to the API requests, an XML response is returned. When unusual situations occur during the clustering procedure, two types of error message are sent back.

```
<b>Parameters missing</b>
<b>Data cannot be found at URL or data syntax error</b>
```

When the correct clustering result is achieved through the clustering procedure, the response is presented in KML format.

```
<?xml version="1.0" encoding="UTF-8"?>
<kml xmlns="http://earth.google.com/kml/2.2">
  <Document>
    <name>GeoClustering Result</name>
  <Snippet> </Snippet>
  <description>
    Total Clusters:4 <br />    <!-- Cluster numbers -->
    Time:<b>0.191354036331</b> Seconds<br /> <!-- Run time -->
  </description>
<!-- Style begin -->
  ...
  <StyleMap id="clustericon1">
    ...
  </StyleMap>
  ...
<!-- Style end -->
<!-- Points info begin -->
  <Folder>
    <Placemark>
      <name>Cluster 1</name>
      <TimeStamp><when></when></TimeStamp>
      <styleUrl>#clustericon1</styleUrl>
      <description>Point0001 2008-09-22T09:00:01-07:00
      </description>
      <point>
        <coordinates>-114.132446,51.079529,0</coordinates>
      </Point>
    </Placemark>
    <Placemark>
      <name>Cluster 1</name>
      ...
    </Placemark>
    <Placemark>
      <name>Cluster 2</name>
    ...
  </Folder>
<!-- Points info end -->
  </Document>
</kml>
```

In the returned KML file, general clustering result information is included in element <description> under <Document>. Number of clusters established and running time are recorded here. The following part is the style information, which defines the icon style for the point in each cluster. When one point needs to use this style, it cites the id of <StyleMap>. All the points in the same cluster have the same cluster id and the same value under <name> and <styleUrl> tags. The original value under <name> and <TimeStamp> of each point will be added into that point's <description> element. In the returned example file, the original point with name "Point0001" becomes the point with the name "Cluster 1" and in the <description> part of this point the original information "Point0001 2008-09-22T09:00:01-07:00" is added.

When there is no cluster found in the dataset following the clustering procedure, an alert response message is returned.

```
<b>Internal error or no Cluster found in the dataset</b>
```

3.2.2 *GUI*

To demonstrate the GeoClustering API, we implemented the GUI as a web page for GeoClustering. The whole web page is implemented by HTML, CSS and JavaScript. To have a better user experience, AJAX (Asynchronous JavaScript and XML) have been used. The third party Web Map APIs are included in the web page. Microsoft Virtual Earth and Google Maps are the two major players in the area of Web Maps. Comparing Google Maps API with Virtual Earth Map Control API, we choose Google Maps as it has a better 3D extension for future development with the Google Earth Plug-in.

From the user's perspective, the web page can be divided into four components: Map Viewer, Data Input Bar, Control Panel and Link Area (Fig. 5). Map Viewer is used for parsing KML files and showing the spatial data layers. To give users a better intuitive feeling for the clustering result, both the convex hull of the cluster, and the markers in each cluster, are shown with the clustering results. 3D viewer is also added to the platform (Fig. 6). Users can switch from the 2D maps smoothly and interact with the 3D map after they install the Google Earth Plug-in. The data Input Bar and Control Panel are responsible for turning user's input into well formatted strings and then calling the GeoClustering API in the form of URLs. Users can also control the display of map layers, conduct clustering, and control the visibility of clustering results through the Control Panel. It also provides some basic information such as the total numbers of points for each cluster. The link area lists three links of the platform. Since datasets with different parameters lead to various results, user can not only view their clustering result history through the link "Gallery", but can also compare the results that show differences between them.

3.3 *Server*

The web server takes the role of communicating with the client and responding to the clustering task requested from the client. Among various web servers and corresponding server

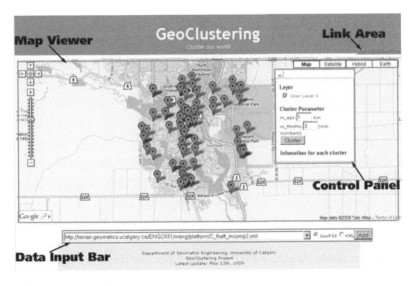

Figure 5. GeoClustering web page.

47

Figure 6. GeoClustering web page—A screenshot view of Calgary crime records clustering result in 3D Map Viewer.

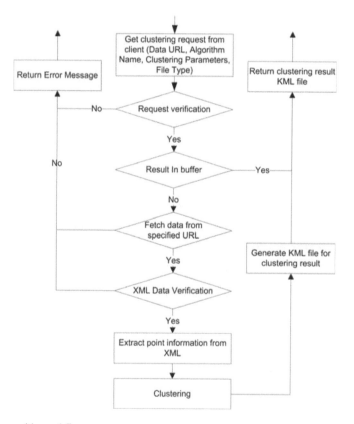

Figure 7. Server-side workflow.

script languages available, Apache + PHP is selected because they are open source software and very stable.

Figure 7 shows the workflow at server side. When the server receives a request, it first verifies whether the request is complete or not. A complete request contains at least

48

three essential parts mentioned in Section 3.2.2.1: `Data URL`, `Algorithm Name` and `Clustering Parameters`. `File Type` is optional. Failure to include any of the three essential parts leads to failure of the verification.

Following verification, the server calls for the function that detects whether there is a target xml results file in the "Buffer". Since clustering is sometimes time consuming, a buffer system is used to save the clustering result for a short period of time. If the same clustering requests are sent a second time, the result can be quickly returned to client. The buffer can speed up the GeoClustering performance, especially when users want to compare the result with previous requests.

If there is no clustering result in the "Buffer", the server will fetch the remote xml file (the user data) through the `Data URL` submitted. If the remote file is unreachable, an error message will be returned. If the server receives the remote data successfully, then the server will verify the completion of the XML files. Any incomplete or incorrect syntax will cause abnormal termination and an error message will be returned. Then, the server extracts all of the points' information from the XML files fetched in the last step and pass the points to the clustering algorithm module in the form of an array. Next, the algorithm module uses the algorithm specified by the request, together with clustering parameters, to conduct the clustering. A new array with the cluster information is then returned to the next module. Finally, in the next module, the server generates KML files from the returned array.

4 EVALUATION

In order to evaluate the prototype version of the GeoClustering platform, two practical application simulations were conducted.

The first application was to identify those areas in Calgary with high intensity of theft crimes. The crime data were gathered from the Calgary Police Online Crime Map. Each crime data record contains a geographical coordinate, the crime type and the time of occurrence. A GeoRSS file was generated based on a sub-dataset ranging from Sept 6 to Sept 19, 2008, with 414 points in total.

For this dataset, when the radius (*Eps*) is set to 1 km, and the minimum number of points is set to 3, 18 clusters are found (See Fig. 8). The largest cluster is centered in and around the downtown area; this contains 79 records. This evaluation was implemented in two different test server environments. On the online testing server (Intel C4 2.4G Hz), average clustering time is 6.4 seconds. On another local test environment (Intel E7200) the average time is less than 3 seconds.

The second application was to quickly identify areas of high earthquake activity around the world. The earthquake data can be found from the USGS National Earthquake Information Center. It provides worldwide earthquake lists, in nearly real-time, in both KML and GeoRSS feeds. The real-time, worldwide earthquake (with magnitude greater than 2.5) list for the past 7 days (June 12, 2009 20:46:55 GMT–June 19, 2009 18:29:48 GMT) is used. The URL address for the earthquake data is http://earthquake.usgs.gov/eqcenter/catalogs/ eqs7 day-M2.5.xml. There are 204 records in total. When the radius (*Eps*) is set to 100 km, and the minimum number of points is set to 5, 4 clusters are found (See Fig. 9). This evaluation was only implemented at the online testing server. Average running time was around 1.1 seconds.

5 CONCLUSIONS AND FUTURE WORK

Geospatial clustering is a very important method in the field of Geographic Knowledge Discovery. It helps people to understand the most prominent and critical patterns and distributions of geographic phenomena. In this chapter, we introduce the design and development of a geospatial clustering web service GeoClustering. GeoClustering can be used to

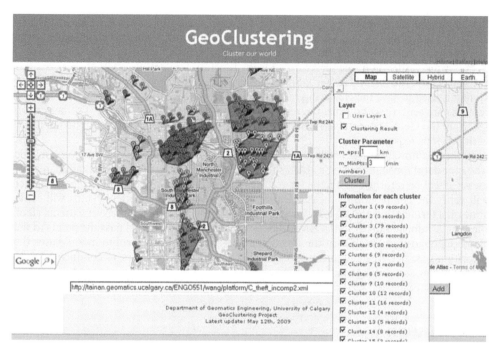

Figure 8.　Clustering result of Crime Data.

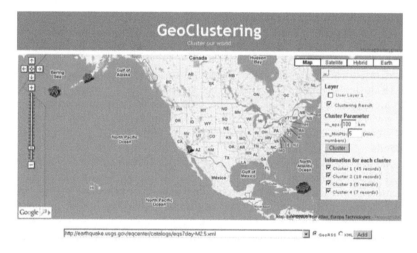

Figure 9.　Clustering result of USGS Earthquake Data.

identify clusters from distributed data sources, with free access at anytime from anywhere. By using open and interoperable service interfaces, users are able to cluster their data online and visualize the clustering patterns on the map easily and conveniently. In the future, we will improve the following aspects of GeoClustering:

First, more algorithms and advanced techniques will be developed and integrated into the platform. The hierarchical clustering methods will be implemented for automated clustering under different zooming levels. To reduce the execution time, clustering, spatial indexing, and data compression techniques will be used to improve network data transmission.

Second, more data types need to be supported by the platform. The current version of GeoClustering only supports static datasets. Dynamic and continuous datasets, such as real time sensor data, have been collected and need to be analyzed as well. Therefore, some new clustering methods and mechanisms dealing with real-time datasets will be added. Since the prototype can only process vector data, in the future, a basic raster data processing module will also be added.

Third, for the above two applications, k-d diagram (Ester et al. 1996) is used to help set parameters. The current prototype cannot provide recommendations for clustering method selection and parameter settings. Clustering methods proposed in spatial data mining is different from the clustering methods in statistics. For many clustering methods, we cannot simply apply "statistic significance" to measure the accuracy of the clustering result. How to select an appropriate clustering method, how to set parameters and constraints, and how to explain the clustering result are not simple, they depend on domain knowledge and the users' goals. Even for the same datasets, different clustering results can be produced based on different users' goals. Some discussions about integrating domain knowledge and user goals into clustering can be found in Gu et al. (2009) and Wang et al. (2010). Additionally, we will investigate how to harness patterns of collective intelligence from our users. In the web 2.0 Era, web users should be treated as participants and contributors rather than simple data consumers. With the implementation of online communication functions in the near future, we aim to enable our users to share and discuss the patterns of clustering results interactively. Such collective intelligence will provide users with a better understanding of spatial data.

Fourth, we will support WPS (Web Processing Service) standard interfaces. OGC approved version 1.0 of the OpenGIS WPS Interface Standard in 2008. The WPS standard defines interfaces that facilitate the publishing of geospatial processes and make it easier to write software clients that can discover and bind to those processes.

REFERENCES

ACME Labs. (2009) [Online] Available from: http://acme.com/javascript/#Clusterer

Agrawl, R., Gehrke, J., Gunopulos, D. & Raghavan, P. (1998) Automatic Subspace Clustering of High Dimensional Data for Data Mining Applications. *ACM SIGMOD Record*, 27 (2), 94–105.

Ankerst, M., Breunig, M.M., Kriegel, H.P. & Sander, J. (1999) OPTICS: Ordering Points to Identify the Clustering Structure. *ACM SIGMOD Record*, 28 (2), 49–60.

Bishr, Y. (1998) Overcoming the Semantic and Other Barriers to GIS Interoperability. *International Journal of Geographical Information Science*, 12 (4), 299–314.

Cheeseman, P., Kelly, J., Self, M., Stutz, J., Taylor, W. & Freeman, D. (1993) AutoClass: a bayesian classification system. In: Bruce, G. Buchanan & David, C. Wilkins (eds.), *Readings in Knowledge Acquisition and Learning: Automating the Construction and Improvement of Expert Systems*, San Francisco, Morgan Kaufmann, pp. 431–441.

ClustrMap. (2009) [Online] Available from: http://clustrmaps.com/

Clusterseer. (2009) [Online] Available from: http://www.terraseer.com/products_clusterseer.php

Ester, M., Kriegel, H., Sander, J. & Xu, X. (1996) A Density-Based Algorithm for Discovering Clusters in Large Spatial Databases with Noise. *Proc. Second International Conference on Knowledge Discovery and Data Mining*, 2–4th August 1996, Portland, Portland, AAAI Press. pp. 226–231

Fisher, D.H. (1987) Knowledge Acquisition via Incremental Conceptual Clustering. *Machine Learning*, 2 (2), 139–172.

Google Maps API Projects. (2009) [Online] Available from: http://googlemapsapi.martinpearman.co.uk

Gu, W., Wang, X. & Ziébelin, D. (2009) An Ontology-Based Spatial Clustering Selection System. In *Proc. the 22nd Canadian Artificial Intelligence Conference* (*CAI 2009*), 25–27 May 2009, Kelowna, Berlin, Springer-Verlag. pp. 215–218.

Guha, S., Rastogi, R. & Shim, K. (1998) CURE: An efficient clustering algorithm for large databases. *ACM SIGMOD*, 27 (2), 73–84.

Han, J., Kamber, M. & Tung, A.K.H. (2001) Spatial Clustering Methods in Data Mining: A Survey. In: Harvey J. Miller & Jawei Han (eds.), *Geographic Data Mining and Knowledge Discovery*. London, Taylor and Francis.

Han, J. & Kamber, M. (2006). *Data Mining: Concepts and Techniques (2nd edition)*. San Francisco, Morgan Kaufmann.

Jacquez, G.M. (2008). Spatial Cluster Analysis. In: John Wilson & Stewart Fotheringham A. (eds.), *The Handbook of Geographic Information Science*. Malden, Blackwell.

Jaffe, A., Naaman, M., Tassa, T. & Davis, M. (2006) Tag Maps, [Online] Available from: http://www.slideshare.net/mor/tag-maps

Karypis, G., Han, E. & Kumar, V. (1999) Chameleon: hierarchical clustering using dynamic modeling. *Computer*, 32 (8), 68–75.

Kaufman, L. & Rousseeuw, P.J. (1990) *Finding Groups in Data: An Introduction to Cluster Analysis*. New York, John Wiley & Sons, Inc.

Kulldorff, M. (2009) SaTScan User Guide. [Online] Available from: http://www.satscan.org/techdoc.html

Lloyd, S.P. (1982) Least Squares Quantization in PCM. *IEEE Transactions on Information Theory*, 28 (2), 129–137.

Maiom. 2009 [Online] Available from: http://www.maiom.com/mappa/

Microsoft Corp. (2009) Virtual Earth Developer Center, MSDN. [Online] Available from: http://msdn.microsoft.com/en-us/library/cc966930.aspx

Miller, H.J. (2008) Geographic Data Mining and Knowledge Discovery. In: John Wilson & Stewart Fotherinham A. (eds.), *The Handbook of Geographic Information Science*, Malden, Blackwell.

Ng, R.T. & Han, J. (2002) CLARANS: A method for clustering objects for spatial data mining. *IEEE Transaction on Knowledge and Data Engineering,* 14 (5), 1003–1016.

OGC. (2006) GeoRSS White Paper, [Online] Available from: http://www.opengeospatial.org/pt/06-050r3

OGC. (2008) [Online] Available from: http://www.opengeospatial.org/standards/kml

Sander, J., Ester, M., Kriegel, H. & Xu, X. (1998) Density-based Clustering in Spatial Databases: the algorithm GDBSCAN and its applications. *Data Mining and Knowledge Discovery*, 2(2), 169–194.

Sheikholeslami, G., Chatterjee, S. & Zhang, A. (1998) WaveCluster: a multi-resolution clustering approach for very large spatial databases. *Proc. International Conference on Very Large Databases (VLDB'98)*, 24–28 August 1998, New York, New York, Springer-Verlag. pp. 428–439.

Tobler, W. (1970) A Computer Movie Simulating Urban Growth in the Detroit Region. *Economic Geography*, 46 (2), 234–240.

W3C. (2004) Web Service Architecture, W3C working Draft. [Online] Available from: http://www.w3.org/TR/2004/NOTE-ws-arch-20040211/

Wang, W., Yang, J. & Muntz, R. (1997) STING: A Statistical Information Grid Approach to Spatial Data Mining. *Proc. the Twenty-third International Conference on Very Large Data Bases*, 25–29 August 1997, Athens, San Francisco, Morgan Kaufmann. pp. 186–195.

Wang, X. & Hamilton, H.J. (2003) DBRS: A density-based spatial clustering method with random sampling. *Proc. the Seventh Pacific-Asia Conference on Knowledge Discovery and Data Mining (PAKDD 2003)*, Apr 30–May 2, 2003, Seoul. pp. 563–575.

Wang, X., Gu, W., Ziébelin, D. & Hamilton, H. (2010) An Ontology-Based Framework for Geospatial Clustering. *International Journal of Geographical Information Science*, 24 (11), 1601–1630.

Yan, H. & Weibel, R. (2008) An algorithm for point cluster generalization based on the Voronoi diagram. *Computers and Geosciences*, 34 (8), 939–954.

Zhang, T., Ramakrishnan, R. & Livny, M. (1996) BIRCH: An Efficient Data Clustering Method for Very Large Databases. *ACM SIGMOD*, 25 (2), 103–114.

APPENDIX

Appendix 1 Pseudocode of DBSCAN (Taken from Ester et al. 1996)

```
DBSCAN (SetOfPoints, Eps, MinPts)//SetOfPoints is UNCLASSIFIED
  ClusterId := nextId(NOISE);
  FOR i FROM 1 TO SetOfPoints.size DO
    Point := SetOfPoints.get(i);
    IF Point.ClId = UNCLASSIFIED THEN
      IF ExpandCluster(SetOfPoints, Point, ClusterId, Eps, MinPts)
    THEN
        ClusterId := nextId(ClusterId)
      END IF
    END IF
  END FOR
END; // DBSCAN
ExpandCluster(SetOfPoints, Point, ClId, Eps, MinPts) : Boolean;
    seeds := SetOfPoints.regionQuery(Point,Eps);
    IF seeds.size < MinPts THEN // no core point
      SetOfPoint.changeClId(Point,NOISE);
      RETURN False;
    ELSE // all points in seeds are density reachable from Point
      SetOfPoints.changeClIds(seeds,ClId);
      seeds.delete(Point);
      WHILE seeds <> Empty DO
        currentP := seeds.first();
        result := SetOfPoints.regionQuery(currentP, Eps);
        IF result.size >= MinPts THEN
          FOR i FROM 1 TO result.size DO
            resultP := result.get(i);
            IF resultP.ClId IN {UNCLASSIFIED, NOISE} THEN
              IF resultP.ClId = UNCLASSIFIED THEN
                seeds.append(resultP);
              END IF;
              SetOfPoints.changeClId(resultP,ClId);
            END IF; //UNCLASSIFIED or NOISE
          END FOR;
        END IF; //result.size >= MinPts
        seeds.delete(currentP);
      END WHILE; //seeds <> Empty
      RETURN True;
    END IF
END; //ExpandCluster
```

Appendix 2 Example of KML Input File

```
<?xml version="1.0" encoding="UTF-8"?>
<kml xmlns="http://earth.google.com/kml/2.2">
  <Document>
    <name>Clustering Example</name>
    <Placemark>
      <name>Point0001</name>
      <TimeStamp>
      <when>2008-09-22T09:00:01-07:00</when>
```

53

```
          </TimeStamp>
          <Point>
            <coordinates>-114.132446,51.079529,0</coordinates>
          </Point>
        </Placemark>
          ...
    </Document>
    </kml>
```

Appendix 3 Example of GeoRSS Input File

```
<?xml version="1.0"?>
<rss version="2.0"
xmlns:geo="http://www.w3.org/2003/01/geo/wgs84_pos#"
xmlns:dc="http://purl.org/dc/elements/1.1">
<channel>
<item>
  <title>Point0001</title>
  <description>2008-09-22T09:00:01-07:00</description>
  <geo:lat>51.079529</geo:lat>
  <geo:long>-114.132446</geo:long>
</item>
</channel>
</rss>
```

Advances in Web-based GIS, Mapping Services
and Applications – Li, Dragićević & Veenendaal (eds)
© *2011 Taylor & Francis Group, London, ISBN 978-0-415-80483-7*

Creating GIS simulation models on a TeraGrid-enabled geospatial web portal: A demonstration of geospatial cyberinfrastructure

Ming Hsiang Tsou & Ick-Hoi Kim
Department of Geography, San Diego State University, California, USA

ABSTRACT: This chapter will introduce a prototype of geospatial web portals, which can enable complicated GIS simulation models in a grid-computing environment (TeraGrid). This prototype demonstrates how complicated GIS simulation models can benefit from geospatial cyberinfrastructure and grid-enabled Internet GIServices. Cyberinfrastructure and grid computing can provide high-performance computing power to overcome the limitations of traditional GIS web services and the constraints of the three-tier Internet GIServices architecture. Complicated geospatial problems and huge volume of geospatial datasets can be effectively processed and calculated under the grid-enabled web portlets. The user-friendly simulation tools on the geospatial web portal will provide geographers and geospatial scientists with a powerful tool to conduct on-line spatial analysis and GIS simulation models efficiently.

Keywords: Cyberinfrastructure, grid computing, GIServices, simulation models, TeraGrid

1 INTRODUCTION

GIS simulation models, such as Cellular Automata (CA) and Multi-Agent Systems (MAS), can provide better understanding and prediction of complex spatial phenomena, including human behavior, urban growth, land use change, and wildfire spread. To facilitate collaborative decision making and active user participation, on-line, web-based simulation models can provide more accessible and interactive GIS analysis functions for multiple users at geographically distributed locations. However, most current web-based GIS simulation models are created using traditional three-tier Internet Geographic Information Services (GIServices). The three-tier architecture with very limited computing power cannot enable complicated GIS simulation models nor process huge volume of data due to hardware constraints of web servers. On the other hand, geospatial cyberinfrastructure and grid computing environments can provide high-performance computing resources that might overcome the limitations of traditional Internet GIServices.

This chapter will introduce a web-based grid computing approach that can enable powerful GIS simulation tools for geographers and geospatial scientists. A prototype was created to demonstrate the feasibility of the grid-enabled web portal and to utilize distributed computing resources in the TeraGrid. TeraGrid (http://www.teragrid.org/) is a nationwide grid computing platform which integrates the most advanced supercomputer resources at geographically distributed sites in the USA, including the San Diego Supercomputer Center (SDSC), the National Center for Supercomputing Applications (NCSA), the Argonne National Laboratory (UC/ANL), and other supercomputer centers (Catlett et al. 2007). The TeraGrid-enabled geospatial web portal can provide supercomputer resources via a common grid computing framework (Open Grid Computing Environment) for GIS simulations.

The current web portal prototype was created within an intranet at San Diego State University for performance and feasibility testing. The geospatial web portal will be open and accessible in the future for scientists and public users and allow them to conduct on-line spatial analysis and web-based GIS simulation models. The web portal will have different user interfaces designed for scientists and public users respectively. For scientists, the geospatial web portal will allow them to build and to experiment with simulation models dynamically. Public users will be able to run and understand spatial phenomena with a simplified user interface through "pre-defined" GIS models and simulation tools for educational purposes.

The focus of GIS simulation functions on this web portal prototype is population growth models with CA forms. Various population growth simulation models can be constructed by using the Decennial Census data and the American Community Survey data. Multiple GIS and census data layers at micro level were analyzed for their spatial relationships in the GIS simulation model.

This geospatial web portal prototype adopted the Open Grid Computing Environments (OGCE) Portal and Gateway Toolkit as the container for Java Specification Request (JSR) 168-compatible portlets. The web portal is scalable by using a Service Oriented Architecture (SOA) for users to select and combine different components in the web portal. All of Web-based GIS simulation models can be powered by the Apache Tomcat Web application container. JavaServer Pages™ (JSP), JavaServer™ Faces (JSF) and JavaScript are the technologies used for developing web interfaces for both portlet and non-portlet applications. Java™ Development Kit (JDK) is the major programming environment for creating these grid-enabled GIS applications and tools. The main research objective is to examine whether grid computing can enhance the performance of GIS simulation models on the web. Ideally, more detailed examination may reveal what type of GIS model (among CA, MAS and others) could be more effective in a grid computing environment (though we did not investigate this here). These findings can encourage more geographers and scientists to utilize geospatial cyberinfrastructure and to understand complicated geographic phenomena.

2 INTERNET GISERVICES AND GEOSPATIAL CYBERINFRASTRUCTURE

Currently, Internet GIServices have played a key role to facilitate the dissemination of geospatial data cost-effectively. In addition, interactive mapping tools allow users to analyze maps over the Internet. Through wired and wireless Internet, Internet GIServices have provided users with flexible and comprehensive Geographic Information Services (Peng & Tsou 2003). The development of Internet GIServices is always energized by innovative Internet technologies. New technologies, such as the Microsoft ASP.NET, Flex, and Ajax, have transformed web-based mapping tools into highly interactive and multimedia GIServices.

In general, Internet GIServices can be categorized into three levels of services: data archive and search (sharing data); information display and query (sharing information); and spatial analysis and GIS modeling services (sharing knowledge) (Tsou 2004). Spatial analysis service is the highest level of Internet GIServices and the most challenging one. There are some successful research projects in developing on-line spatial analysis and GIS simulation modeling services. Anselin et al. (2003) developed web-based spatial data exploration tools, and Tsou (2004) implemented Java-based image processing tools. With respect to simulation modeling, Semboloni et al. (2004) presented an interactive web multi-agent simulation model for the development of a city. In the simulation model, human users can play with the artificial agents and experiment urban simulation. Yassemi & Dragicevic (2008) showed a web-based forest fire modeling tool in order to help experts and decision makers in emergency situations.

Although these applications revealed the great potential of Internet GIServices for GIS models and simulations, most traditional desktop GIS analysis functions have not yet been transformed into three-tier Internet GIServices. In addition, computational limitations become a major obstacle when users need to implement more complex and large scale simulation models (Lei et al. 2005, Semboloni et al. 2004, Stevens & Dragicevic 2007, Tang & Wang 2009). We need to develop a new approach to enhance the spatial analysis functions in Internet GIServices. Distributed computing can be an alternative method to overcome these computational limitations (Lei et al. 2005, Tang & Wang 2009, Yang & Raskin 2009). In addition,

Distributed Geographic Information Processing (DGIP) frameworks (Yang & Raskin 2009) will facilitate the establishment of geospatial cyberinfrastructure and grid-enabled GIS analysis functions.

The geospatial cyberinfrastructure, which integrates high-performance computing resources, can overcome the constraints of the traditional three-tier Internet GIServices architecture. Cyberinfrastructure is an integrated information infrastructure connecting distributed computing resources (hardware and software), virtual organizations (individuals and multiple organizations) and knowledge bases for the advancement of scientific research and education (NSF 2007, Wang & Liu 2009, Zhang & Tsou 2009). Through cyberinfrastructure, researchers can share data, analysis tools, and research outcomes over the Internet. As for geography and GIScience research, Zhang & Tsou (2009) used the term, "*geospatial cyberinfrastructure*", to indicate the cyberinfrastructure created for the advancement of geospatial technologies and GIScience.

Geographic analysis and GIS simulation largely require intensive computing power and handle huge data volumes. This research will adopt TeraGrid for the grid computing framework in order to meet these requirements. TeraGrid is widely utilized by US scientists as high-end experimental facilities to solve complicated scientific problems, such as climate change and molecule simulation. TeraGrid also offers a variety of gateway web portals for various scientific research, such as the Geosciences Network (GEON 2010), the Science Environment for Ecological Knowledge (SEEK 2010), and the Linked Environments for Atmospheric Discovery (LEAD 2010). These web portals enable users to conduct analysis and to share knowledge over the Internet. An early example of spatial Web portal prototypes was introduced by Zhang & Tsou (2009). The prototype focused on generic grid-computing performance enhancement issues by using a simplified transportation accessibility analysis model. Different from this early example, this paper focuses on more complicated GIS models (cellular automata and dasymetric mapping) with a realistic geographic problem (population growth in the County of San Diego). The research findings of this new prototype can provide more valuable experiences and suggestions for developing high performance GIS simulation tools.

In addition to high performance computing power, a grid-enabled web portal can facilitate research collaboration. Researchers can upload spatial data and analysis tools on the web portal and share them with other scientists. Through research collaboration, Problem Solving Environments (PSEs) can be established. PSEs utilize high computing environments to provide comprehensive computing solutions (algorithms, visualization, models, etc) to complicated problems, such as geographic questions and models (Gallopoulos et al. 1994). Researchers can collaborate together to solve target problems. The geospatial web portal can also help us to construct geography-oriented Virtual Organizations (VOs). To facilitate the collaboration of multiple researchers who might have different research methods and computing skills, a good PSE needs to provide easy-to-use user interfaces and accessible web tools (Takatsuka & Gahegan 2001, Zhang & Tsou 2009).

In order to broaden and enhance geographers' knowledge and spatial models, loose-coupling GIServices can be efficient (Clarke & Gaydos 1998, Zhang 2007). The Service Oriented Architecture (SOA) is a service design architecture to share Internet services among scientists. Distributed services can be invoked separately and combined easily. The SOA can be implemented on the TeraGrid using the OGCE (Open Grid Computing Environment) toolkit. Users of geospatial web portals can request each service and combine necessary GIS services to fulfill their needs.

3 OVERVIEW OF GIS SIMULATION

Spatial phenomena on the Earth can be modeled with GIS simulations. Urban growth, land use change, population migration, vegetation and wild animal habitat, and wildfire spreads can be simulated and projected through GIS simulation models. There are three types of spatial simulation methods: Cellular Automata (CA) (Clarke et al. 1997, Couclelis 1985, O'Sullivan 2002, Tobler 1979); Multi-Agent System (MAS) (Benenson & Torrens 2004, Parker et al. 2003); and Geographic Automata System (GAS) (Benenson & Torrens 2004, Torrens 2007). CA defines spatial relationships in continuous fields. MAS and GAS define spatial phenomena as individual objects. These models complement each other and provide various perspectives of spatial relationships and associations. According to Benenson & Torrens (2004), GAS (Geosimulation) indicates a research domain for developing object-based high resolution geographic simulation models.

CA was invented by Ulam and Neumann in the mid-1940s and it was adopted in Geography in the late 1950s. CA models were widely used in the late 1980s and became one of main tools to explore geographic modeling (Benenson & Torrens 2004). Since Tobler (1979) introduced a geographic model based on a von Neumann neighborhood, his paper has inspired several urban geographers to adopt CA models, including Helen Couclelis (1985), Michael Batty (1997), and Keith Clarke et al. (1997). Today CA models are very popular for urban growth simulation (Benenson & Torrens 2004).

CA model is composed of four components, including contiguous or adjacent cells, states of cells, transition rules, and neighborhoods (Batty 1997). Most of CA data are in a grid-based raster format. The raster format is suitable to represent regular neighborhoods. However, many spatial phenomena are adjacent irregularly. For example, each land parcel does not have the same sizes or shapes. Some revised CA models adopt vector data format to address the irregularity of spatial entities such as land parcels and houses (Benenson et al. 2002, O'Sullivan 2002). Graph structures (O'Sullivan 2001, 2002), Delaunay triangulations (Semboloni 2000), Voronoi polygons (Shi & Pang 2000), and vector data (Stevens & Dragicevic 2007) were exploited to overcome the irregularity of spatial phenomena in CA modeling. In a graph structure, entities are represented as vertices and they are connected by edges. The neighborhood of vertex is the set of vertices joined by edges. A Delaunay triangulation, similar to the graph structure, defines nodes as the centroids of entities and they are connected by networks. In a Voronoi diagram, nodes are centroids of entities and they are connected by links. The neighborhoods are entities sharing edges of Voronoi polygon edges. Although vector-based CA can deal with irregular spatial entities, there are still constraints to represent spatial entities. In particular, CA is limited to process non-fixed objects. The insufficiency of the CA to deal with mobile objects has led to the advancement of agent automata designed to represent socioeconomic characteristics (Benenson & Torrens 2004).

Unlike CA, MAS focuses on agent behaviors. Most MAS models can handle both raster and vector formats. Agents are automata that can process and exchange information with other agents (Benenson & Torrens 2004). CA models focus on infrastructure and environments such as land parcels, land-use, slopes and roads whereas MAS considers interaction with other moving agents such as householders, cars and pedestrians. For example, MAS can simulate household's migration behaviors by considering their decision making criteria or simulate pedestrians walking on the streets. Benenson & Torrens (2004) highlighted the advantages of MAS as four folds: 1) agents can reflect directly human behavior; 2) agent-based models do not require comprehensive knowledge; 3) behavior rules can reveal the gaps in our knowledge of urban process; and 4) self-organizing of human behavior can be explored.

GAS combines the concepts of CA and MAS to address weaknesses of CA and MAS to represent spatial relationships. Since GAS unites CA and MAS, GAS consists of fixed and non-fixed automata (Benenson & Torrens 2004). Fixed automata may be infrastructure such as roads, buildings and lands, but non-fixed automata can be persons or cars. Time scale of each automata depends on their movements. If a simulation model focuses on pedestrian movements, a time scale is seconds or minutes. If the model is related to land use change, the time scale could be years. Neighbors of non-fixed automata in GAS are different from those in CA. In CA, neighbors are defined as cells adjacent to a cell. However, agents with similar characteristics of an agent can be neighbors in GAS. That might be a more realistic concept of neighbors but it is much complicated to match neighbors in computational models.

4 PROTOTYPE OF WEB PORTAL ON THE TERAGRID

This chapter introduces a prototype of geospatial web portals, which can provide high performance grid-enabled GIServices. The prototype adopted the OGCE toolkit as a main container to provide comprehensive Internet GIServices, including data loading and processing, spatial analysis, grid computing, simulation, 2D mapping, and 3D mapping. Figure 1 illustrates a screen shot of the prototype with its major functions (portlets) in the menu. Each Internet GIService (portlet) is independent but sharable based on the SOA. The SOA is implemented with the Java 168 Portlet Specification to offer interoperability on the web portal. Users will be able to combine multiple portlets for their research and can visualize the GIS outcomes on the web portal. The portlets are loosely coupled to provide independent Internet GIServices.

As for the development of geospatial web portals, this research will utilize a four-tier Internet GIService architecture (Zhang 2007). The first tier is the presentation tier. The presentation tier provides user interfaces so that users can access distributed Internet GIServices. The second tier is the logic tier. The logic tier defines a set of geospatial problem semantics. The geospatial problem semantics formalize problem solving procedures with several sub tasks. The logic tier connects the user interfaces in the presentation tier to the grid-enabled GIServices in the service tier. The third tier is the service tier. The service tier is to host distributed grid-enabled GIServices defined in the logic tier. Through the service tier, users can utilize data processing, analysis and visualization. The last tier is the grid tier. The grid tier is essential to advance the Internet GIServices. The grid tier provides grid-based resources, such as grid network and high performance computing nodes.

The four-tier framework can overcome the limitations of the traditional three-tier framework of the Internet GIServices. The grid tier, in particular, provides high performance computing and allows users to solve complex geographic problems. The OGCE toolkit will provide user interfaces as the presentation tier. Users can define their problems and integrate portlets on the Web portal. The OGCE toolkit is the combination of tools for grid-enabled Internet services, including the Apache Tomcat, the JSP, the JSF, and the GridSphere. The Apache Tomcat contains diverse Web applications. The server side script of the Apache Tomcat is the JSP. The JSF and JavaScript are also used to implement Web interfaces for both portlet and non-portlet applications. The JDK is used for creating GIS applications and tools

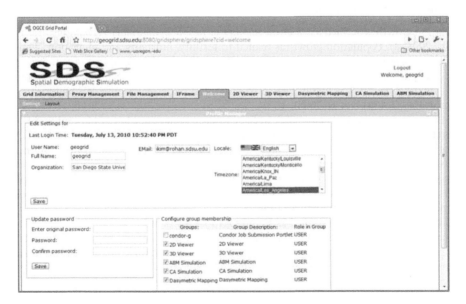

Figure 1. The screen shot of the Spatial Demographic Simulation (SDS) web portal prototype.

59

to access grid computing. Data and parallel programs are loaded on the TeraGrid through portlets on the Web portal. Then, the Java applications invoke parallel applications on the TeraGrid. Grid FTP can enable data and applications to transfer to other sites.

The Globus Toolkit is also important for the OGCE toolkit as a grid-enabled Web portal. Grid programming is implemented through the Globus Toolkit. The Globus toolkit is an open source software for building grids. Through the Globus Toolkit libraries, the Java application can connect to the grid resources. In addition, the portal framework is implemented with the GridSphere, which is an open source portlet-based web portal framework.

5 CASE STUDY—POPULATION GROWTH SIMULATION MODEL

The geospatial web portal prototype used a population growth simulation model as a case study to demonstrate the feasibility and performance enhancement of geosimulation functions on the TeraGrid. The required GIS operations in this case study include land use change modeling, dasymetric mapping, and CA population growth modeling. We implemented the dasymetric mapping functions and the CA population growth model by using parallel algorithms with Message Passing Interfaces on the TeraGrid. Necessary datasets were uploaded through the File Manager on the TeraGrid User Portal. To utilize the TeraGrid, we need to convert these GIS operations into parallel programming algorithms. Through parallel processing, datasets are divided and processed separately by multiple processors on the TeraGrid. Since one dimensional array parallel algorithms are easier to implement in the TeraGrid, several two dimensional datasets in the GIS models, such as land use and population, were converted into one dimensional array for parallel processing. Detail algorithms will be discussed later.

Because of limited funding and resources, several GIS operations in the population simulation are not implemented on the TeraGrid. Basic land use and census data processes were executed separately by the ESRI ArcGIS Desktop locally, and then we uploaded the processed data into the web portal for dasymetric and CA models. Markov chains and logistic regression analysis in land use change modeling were also conducted by the GIS software, IDRISI, locally.

5.1 *Overview of population growth*

Population growth is defined as "*how the number of people in that place is changing over time*" (Weeks 2008: 3). Population growth is related to fertility, mortality and migration. Thus, to simulate population growth, a wide variety of factors of the demographic transition need to be considered. Some factors are more dominant in the model. For example, recent population growth in the USA is highly related to migration patterns because of low fertility and low mortality (Johnson et al. 2005).

In the USA between 1995 and 2000, over 22 million Americans have migrated to different states (Franklin 2003). An examination of these numbers can reveal the reasons for particular migration patterns and contribute to understanding of the redistribution of population. It is necessary to note that migration patterns are very dynamic. Some states lose populations, whereas other states gain in-migrants. Whether a person decides to migrate or not depends on his or her personal circumstances such as marital status, gender, family size, household income, children and unemployment rates.

To build population migration, MAS and GAS might be more suitable than CA models because the first two models can define human behavior and implement objects' movements. However, the simulation model in the web portal prototype adopted a CA model because it is the easiest model to be created in a grid computing framework. We hope that the CA population growth model can demonstrate the feasibility of the grid-enabled web portal to conduct complicated analysis of geographic phenomena. The population growth model in this case study will project population of San Diego County in 2010. One thing to note is that we did not select the most comprehensive and complicated population growth model since we are not focusing on the accuracy of the population growth model. Instead, the goal of this prototype is to illustrate the feasibility and potentials of GIS simulation models and geospatial web portals on a grid computing environment.

5.2 *Local data processing*

The case study of this research simulates the population growth in the County of San Diego. To build the ideal population growth model, the first step was to collect and compile necessary GIS datasets. The Decennial Census datasets from the US Census Bureau were used for basic input data, including total population, occupancy, and vacancy. The Decennial Census data was the basis of the population growth models and provided the population growth patterns in details. In addition, the American Community Survey datasets (also from the US Census Bureau) can be used for building human behavior rules. Analyzing the American Community Survey data is expected to offer individual migration behavior rules. In particular, the PUMS (the Public Use Microdata Sample) files in the American Community Survey can describe more detailed information about population patterns. However, the American Community Survey datasets do not cover all counties but several counties per each State. Thus, we only used occupancy and vacancy rates from the Decennial Census datasets to estimate individual patterns and to construct transition rules.

The TIGER spatial datasets from the US Census Bureau provide the fundamental boundaries for census attribute datasets. To derive population growth, we needed to compare datasets of different decennial years. However, the boundary datasets of the TIGER in 1990 and in 2000 are not matched to each other. Therefore, we adopted the NHGIS (National Historical Geographic Information System 2010) datasets, which provided census attribute datasets and well matched spatial datasets of different years. In addition, since the NHGIS provides historical datasets, it will be better to project long-term population growth based on the historical census datasets.

In addition to the census datasets, land use datasets were also considered in the population growth models. The USGS LULC (the US Geological Survey Land Use and Land Cover) data might be used as the land use datasets in our models. Non-residential areas such as lake and natural habitat areas should be excluded for population growth modeling. Dasymetric mapping was adopted for the prediction of population distribution in residential and other land-use areas.

The local governments may provide more detail land use datasets, and more accurate population growth modeling can be created. The case study of this research utilized the SANDAG (the San Diego Association of Governments) website (http://www.sandag. org/) to download vector-based land use datasets in 1986, 1990, 1995, 2000, 2004 and 2008. To compare the decennial population growth, the 1990 and 2000 datasets were utilized. Then, the land use datasets were converted to the ESRI ASCII format by the ESRI ArcGIS Desktop software with 100 ft cells. As there were over one hundred land use types, the land use codes were reclassified and simplified.

5.3 *Local land use change modeling*

Population growth is closely related to urban growth and land use change. Population growth is the main reason of urban growth and, at the same time, urban growth influences population increase. Thus, land use change modeling is also related to the population growth modeling. As for the land use change modeling, the Markov chains and the logistic regression were conducted. The land use types of 1990 and 2000 datasets were classified as residential, commercial, industry, office, agriculture and vacant areas. Due to the constraints of funding and resources, we utilized the IDRISI (a raster-based GIS software) to conduct the

Markov chain analysis (rather than developing a new Markov chain analysis portlet for the prototype). Table 1 is the transition area matrix of Markov chains.

Next, logistic regression analysis about land use change was conducted. The logistic regression analysis calculated probabilities for residential areas. The dependent variable was the residential areas in 2000. The residential areas were extracted from the SANDAG land use data of 2000. The independent variables were DEM, Slope, and distances from major roads. We used the logistic regression analysis to predict land use change probability, and then applied this result (Fig. 2) to the Markov chain model for the 2010 land use prediction. The equation of the logistic regression was "$Logit(residential) = -0.8134 - 0.000412*DEM + 0.000206*Slope - 0.000232*distances from major roads$". Pseudo R^2 was 0.1929. The cut-off value was 0.2465.

The following step was to combine Markov chains and logistic regression analysis to project land use change in 2010. Based on probabilities of the logistic regression model, the transition areas were allocated to each land use type. Figure 3 shows the land use simulation results.

5.4 *Grid-enabled dasymetric mapping*

Dasymetric mapping is a method to refine choropleth maps and disaggregate the aggregated data. Since census datasets are aggregated within enumeration units (such as census tracts), population is evenly distributed in non-residential areas, such as rivers, roads, industry areas and forests. To build the population growth model, the first step was to allocate the people to the residential areas. Therefore, we used dasymetric mapping to refine the location of residential areas and the population growth regions.

Table 1.   Transition area matrix of Markov chains. The number is total cells of each land use type.

Land use	Residential	Industry	Commercial	Office	Agriculture	Vacant
Residential	1049602	236	913	1706	12614	25707
Industry	756	87431	716	1029	21	3637
Commercial	185	242	49716	535	7	488
Office	468	517	170	662850	48	1540
Agriculture	29280	1510	2022	2344	656455	67061
Vacant	99871	16773	6878	8699	20547	4285760

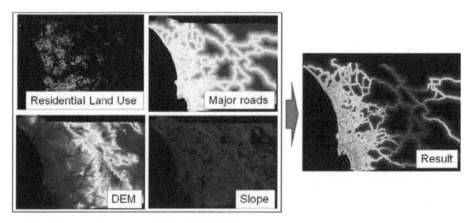

Figure 2.   Logistic regression modeling. The logistic regression analysis was conducted by the IDRISI software locally.

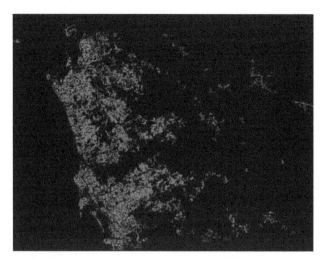

Figure 3. The land use change simulation result. The grey colored cells are projected as residential areas in 2010.

The dasymetric mapping is based on the weights of land use types. Then, population is allocated based on the weights (Equation 1):

$$pop_i = \frac{w_i \times pop}{\sum w} \tag{1}$$

where pop_i is the population of a cell. The pop denotes the total population of the enumeration unit that the cell belongs. The population of the enumeration unit is divided by total weights and then multiplied by the weight of the cell. The main challenge of dasymetric mapping is to calculate the weight factors for each land use type. Although the weights can be derived from scientific methods, such as regression models, users can also assign the weights manually in the model (Fig. 4).

In this research, we developed specialized parallel programming algorithms for dasymetric mapping and implemented them on the TeraGrid (Fig. 5). Parallel processing algorithms are more effective than regular algorithms on the TeraGrid because we can utilize multiple central processing units at the same time. Figure 5 is the pseudo codes for parallel processing of dasymetric mapping. This parallel algorithm of dasymetric mapping is very simple. The two dimensional lattice (land use coverage) is represented by one dimensional array. Parts of arrays are allocated to each processor by MPI_Bcast function. Then processing outcomes are collected by MPI_Gather function. Figure 6 illustrates the input data for dasymetric mapping and its result.

5.5 Grid-enabled cellular automata population growth simulation

The final step of the population growth models was to create a grid-enabled cellular automata (CA) model with the outcomes of dasymetric mapping, Markov chains, and logistic regression results. The contiguity filter was defined as the Moore neighborhood. The range of neighborhood is 11×11. Since most of immediate neighbors of cells were roads, wider neighborhoods were required. The range of 11 neighboring cells was 1100 feet and could cover the immediate block groups in most of census blocks.

Figure 7 illustrates the transition rules in the population growth CA model. When land use of a center cell changes from other types to residential, the population of the center cell in 2010 will become the median population of land use (changed) cells among 11×11

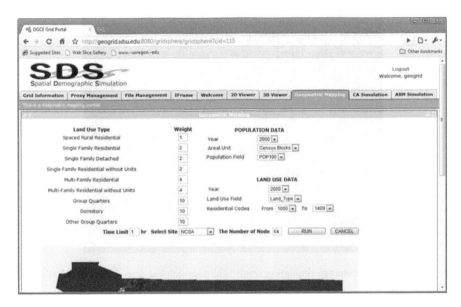

Figure 4.   The design of the dasymetric mapping portlet (under developing).

```
//Algorithm for dasymetric mapping

Read land use type, census block ID, population datasets

Calculate the weights of the cells based on land use types.

MPI_Init()
MPI_COMM_rank (MPI_COMM_WORLD, &rank)
MPI_COMM_size (MPI_COMM_WORLD, &size)

blocksize = total number of cells / (size - 1)

MPI_Bcast(Land use type, census block ID, population)

For (i = rank * blocksize to (rank +1) * blocksize) {
     Population of the cell [i] = the weight of the cell * total population of the census block
     / total weights of the census block
}

If (rank == 0) MPI_Gather (computation results from every processor)

MPI_Finalize()
```

Figure 5.   Pseudo parallel programming codes of dasymetric mapping.

neighboring cells. If there are no neighboring cells of which land use types change, the average value of total populations of all land use change cells in San Diego will be assigned as the
population of the center cell.

If the land use type of the center cell was not changed, vacant rates were compared. In
that case, comparing vacant rates reflected the attraction of the cell. The CA population
growth model was also implemented by parallel processing algorithms on the TeraGrid. All
datasets were casted to multiple processors by MPI_Bcast function (Fig. 8). Then the transition rules were processed separately by multiple processors. The parallel algorithm of the
CA population growth model was similar to that of dasymetric mapping. However, the CA
model needed to check the status of neighboring cells. Figure 8 is the pseudo codes of the
CA population growth model.

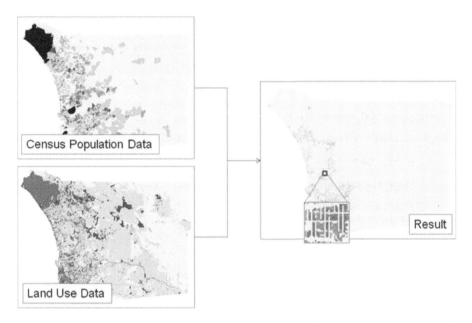

Figure 6.   Processing aggregated population data and the dasymetric mapping result on the TeraGrid.

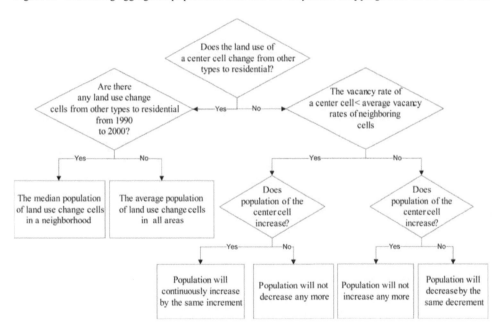

Figure 7.   Transition rules of the population growth simulation model.

5.6 *Performance of Grid-enabled GIS simulation models*

After we implemented the dasymetric mapping algorithms and the CA population growth model for the web portal on the TeraGrid, we did performance testing and calculated the run-time for the GIS simulation models in different TeraGrid settings. Table 2 and Figure 9 illustrate the computation times of dasymetric mapping from one node to 30 nodes. As the number of nodes increase, the computation time decreases. Each node on the TeraGrid has

```
//Algorithm for the CA population growth model

Read land use type, census block ID, population, occupancy and vacancy datasets.

MPI_Init()
MPI_COMM_rank (MPI_COMM_WORLD, &rank)
MPI_COMM_size (MPI_COMM_WORLD, &size)

blocksize = total number of cells / (size - 1)

MPI_Bcast(Land use type, census block ID, population, occupancy and vacancy)

For (i = rank * blocksize to (rank +1) * blocksize) {

        Check the land use change of the center cell

        //Explore neighboring cells.
        For (m = -5; m < 6; m++)
        {
            For (n = -5; n < 6; n++)
            {
                Neighboring cells = i + m + n * total number of columns;
                Calculate vacancy rates of neighborhoods;
            }
        }
        Compare the vacancy rate of the center cell with neighboring cells;
        Calculate population in 2010;
}

If (rank == 0) MPI_GATHER (computation results from every processor)

MPI_Finalize()
```

Figure 8.   Pseudo parallel programming codes of the CA population growth model.

Table 2.   Dasymetric mapping computation times on the TeraGrid.

Number of nodes	1	2	3	4	5	10	15	20	30
Time (sec.)	3426	1356	944	766	667	486	429	402	375

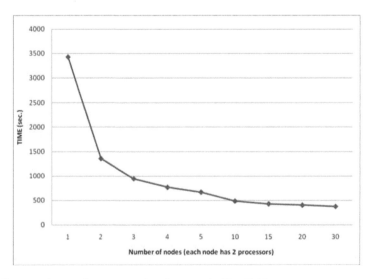

Figure 9.   Dasymetric mapping computation times on the TeraGrid.

Table 3. CA population growth model computation times on the TeraGrid.

Number of nodes	1	2	3	4	5	10	15	20	30
Time (sec.)	154	112	105	100	98	91	89	90	87

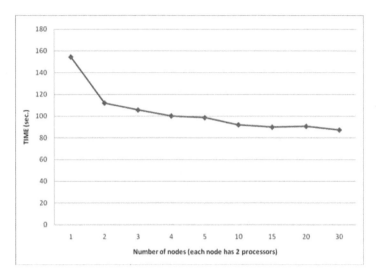

Figure 10. CA population growth model computation times on the TeraGrid.

two high-performance processors. This result shows how significantly the grid-computing framework can improve the performance of the GIS simulation model (from 3426 seconds with one node to 375 seconds with thirty nodes). We also tested the same model in a local powerful computer. It took 3420 seconds (similar to the one node in the TeraGrid) to complete the GIS simulation.

Table 3 and Figure 10 illustrate the computation times of the CA population growth model. The result shows a similar pattern with the dasymetric mapping computation times. The performance testing indicates that TeraGrid and parallel programming algorithms can reduce GIS simulation computation times significantly and provide more efficient distributed GIS modeling services via the geospatial web portals.

6 DISCUSSION AND CONCLUSION

This chapter discusses the possibilities of GIS simulation models established on a grid computing environment (TeraGrid) and GIS computation performance testing results in different TeraGrid settings. The OGCE toolkit was utilized for the framework of the geospatial web portal. Some major GIS functions, such as data processing and land use prediction, have not been implemented yet, but are under development. This web portal illustrates a great potential framework for next generation Internet GIServices.

Along with the advancement of remote sensing technology and web GIS tools, geospatial datasets for GIS models have become much larger, with higher spatial resolutions and temporal resolutions. In our case study, it took 57 minutes to process dasymetric mapping in a local GIS workstation, but the same model took only 6.3 minutes on the TeraGrid (with 30 nodes). With multiple user requests, the grid-enabled Internet GIServices can be more efficient to compute complicated GIS simulation models and prevent crashes or overloads of web servers. This case study only addressed the parallel processing algorithms in grid computing. Future studies will need to consider data input and output settings with various formats in the TeraGrid. We might compare different GIS databases for grid-enabled models and seek the best setting of GIS data format (representation) in a grid computing environment.

Grid-enabled Internet GIServices with high performance computing capability can facilitate the advancement of GIScience research, provide a problem solving environment (PSE), and formalize a virtual organization (VO). All computing resources for the GIScience research can be provided through the user-friendly web portal. Many researchers can collaborate to solve a challenging geography problem on the web portal. They can conduct GIS analysis and share the outcomes directly on the TeraGrid. Moreover, the grid-enabled web portal can provide educational tools to the public. Simulation model outcomes can convey knowledge about spatial phenomena effectively to the public and students. Users can easily manipulate the GIS simulation models (changing the prediction year or weight factors) and conduct the GIS model by themselves.

This study did not adopt the most accurate simulation model for population growth, but focused on the potentials of TeraGrid and the development of grid-enabled Internet GIServices. More comprehensive GIS simulation models will be required to provide better prediction results in the future. Besides CA models, the population growth simulation models of MAS and GAS should be built and tested on the TeraGrid. Traditional CA models are fundamentally based on the raster-based approach. Although grids of raster datasets are very suitable for parallel processing, the raster-based approach has many limitations to represent spatial phenomena. Instead, MAS and GAS might be a better model for grid computing because it combines both vector and raster based approaches. Migration of people is an example why a vector-based approach is needed. CA models have limitations to describe the movement of objects. Raster models have difficulties to represent accurate locations of objects. The vector-based approach is more suitable for the migration of people. Thus GAS may be more appropriate to build a more practical population growth simulation model. However, parallel processing algorithms of GAS can be more difficult in developing population growth simulation models on the TeraGrid.

In summary, this study provides a successful example of grid-enabled Internet GIServices for executing population growth simulation models. We hope that the GIScience community will continue this effort and develop more grid-enabled GIS models in various domains, such as hydrological models, weather change models, and traffic models. The development of new geospatial web portals and grid-enabled GIServices will be the major components of the next generation of Internet GIServices.

ACKNOWLEDGEMENT

This research was supported in part by the National Science Foundation ATE program, National Geospatial Technology Center (DUE #0801893) and the National Science Foundation through TeraGrid resources provided by San Diego Supercomputer Center.

REFERENCES

Anselin, L., Kim, Y.W. & Syabri, I. (2003) Web-based analytical tools for the exploration of spatial data. *Journal of Geographical Systems*, 6 (2), 197–218.
Batty, M. (1997) Cellular automata and urban form: A primer. *Journal of the American Planning Association*, 63 (2), 266–275.
Benenson, I. & Torrens, P.M. (2004) *Geosimulation: Automata-based modeling of urban phenomena.* West Sussex, UK, John Wiley & Sons Ltd.
Benenson, I., Omer, I. & Hatna, E. (2002) Entity-based modeling of urban residential dynamics: The case of Yaffo, Tel Aviv. *Environmental and Planning B: Planning and Design*, 29, 491–512.

Catlett, C. et al. (2007) TeraGrid: Analysis of Organization, System Architecture, and Middleware Enabling New Types of Applications, HPC and Grids in Action, Ed. Lucio Grandinetti, IOS Press 'Advances in Parallel Computing' series, Amsterdam.

Clarke, K.C. & Gaydos, L.J. (1998) Loose-coupling a cellular automaton model and GIS: Long-term urban growth prediction for San Francisco and Washington/Baltimore. *International Journal of Geographical Information Science*, 12 (7), 699–714.

Clarke, K.C., Hoppen, S. & Gaydos, L. (1997) A self-modifying cellular automaton model of historical urbanization in the San Francisco bay area. *Environment and Planning B: Planning and Design*, 24 (2), 247–261.

Couclelis, H. (1985) Cellular worlds: A framework for modeling micro-macro dynamics. *Environmental and Planning, A17*, 585–596.

Franklin, R.S. (2003) *Domestic migration across regions, divisions, and states: 1995 to 2000.* US Census Bureau.

Gallopoulos, E., Houstis, E. & Rice, J.R. (1994) Problem-solving environments for computational science. *IEEE Computational Science and Engineering*, 1, 11–23.

Johnson, K.M., Voss, P.R., Hammer, R.B., Fuguitt, G.V. & McNiven, S. (2005) Temporal and spatial variation in age-specific net migration in the United States. *Demography*, 42 (4), 791–812.

Lei, Z., Pijanowsk, B.C. & Olson, J. (2005) Distributed modeling architecture of a multi-agent-based behavioral economic landscape (MABEL) Model. *Simulation*, 81 (7), 503–515.

NSF (National Science Foundation) Cyberinfrastrucutre Council. (2007). *Cyberinfrastructure vision for 21 st century discovery.* Arlington, VA, National Science Foundation.

O'Sullivan D. (2001). Graph-cellular automata: a generalized discrete urban and regional model. *Environment and Planning B: Planning and Design*, 28, 687–705.

O'Sullivan, D. (2002) Toward micro-scale spatial modeling of gentrification. *Journal of Geographical Systems*, 4, 251–274.

Parker, D.C., Manson, S.M., Janssen, M.A., Hoffman, M.J. & Deadman, P. (2003) Multi-agent systems for the simulation of land-use and land-cover change: a review, *Annals of the Association of American Geographers*, 93 (2), 314–337.

Peng, Z.R. & Tsou, M.H. (2003) *Internet GIS: Distributed geographic information services for the internet and wireless network.* New Jersey, John Wiley & Sons.

Semboloni, F. (2000) The growth of an urban cluster into a dynamic self-modifying spatial pattern. *Environmental and Planning B: Planning and Design*, 27, 549–564.

Semboloni, F., Assfalg, J., Armeni, S., Gianassi, R. & Marsoni, F. (2004) CityDev, an interactive multi-agents urban model on the web. *Computers, Environment and Urban Systems*, 28, 45–64.

Shi, W.Z. & Pang, M.Y.C. (2000) Development of Voronoi-based cellular automata-an integrated dynamic model for Geographical Information Systems. *International Journal of Geographical Information Science*, 14 (5), 455–474.

Stevens, D. & Dragicevic, S. (2007) A GIS-based irregular cellular automata model of land-use change. *Environment and Planning B: Planning and Design*, 34, 708–724.

Takatsuka, M. & Gahegan, M. (2001) Sharing exploratory geospatial analysis and decision making using GeoVISTA studio: From a desktop to the web. *Journal of Geographic Information and Decision Analysis*, 5 (2), 129–139.

Tang, W. & Wang, S. (2009) HPABM: A hierarchical parallel simulation framework for spatially-explicit agent-based models. *Transactions in GIS*, 13 (3), 315–333.

Tobler. W. (1979) Cellular geography. In: Gale, S. & Ollson, G. (eds.) *Philosophy in Geography*, Dordrecht, D. Reidel Publishing Company. pp. 379–386.

Torrens, P.M. (2007) A geographic automata model of residential mobility. *Environment and Planning B: Planning and Design*, 34 (2), 200–222.

Tsou, M.H. (2004). Integrating web-based GIS and image processing tools for environmental monitoring and natural resource management. *Journal of Geographical Systems*, 6 (2), 155–174.

Wang, S. & Liu, Y. (2009) TeraGrid GIScience gateway: Bridging cyberinfrastructure and GIScience. *International Journal of Geographic Information Science*, 23 (5), 631–656.

Weeks, J.R. (2008) *Population*. Belmont, CA, Thomson Wadsworth.

Yang, C. & Raskin, R. (2009) Introduction to distributed geographic information processing research. *International Journal of Geographical Information Science*, 23 (5), 553–560.

Yassemi, S. & Dragicevic, S. (2008) Web cellular automata: a forest fire modeling approach and prototype tool. *Cartography and Geographic Information Science*, 35 (2), 103–115.

Zhang, T. (2007) Developing grid-enabled internet GIService to support geospatial cyberinfrastructure: A pilot study in accessibility. Doctoral dissertation. San Diego State University.

Zhang, T. & Tsou, M.H. (2009) Developing a grid-enabled spatial web portal for internet GIServices and geospatial cyberinfrastructure. *International Journal of Geographic Information Science*, 23 (5), 605–630.

WEBSITES

Geoscience Network (GEON). [Online] Available from: http://www.geongrid.org [Accessed 13th July 2010].

Linked Environments for Atmospheric Discovery (LEAD). [Online] Available from: https://portal.leadproject.org [Accessed 13th July 2010].

National Historical Geographic Information Systems (NHGIS). [Online] Available from: http://www.nhgis.org [Accessed 13th July 2010].

Science Environment for Ecological Knowledge (SEEK): [Online] Available from: http://seek.ecoinformatics.org/ [Accessed 13th July 2010].

*Advances in Web-based GIS, Mapping Services
and Applications – Li, Dragićević & Veenendaal (eds)*
© 2011 Taylor & Francis Group, London, ISBN 978-0-415-80483-7

An OGC Web Processing Service for automated interpolation

Jan Dürrfeld, Jochen Bisier & Edzer Pebesma
Institute for Geoinformatics, University of Muenster, Muenster, Germany

ABSTRACT: For decision support in disaster management, maps of interpolated values, based on measurements from sensor networks, are indispensable. It should be possible to get such maps automatically in near real-time. This chapter describes an automated interpolation service that was developed in the context of the INTAMAP project. The OGC Web Processing Service and *UncertML*, an XML grammar for encapsulating uncertainties, are introduced and explained, and the actual implementation status is reported. In addition, the requirements that have to be considered when automating interpolation are described as well as several use-cases. Finally, the value of such a service and its generic architecture are discussed from a remotes sensing workflow development perspective.

1 INTRODUCTION

Over the last few years there has been a strong development in the domain of sensor networks and geospatial web services. Several specifications have been released by the Open Geospatial Consortium (OGC) for interfacing and exchanging geospatial data. Examples include the Web Processing Service (WPS) (Open Geospatial Consortium Inc. 2007e), Sensor Observation Service (SOS) (Open Geospatial Consortium Inc. 2007c), and Observations and Measurements (O&M) (Open Geospatial Consortium Inc. 2007a). In addition, several applications that implement these standards have been developed (Raape et al. 2009, Stasch et al. 2008), and provide building blocks necessary for the chaining of web services. In this manner, Geospatial Data Infrastructures (GDI) are realised using Service Oriented Architectures (SOA), so that they can combine relatively static information such as soil types or administrative boundaries with dynamic sensor information like temperature or rainfall measurements, or remote sensing imagery. This combination at the technical level provides interesting challenges for data analysis workflows. This chapter describes the development of an automated interpolation web service.

In case of an emergency and crisis management, maps of interpolated values derived from sensor data are required for decision making, e.g. because decisions relate to areas rather than to punctual measurements. For example, it may be necessary to make a decision, based on observations of monitoring stations, as to which areas have to be evacuated. For such an instance, an automated interpolation service is useful. Another motivation for developing such a service is that end users who have no strong background in interpolation methods should be able to use advanced interpolation methods without interacting with expensive or complicated software.

Automated interpolation raises a number of questions that lead to several requirements that have to be considered before such a service can be implemented. Which of the supported interpolation methods should be automatically chosen? Which observations should be used for an appropriate interpolation result? How do we deal with anisotropy? How do we deal with extreme value distributions? How do we deal with interpolation errors and their propagation through the workflow? One of the major aims of the INTAMAP (INTAMAP 2009) specific targeted research project (STREP), funded under the sixth framework program of the European Commission, was the development of an automated interpolation service. This chapter describes the architecture, concepts, and applications of that service and discusses its relevance from a remote sensing perspective. Details about the actual interpolation methods implemented, and the criteria and decision tree used to choose between them are given in Pebesma et al. 2011.

The outline of this chapter is as follows. In the next section the actual status of the implementation of the automated interpolation web service is described. The system architecture, based on the OGC WPS specification and on *UncertML* is introduced. Then, choosing interpolation methods and estimating parameters without human interaction will be discussed in Section 3. Section 4 explains some use cases for which we believe automated interpolation is meaningful and the last section contains conclusions and a discussion, and addresses how this service could be useful for remote sensing applications, or how the generic architecture might be used, for example for developing an automated classification service.

2   A SERVICE FOR AUTOMATED INTERPOLATION

The automated interpolation service is implemented as an OGC WPS. It consists of two components: a WPS interface and an R environment with several R packages as an interpolation back-end. For the WPS, the Java implementation of 52° North (52North 2010) has been used. The WPS interface is responsible for the communication between a client and the interpolation service instance, and for parsing requests and extracting the required information such as the interpolation input data. R is an environment for statistical computing that is used for implementing different interpolation methods available in R packages. To connect the WPS interface with the R environment, the R package Rserve, which runs R in listening mode, is used to allow R commands to be sent over the TCP/IP protocol to the R environment and to retrieve computed results.

This section gives an overview of the system architecture and is divided into three subsections. The first gives a brief introduction of the WPS specification and the second introduces *UncertML*, a markup language for encoding probabilistic information. The last subsection describes the system architecture and the workflow.

2.1   *WPS specification*

The WPS standard specifies an interface that provides web services for the processing of geospatial data. For instance, the processing could be either a simple calculation like the determination of NDVI (Normalised Differenced Vegetation Index) or a more complex methodology like image classification or running a dynamic spatial model. Therefore, the general form of input and output parameters, methods for addressing input as well as output data, and the communication between the server and the client are standardised. This is achieved by defining XML schemata and request operations which have to be applied in a WPS compliant Processing Service.

The interface requires the implementation of three mandatory request operations: `getCapabilities`, `describeProcess`, and `executeProcess` as shown in Figure 1. First, a client sends a `getCapabilities` request to retrieve an XML document providing general information about the server instance. This contains a list of processes provided by the service and can contain additional metadata like contact information of the provider or keywords. In a second step, the client requests detailed information about a certain service. Finally, the client sends an execute request to start a specific process by posting an XML document which contains the input data and the requested output parameters to be returned. As presented in Figure 2, the `getCapabilities` request is inherited from the Web Service Common specification (WS-Common) (Open Geospatial Consortium Inc. 2007f) and serves users with general information about the service instance. According to

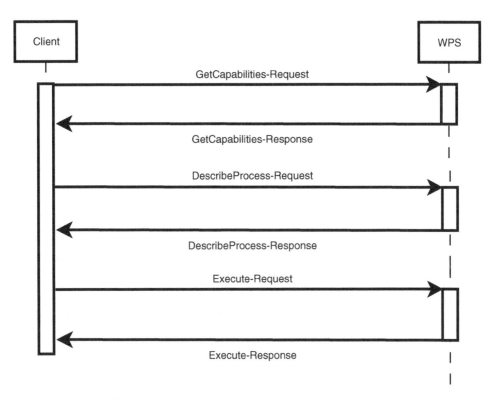

Figure 1. Sequence diagram of WPS.

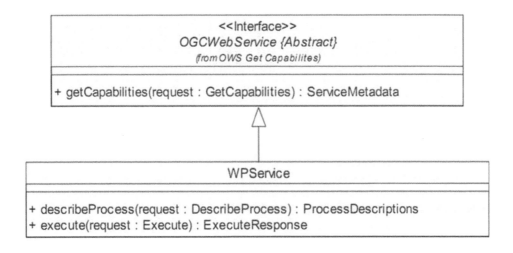

Figure 2. UML class diagram of the WPS interface.

73

the specification this operation must be implemented as a HTTP GET method with Key Value Pair (KVP) encoding. An exemplary request is:

```
http://wps.intamap.server:8080/WebProcessingService?
Service=WPS&
Request=getCapabilities&
Version=1.0.0
```

The request consists of several KVPs: Request defines the request method (e.g. getCapabilities); Version ascertains the version number of the applied service specification (e.g. 1.0.0); and Service identifies the service that is used (here WPS). The response should be a capabilities document that extends the capabilities document of the WS-Common by a ProcessOffering part (see Listing 1). This part lists the available processing services whereby each process has an identifier (Listing 1, line 68) and a title (Listing 1, line 69). Besides these particulars, processes can have: an abstract, a profile and a link to a Web Service Description Language (WSDL) file. The profile has the purpose to classify different web processing services for generic use (not contained in the Listing). The WSDL file is required if, for example, a BPEL (Business Process Execution Language) engine is used to chain services.

The describeProcess request returns detailed information about a process or processes. The request operation must be implemented as a HTTP GET method with KVP encoding. The request must contain the KVP as listed in Table 1. An example request is as follows:

```
http://wps.intamap.server:8080/WebProcessingService?
Service=WPS&
Request=describeProcess&
Version=1.0.0&
Identifier=org.intamap.wps.Interpolate
```

The response document extends the information about a specific service part by the description of the input (DataInputs) and output (ProcessOutputs) parameters, whereby the number of input and output parameters is not limited. As presented in Listings 2 and 3, each input and output parameter is referenced by an identifier (Listing 2, line 8 and Listing 3, line 195) and has a title (Listing 2, line 9 and Listing 3, line 196) like different processes in the getCapabilities

```
66    <wps:ProcessOfferings>
67        <wps:Process wps:processVersion="1">
68            <ows:Identifier>org.intamap.wps.Interpolate</ows:Identifier>
69            <ows:Title>INTAMAP Automatic Interpolation</ows:Title>
70        </wps:Process>
71    </wps:ProcessOfferings>
```

Listing 1.   Excerpt of a getCapabilities response document.

Table 1.   KVP for the describe Process operation.

Key	Value
Service	WPS
Request	describeProcess
Version	1.0.0
Identifier	name of process

74

document and could contain an abstract (Listing 2, line 10 and Listing 3, line 197). Moreover, their data types are listed. Three different data types are defined in the specification and could be assigned: `LiteralValue`, `ComplexValue`, and `BoundingBoxValue`. The `Literal-Value type` (Listing 2, lines 16–21) represents simple data like integer values or strings. One can define values (Listing 2, line 17), a range of values, a default value, and/or the unit of measure (UOM). The `ComplexValue` type is for the representation of complex data structures like Geography Markup Language (GML) (Open Geospatial Consortium Inc. 2007b) features or images (Listing 3, lines 198–211). It is described by a MIME type (Listing 3, line 201). In addition, the encoding and a URI to a schema can be specified (Listing 3, line 202).

The `executeProcess` request starts a process with the input data provided. According to the specification, the operation must be implemented as a HTTP POST method with XML encoding. The XML document, which is uploaded to the server via HTTP POST, contains the values assigned to the input and output parameters. The input data is either encoded in the XML request document or is a Uniform Resource Identifier (URI) to a web accessible resource. The response depends on the value of the `ResponseForm` parameter in a request. The output is either embedded in the XML response document or it contains a reference to a web accessible online resource. Due to the fact that geoprocessing could take a

```
7     <Input minOccurs="0" maxOccurs="1">
8         <ows:Identifier>SOSURL</ows:Identifier>
9         <ows:Title>A URL pointing towards a fully functioning Sensor Observation
            ... Servce</ows:Title>
10        <ows:Abstract>
11            Using this URL you can specify a full SOS GET request, or just
                ...provide a URL that will be used to send
12            the 'SOSRequest' parameter to. If the SOSRequest parameter is
                ...empty and the URL is not a valid GET request
13            then an exception will be raised. Also, we only support the
                ...Observations and Measurements 'Measurement' type
14            as a result from a SOS request.
15        </ows:Abstract>
16        <LiteralData>
17            <ows:DataType xmlns:xs="http://www.w3.org/2001/XMLSchema"
                ...ows:reference="xs:string"/>
18            <ows:AllowedValues>
19                <ows:Value>Any valid SOS GET request or a URL pointing to
                    ...a SOS</ows:Value>
20            </ows:AllowedValues>
21        </LiteralData>
22    </Input>
```

Listing 2.   Excerpt 1 of a DescribeProcess response document.

```
194   <Output>
195       <ows:Identifier>PredictedValues</ows:Identifier>
196       <ows:Title>The predicted values</ows:Title>
197       <ows:Abstract>This output contains the interpolation result encoded as
            ...UncertML</ows:Abstract>
198       <ComplexOutput>
199           <Default>
200               <Format>
201                   <MimeType>text/XML</MimeType>
202                   <Schema>http://schemas.uncertml.org/1.0.0/UncertML
                        ....xsd</Schema>
203               </Format>
204           </Default>
205           <Supported>
206               <Format>
207                   <MimeType>text/XML</MimeType>
208                   <Schema>http://schemas.uncertml.org/1.0.0/UncertML
                        ....xsd</Schema>
209               </Format>
210           </Supported>
211       </ComplexOutput>
212   </Output>
```

Listing 3.   Excerpt 2 of a DescribeProcess response document.

```
10    <om:ObservationCollection xmlns:om="http://www.opengis.net/om/1.0" xmlns:gml="http://www.
      ...opengis.net/gml" xmlns:xlink="http://www.w3.org/1999/xlink" xmlns:sa="http://www.
      ...opengis.net/sampling/1.0">
11     <om:member>
12      <om:Observation gml:id="MEUSE1">
13       <om:samplingTime />
14       <om:procedure />
15       <om:observedProperty />
16       <om:featureOfInterest>
17        <sa:SamplingPoint>
18         <sa:sampledFeature />
19         <sa:position>
20          <gml:Point>
21           <gml:pos>181072.0 333611.0</gml:pos>
22          </gml:Point>
23         </sa:position>
24        </sa:SamplingPoint>
25       </om:featureOfInterest>
26       <om:result>11.7</om:result>
27      </om:Observation>
28     </om:member>
```

Listing 4. Part of an ExecuteProcess request document where a single point observations is encoded using O&M.

```
2840 <wps:Input>
2841  <ows:Identifier>Domain</ows:Identifier>
2842  <wps:Data>
2843   <wps:ComplexData>
2844    <gml:Polygon gml:id="POLY1" xmlns:gml="http://www.opengis.net/gml" xmlns:xsi="http://www.w3.
         ...org/2001/XMLSchema-instance">
2845     <gml:exterior>
2846      <gml:LinearRing>
2847       <gml:posList>181540 333140 180700 330100 179380 329660 179220 329620 178900 329620
          ...178780 329660 178660 329740 178540 329860 178500 329980 178460 330140 178460
          ...330180 178500 330420 178540 330500 179580 332100 180940 333500 181180 333740
          ...181220 333700 181300 333580 181540 333180 181540 333140</gml:posList>
2848      </gml:LinearRing>
2849     </gml:exterior>
2850    </gml:Polygon>
2851   </wps:ComplexData>
2852  </wps:Data>
2853 </wps:Input>
```

Listing 5. Excerpt 2 of a ExecuteProcess request document. The sequence of numbers indicates the coordinates of the linear ring enclosing the interpolation locations.

long time, a further parameter allows a request for asynchronous messaging. After initialising a specific process a status message is sent to the client with information about the start time. Furthermore, it is possible to request the storage of the output on a server. Listings 4 and 5 are excerpts of a request document. They show the assignment of input parameters for the INTAMAP WPS. Listing 4 shows an observation corresponding to the O&M specification, and Listing 5 an assignment of a polygon as domain.

2.2 *UncertML*

Interpolated values are always subject to interpolation errors that can be described in terms of probability distributions. Existing data encoding formats do not support probability distributions. Therefore, the XML grammar *UncertML* (Williams et al. 2009) has been developed in the context of the INTAMAP project. *UncertML* is a conceptual model that is designed for encapsulating probabilistic uncertainties and that is realised by defining XML schemata. The use of these schemata leads to an interoperable framework for quantifying and exchanging uncertainties. This allows uncertainty to be propagated through processing chains and thus providing a more realistic representation of knowledge. The design of *UncertML* is generic and not confined to spatial, temporal or spatio-temporal data.

Uncertainty can be expressed in different ways using *UncertML*. For this purpose three methods were declared: `realisations`, `statistics`, and `distributions`. Below is an overview for each of these methods for describing uncertainty including the common elements. `Realisations` are used if uncertainty is characterised by a sample. This could be a sample for a single variable or a set of spatial fields. It can be used to characterise complex distributions and multivariate correlation structures, typically output or part of a Monte Carlo experiment, and represents uncertainty distributions by a finite sample approximation. In opposition to `realisations`, `distributions` can be used to encode uncertainty if distributions of data are clearly understood. The elements listed below are specifically designed to allow a concise encapsulation of all parametric distributions without sacrificing the simplicity of *UncertML*: `Distribution`, `DistributionArray`, `Mixturemodel`, and `MultivariateDistribution`. `Statistics` are used to provide a summary of a variable ranging from measures of location (e.g. mean, mode, or median) to measures of dispersion (e.g. range, standard deviation, or variance). There is a large number of options available in *UncertML* for describing such statistics. While certain statistics do not provide any information about uncertainty they are often used in conjunction with other statistics to provide a concise but detailed summary. The following elements are available: `Statistic`, `Probability`, `DiscreteProbability`, `Quantile`, `Moment`, `StatisticsArray`, and `StatisticsRecord`. For a complete overview of how these elements can be used to describe various statistics the reader is referred to the encoding specification in an OGC discussion paper (Williams et al. 2009). A follow-up project of INTAMAP, called UncertWeb, uses and further develops *UncertML* to realize the uncertainty enabled model web.

2.3 *Implementation*

For the **INTAMAP** WPS the Java based implementation of the WPS from 52° North was used and extended. Figure 3 shows an overview of the system architecture. To serve as a frontend to the R statistical computing environment, a WPS to R interface has been developed. The Java package `org.rosuda.REngine` was used to communicate with an R environment over TCP/IP. The interpolation process is done in R by calling appropriate functions which have been developed in different R packages (Pebesma 2004). The supported input and output parameters are defined in an XML document. Table 2 summarises the defined input parameters and their purposes. The `ObservationCollection` parameter includes the measured values

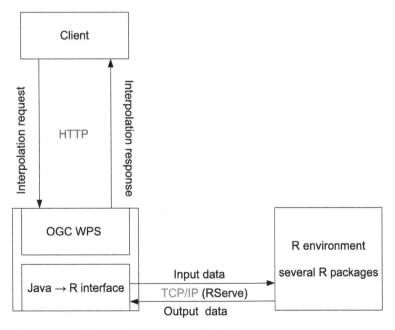

Figure 3.   The system architecture of the INTAMAP project.

Table 2. Overview of the defined input parameters.

input parameter	data type	mandatory	description
SOSURL	LiteralValue	true	A URL pointing towards a fully functioning Sensor Observation Service.
SOSRequest	ComplexValue	true	This parameter is a full valid SOS GetObservation request. The result of this request once sent to the URL contained within the SOSURL should be an ObservationCollection containing Measurement elements.
Observation Collection	ComplexValue	true	You can supply a series of ObservationCollection types to group observations. This is useful for bias correction. Currently only a single ObservationCollection is parsed.
domain	ComplexValue	false	The prediction domain specifies the locations at where you wish the algorithm to predict, it can be a GML Point, Polygon, MultiPoint, MultiPolygon or RectifiedGrid.
Prediction Type	ComplexValue	false	This parameter allows you to specify the exact type you want to be returned (encoded in *UncertML*). Currently the supported types are: Mean, Variance, Quantile and Probabilities—they can be grouped using a StatisticsRecord.
Method Parameters	ComplexValue	false	A collection of key/value pairs allowing customisation over the interpolation methods.
TUCAnisotropy	LiteralValue	false	Use the TUC method for anisotropy detection. The default is set to true.
BiasCorrection	LiteralValue	false	Apply bias correction to observations.
numberOf Predictions	LiteralValue	false	When supplying polygons as a domain you can specify the number of desired prediction locations as represented using this parameter.
maximumTime	LiteralValue	false	You can specify the maximum amount of time you are willing to allow for prediction. If the method is not able to complete within the required time an exception is raised.
MethodName	LiteralValue	false	You can request a specific method to be used for interpolation. Currently there is only 1 supported method: 'automap'.

corresponding to the O&M specification, which form the input of the interpolation. As an alternative, one can assign a SOS GetObservations request with the SOSRequest parameter and to use the response as an ObservationCollection. The PredictionType parameter specifies the type of statistical characteristics that should be specified in *UncertML*. Currently, the supported types are: Mean, Variance, Quantile and Exceeding Probabilities for the predictions. The domain parameter defines the prediction locations and is one of the listed geometry types according to the GML specification: grid, multipoint, polygon and multipolygon. This parameter is optional and in the case that it is not assigned in the request a grid will be generated for the prediction locations. In case of a polygon, the polygon is interpreted as the outer boundary of a gridded region. The maximumTime parameter allows the specification of the maximum allowed amount of time for the interpolation. If the method is not able to complete within the required time an exception is raised. The maximumTime parameter was introduced for time-critical application, when a result should be accessible within a certain amount of time. The motivation for this is that certain interpolation methods tailored to deal with extreme value distributions have extremely long run times for moderately large data sets.

The defined output parameters are shown in Table 3. Like the request document it is a composition of SWE, GML, and *UncertML* elements. The prediction locations (Domain) are encoded in GML and the prediction types (Predicted-Values) in *UncertML*. As an alternative to embedding the interpolation results in the response document, the prediction values can be saved as georeferenced raster data in the GeoTIFF format and served over a Web Coverage Service (WCS) (Open Geospatial Consortium Inc. 2007d) server instance so that they are accessible through a more usual interface for raster data. Besides providing information about the Domain and the PredictedValues a prediction report is generated that embodies information about the pre-processing, processing and post-processing of the data used.

The different interpolation methods were implemented as R packages automap, spatialCopulas, and astonGeostats. The first one is for ordinary kriging and is a front-end to the functionality provided by the R package gstat (Pebesma 2004). It is applied if the observations are close to normally distributed and measurement errors are specified. The second package provides a Maximum Likelihood estimation that uses copula

Table 3. Overview of the defined output parameters.

output parameter	description
PredictedValues	This output contains the interpolation result encoded as *UncertML*.
Domain	This output contains the prediction domain encoded as GML.
PredictionReport	A full report of the pre-processing, processing and post-processing applied to the data.
MethodParameters	You can use this output to emulate the exact conditions used in this request in future requests (by using the MethodParameters input).
RasterOutput	A list of URLs that point to GeoTiff files generated by the interpolation service.

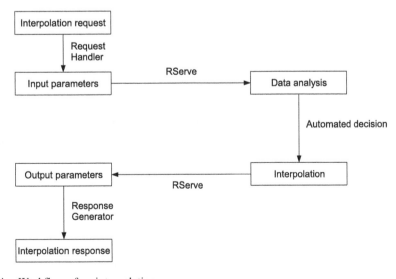

Figure 4. Workflow of an interpolation.

functions and that is applied if the observations are extreme value distributed and no errors are encoded. If the measurements contain a specified error distribution, a Maximum Likelihood estimation with a projected sparse Gaussian process is used (Ingram et al. 2007).

The workflow of an interpolation is shown in Figure 4. First, a user sends an execution request to the WPS. This request contains either the observations or a SOS GetObservation request is assigned and the observation are directly sent from the SOS instance to the WPS. The WPS then extracts the assigned parameters and creates the appropriate R commands. Via TCP/IP these commands are sent to the R environment, where the input data is analysed and the interpolation method is chosen and applied. The results of the interpolation are then sent back to the WPS and the response document is generated.

3 AUTOMATING INTERPOLATION

The phrase "automatic interpolation" suggests that interpolation can be done in an automatic fashion. Although some people may refute this opinion, one should realise that for example the inverse distance weighted interpolation algorithm (Burrough and McDonnell 1998) always returns a more or less sane value, regardless of the data. More interesting forms of interpolation are those where the variability of the data is captured by some form of a statistical model for spatial variation, such as is the case for geostatistical models. For such models, there is always the requirement that (i) the form of the model should be capable of accommodating the characteristics of the data and (ii) it should be possible to automatically fit the parameters of the model from the observation data. Given a choice for a class of models, it is always possible to come up with data for which either (i) or (ii) fail.

Having said that, we should acknowledge that automatic interpolation algorithms that always succeed are out of reach. One alternative to continue would be to leave statistical interpolation in the hands of the professional geostatisticians, meaning that their tools will not be used in cases where their service is not available or takes too long, e.g. in emergency management situations. The challenge then becomes to develop and choose algorithms that are versatile and succeed often.

Users of the interpolation service should realise that they assume that the data submitted to the automatic interpolation can be interpolated. As with any software that does data analysis, some of the responsibility will always remain with the user.

Issues commonly encountered when interpolation problems become difficult are: (i) outlier detection and removal, (ii) dealing with extreme value distributions, (iii) anisotropy, (iv) dealing with measurement errors that have non-constant variance, and (v) change of support. These are discussed in the following subsections.

3.1 *Outlier detection*

Outliers are considered extreme observations that clearly behave differently to the bulk of the observations. They can result from plain errors (typos, digital noise in data loggers) or may be true values but arising from a process different from the process that generated the majority of the observations. In the case of outliers that are erroneous data, they should be removed before interpolation. Outlier removal is not implemented in the automated interpolation method developed within INTAMAP because it cannot be decided automatically whether extreme values are erroneous or whether they may indicate actual extreme values. Instead, in case of extreme and outlying values, the interpolation will try hard to deal with them. In many cases outlier removal will be manual and happens interactively in a client environment rather than directly in an interpolation service.

3.2 *Extreme value distributions*

In cases where data with extremes are offered to an interpolation service, usual linear geostatistical approaches fail because the modelling of the spatial correlation is difficult to perform.

Recent developments (Pilz and Spöck, 2008) have suggested to use Box-Cox transforms of the data, but experiments have shown that these methods are outperformed by interpolation methods based on the Copula method (Pilz et al. 2008). In the present interpolation service, based on the marginal distribution of the observation data, an automatic decision is taken whether a linear ("classical") geostatistical approach is taken, or whether the (Gaussian) Copula-based interpolation method is used.

3.3 *Anisotropy detection*

Anisotropy, the dependency of spatial correlation on direction, may be present and sufficiently strong that taking it into account will improve interpolation. The automated interpolation algorithm does take anisotropy into account using a novel method developed by Chorti and Hristopulos (2008). It allows for a non-parametric estimation of two-dimensional, geometric anisotropy and gives standard errors for the parameters. Only in cases where the data indicate significant anisotropy, will it be corrected for this by rescaling the coordinates. In the next step, in an isotropic coordinate system, the spatial correlation will be modelled.

3.4 *Measurement errors*

Although all measurement data are subject to measurement errors, linear geostatistical methods can only deal with the case where this error has a constant variance by way of fitting a nugget variance. In cases where, for example, data from different sources are merged, one may have knowledge about specific measurement error distributions that could, for example, have strongly differing variances. In such a case, interpolation needs to incorporate a different weighting, depending on the magnitude of these errors. A specific method has been developed (Ingram et al. 2007) called the *projected sparse Gaussian process*. When observations are provided with specific, varying error distributions, this method is chosen as the default interpolation method.

3.5 *Change of support*

In specific cases, interpolated values are not needed for elements that have the physical size of a measurement (typically centimetres to metres) but for aggregates (e.g. averages) of elements over larger areas–one may think of the average concentration over a city, over a county or country, or over a remote sensing pixel of 30 m × 30 m. In such cases, interpolated values and their standard error (or error distribution) have to be integrated over this area. In classical geostatistics this is performed with *block kriging*; the INTAMAP interpolation service provides an estimation of aggregated values for various aggregation methods (e.g. areal mean, areal fraction above a threshold) and flexible target supports (e.g. square blocks, irregular polygons).

4 USE-CASES

This section gives a short introduction of two scenarios for which the automatic interpolation service provided by IN-TAMAP was applied, namely, radiological emergencies and human exposure to air pollutants.

4.1 *Radiological emergencies*

The primary use-case leading to the development of INTAMAP concerns automatic mapping of radiation levels over Europe. It is based on measurements provided by more than 4200 stations distributed across 33 countries participating in the European radiological data exchange platform (EURDEP). These measurements (Fig. 5a) are provided on a daily basis and on a 30 minutes basis in emergency mode. The INTAMAP service can provide maps

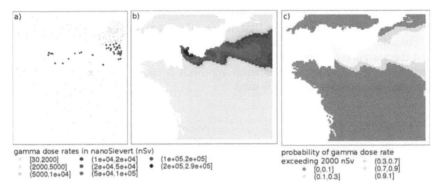

gamma dose rates in nanoSievert (nSv)
- [30,2000]   • (1e+04,2e+04]   • (1e+05,2e+05]
- [2000,5000]   • (2e+04,5e+04]   • (2e+05,2.9e+05]
- (5000,1e+04]   • (5e+04,1e+05]

probability of gamma dose rate
exceeding 2000 nSv   • (0.3,0.7]
- [0,0.1]   (0.7,0.9]
- (0.1,0.3]   (0.9,1]

Figure 5.   Simulation of a radiological emergency in northern France, at (a) monitoring network locations, (b) interpolated map based on simulated measurements (b), and (c) exceedance probabilities based on interpolation error distributions.

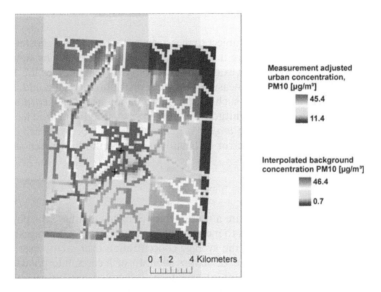

Measurement adjusted
urban concentration,
PM10 [µg/m³]
45.4
11.4

Interpolated background
concentration PM10 [µg/m³]
46.4
0.7

0 1 2   4 Kilometers

Figure 6.   Modelled concentration of air pollutants in the city of Muenster in Germany (fine grid) and background air pollution in the surrounding area (course grid).

of interpolated radiation levels all over Europe (Fig. 5b) with or without little intervention by humans. It also provides maps of uncertainties associated with these radiation levels, for example, the probabilities of the radiation level exceeding a critical threshold at a certain location (Fig. 5c). This information, combined with information on population densities, can be used to assess the population at risk.

4.2   *Human exposure to air pollutants*

Another project that is currently using the INTAMAP service is dealing with in-situ air quality data gathered by the European Environmental Agency (EEA) to estimate human exposure to air pollutants (Henneböhl et al. 2009). The data used for interpolation are concentrations of air pollutants such as particulate matter (PM10), ozone (O_3) or nitrogen dioxide (NO_2) measured at background stations throughout Europe. These measurements can be retrieved from the European Air quality data base. Based on those measurements the INTAMAP service can be used to compute maps of background air pollution. The actual concentration of air pollutants at a specific location is modelled via dispersion models that describe the distribution of air pollutants emitted from local sources, measurements from urban stations provided by a Sensor Observation Service (SOS) and background air pollution maps (Figure 6). Given a human's trajectory through space and time, the concentrations can be aggregated to estimate the amount of air pollutants a human is exposed to.

5 DISCUSSION AND CONCLUSIONS

INTAMAP has developed an interoperable automated interpolation service by applying OGC standards to provide standardised mechanisms, thus allowing reuse and ensuring interoperability in a SOA. In addition, *UncertML* has been developed and used to encapsulate uncertainties, thus facilitating the quantified analysis and propagation of errors through workflows.

Interpolation of measured, continuously varying phenomena is not the most typical problem encountered in remote sensing research. There are a number of other areas where interpolation does, or may, play a role. One of the most obvious is that of interpolating pixel values for cloud removal. Another example is the estimation of biomass, based on ground truth measurements and hyperspectral imagery (De Jong et al. 2003). In this latter example, regression was used to calibrate the imagery data to biomass estimates, and residual spatial correlation motivated the interpolation of the regression residuals. Applications of this kind of analysis, where one or more grid coverages are part of the interpolation request, have already been realised within the INTAMAP project. In general, when a global calibration function is used to translate image values based on ground truths of a continuously varying feature, residuals are usually spatially correlated and residual interpolation will improve the spatial prediction or mapping of the feature (Pebesma 2006).

When one is willing to look at the INTAMAP project without focussing on the interpolation problem as such, remote sensing data processing workflows could benefit from several of its components:

1. an OGC-compliant, service-oriented workflow to enable interoperability, scalability, reuseability and chainability of components,
2. the current setup, where the input is point data and the output a grid coverage, extended to the case where the input includes multiple (links to WCS for) coverages, and can serve as a blueprint for an automated classification service,
3. *UncertML* to allow interoperable, integral encoding and propagation of errors, including classification errors for categorical variables, and
4. interfacing the R open source statistical environment through an OGC-compliant WPS to further expose the feature-rich machine learning and image analysis algorithms available in the R environment (e.g. Sluiter and Pebesma 2010) to automated or non-automated remote sensing workflows.

Although several alternative workflow systems are currently available (e.g. Ludäscher et al. 2006, Oinn et al. 2004), the OGC framework at least tries to take care of the spatial aspect of data correctly, as well as the different types of spatial data (vector, raster) that are commonly used. For building workflows, WPS can be a vital component. Building, cataloging, and chaining services into model webs are current topics under research.

ACKNOWLEDGEMENTS

This work has been funded by the European Commission under the Sixth Framework Programme by the Contract N. 033811 with the DG INFSO, action Line IST-2005-2.5.12 ICT for Environmental Risk Management. The views expressed herein are those of the authors and are not necessarily those of the European Commission.

REFERENCES

52 north web site. http://www.52north.org. [Online; Accessed 01st Jan 2011].

Burrough, P.A. & McDonnell, R.A. (1998) *Principles of Geographical Information Systems*. Oxford: Oxford University Press.

Chorti, A. & Hristopulos, D. (2008) Systematic identification of anisotropic correlations in spatially distributed data sets. *IEEE Transactions on Signal Processing*, 4738–4751.

De Jong, S.M., Pebesma, E.J. & Lacaze, B. (2003) Above-ground biomass assessment of mediterranean forests using airborne imaging spectrometry: the dais peyne experiment. *International Journal of Remote Sensing*, (7), 1505–1520.

Henneböhl, K., Gerharz, L.E. & Pebesma, E.J. (2009) An OGC web service architecture for near real-time interpolation of air quality over europe. In *Proceedings of StatGIS 2009, Jun 17–19 2009, Milos, Greece, G. Dubois (Ed.)*.

Ingram, B., Cornford, D. & Evans, D. (2007) Fast algorithms for automatic mapping with space–limited covariance functions. *Stochastic Environmental Research and Risk Assessment*, (5), 661–670.

Intamap, interoperability and automated mapping. http://www.intamap.org [Online; Accessed 01st Jan 2011].

Ludäscher, B., Altintas, I., Berkley, C., Higgins, D., Jaeger-Frank, E., Jones, M., Lee, E., Tao, J. & Zhao, Y. (2006) Scientific workflow management and the kepler system. *Concurrency and Computation: Practice & Experience*, (10), 1039–1065.

Oinn, T., Addis, M. & Ferris, J. (2004) Taverna: a tool for the composition and enactment of bioinformatics workflows. *Bioinformatics*, 3045–54.

Open Geospatial Consortium Inc. (2007a) Observations and measurements. OpenGIS implementation specification, OGC 07–022r1. Technical report.

Open Geospatial Consortium Inc. (2007b) Opengis geography markup language (gml) encoding standard. OpenGIS encoding specification, OGC 07–036. Technical report.

Open Geospatial Consortium Inc. (2007c) Sensor observation service. OpenGIS implementation specification, OGC 06–009r6. Technical report.

Open Geospatial Consortium Inc. (2007d) Web coverage service. OpenGIS implementation specification, OGC 07–067r5. Technical report.

Open Geospatial Consortium Inc. (2007e) Web processing service. OpenGIS implementation specification, OGC 05–007r7. Technical report.

Open Geospatial Consortium Inc. (2007f) Web service common. OpenGIS implementation specification, OGC 06–121r3. Technical report.

Pebesma, E., Cornford, D., Dubois, G., Heuvelink, G.B.M., Hristopoulos, D., Pilz, J., Stöhlker, U., Morin, G. & Skøien, J.O. (2010) INTAMAP: The design and implementation of an interoperable automated interpolation web service. *Computers & Geosciences, in press*.

Pebesma, E.J. (2004) Multivariable geostatistics in S: the gstat package. *Computers & Geosciences*, (7), 683–691.

Pebesma, E.J. (2006) The role of external variables and gis databases in geostatistical analysis. *Transactions in GIS*, (4), 615–632.

Pilz, J., Kazianka, H. & Spöck, G. (2008) Interoperability – spatial interpolation and automated mapping. In *HAICTA 2008 - 4th International Conference on Information and Communication Technologies in Bio and Earth Sciences*, pp. 18–20.

Pilz, J. & Spöck, G. (2008) Why do we need and how should we implement bayesian kriging methods. *Stochastic Environmental Research and Risk Assessment*, 621–632.

Raape, U., Teßmann, S., Wytzisk, A., Steinmetz, T., Wnuk, M., Hunold, M., Strobl, C., Stasch, C., Walkowski, A.C., Meyer, O. & Jirka, S. (2009) Decision support for tsunami early warning in indonesia: The role of standards. In *Cartography and Geoinformatics for Early Warning and Emergency Management*.

Sluiter, R. & Pebesma, E.J. (2010) Comparing techniques for vegetation classification using multi- and hyperspectral images and ancillary environmental data. *International Journal of Remote Sensing*, (23), 6143–6161.

Stasch, C., Walkowski, A.C. & Jirka, S. (2008) A geosensor network architecture for disaster management based on open standards. In: Ehlers, M., Behncke, K., Gerstengabe, F.W., Hillen, F. Koppers, L.L. Stroink, L. & Wächter, J. (eds.), *Digital Earth Summit on Geoinformatics 2008: Tools for Climate Change Research*, pp. 54–59.

Williams, M., Cornford, D., Bastin, L. & Pebesma, E. (2009) Uncertainty Markup Language (UncertML). *OGC Discussion Paper, Document Number: 08–122r1*.

Advances in Web-based GIS, Mapping Services and Applications – Li, Dragićević & Veenendaal (eds)
© *2011 Taylor & Francis Group, London, ISBN 978-0-415-80483-7*

GeoGlobe: A Virtual Globe for multi-source geospatial information integration and service

Jianya Gong, Longgang Xiang, Jin Chen & Peng Yue
State Key Laboratory of Information Engineering in Surveying, Mapping and Remote Sensing, Wuhan University, Wuhan, Hubei, China

Yi Liu
School of Geodesy and Geomatics, Wuhan University, Wuhan, Hubei, China

ABSTRACT: The emergence of Virtual Globe software systems revolutionizes the traditional way of using geospatial information, making it usable by the general public instead of only domain experts. However, it is still a challenge to share and serve multi-source heterogeneous geospatial information over the Web. This chapter discusses several key technologies for multi-source geospatial information sharing and serving through Virtual Globe. First, a global seamless data model that supports multiple sources and multiple scales is proposed. Second, a balanced scheduling method for distributed multiple servers in a wide area network (WAN) to enable fast transmission and multi-user concurrent access of geospatial data is described. Third, a peer-to-peer (P2P) transmission strategy is designed to support more efficient transmission of geospatial data. Finally, a network based Virtual Globe software system, named GeoGlobe, is introduced and illustrated for applicability of the proposed approaches.

Keywords: Virtual globe, multi-resolution pyramid, load-balancing, GeoOD-P2P, GeoGlobe

1 INTRODUCTION

Digital Earth, a vision firstly proposed by the former U.S. Vice-President Al Gore in 1998, aimed at harnessing the world's data and information resources to develop a virtual three dimensional (3D for short) model of the Earth in order to monitor, measure, and forecast natural and human activities on the planet. The recent popular Virtual Globe software systems such as Google Earth and World Wind bring Gore's dream into reality (Declan 2006). A Virtual Globe is a 3D software model or representation of the Earth or another world, which provides the user with the ability to freely move around in the virtual environment by changing the viewing angle and position (Wikipedia 2010a). Since Google Earth and World Wind brought the concept of Virtual Globe into the general public's consciousness, our concept of how to view the planet we live on has permanently changed, similar to the way the Internet changed the way we store, access and sort information. With a virtual globe product, users can freely fly anywhere on a virtual earth, with different views of the Earth, such as satellite imagery, geographical features, terrain, 3D building, and advanced stars, atmosphere or sunlight effects. Thus, Virtual Globe revolutionizes the traditional way of using geospatial information easily accessible and usable by the general public, instead of only domain experts.

Since the technology of Virtual Globe effectively communicates the research of earth scientist to both other scientists and the general public, millions of people across the world are using these software systems in their spatial exploration. For example, Google Earth was downloaded over 100 million times in the first 15 months of its release (Spano 2006). Former U.S. President George W. Bush has said that he uses Google Earth to look at his Texas ranch (Glaister 2006). Moreover, the phenomenon has even inspired a Nature news article, in which the author writes: "to the casual user ... the appeal of Google Earth is the ease with which you can zoom from space right down to the street level" (Declan 2006). Thanks to Virtual Globe, the overhead associated with accessing global archives of satellite imagery can be reduced vastly, and therefore scientists enjoy experimenting with Virtual Globes to show their research to the general public, resulting in many virtual globe based applications in various fields. For examples, the World Wind was used for interactive Internet visualization of global MODIS burned area product (Boschetti et al. 2008), the Google Earth was used to share and visualize environmental data (Blower et al. 2007), and to rapidly deliver weather products (Smith & Lakshmanan 2010). Schöning et al. (2008) developed a multi-touch virtual globe to facilitate the asking and answering of simple "why" questions, etc.

With the advancement of technologies in Earth observation, Geographic Information System (GIS) and image processing, it is possible now to collect global data with multiple dimensions, multiple time phases and multiple scales. The quality and quantity of spatial or geographically-referenced date have been improved and increased dramatically in the past few decades, and this trend will most likely continue in the foreseeable future (Goodchild 2007, Wang 2008). The collected data has gained wide applications in the fields of meteorology, agriculture, forestry, etc. Usually, these applications are implemented by using different software platforms, which lack of inter-systems sharing and interoperability. While the rapid development of Virtual Globe shows some promises of providing hypermedia-supported, interactive and distributed geospatial information services to the worldwide users, how to provide solutions for high efficient organization, management, transmission, and visualization of multi-source heterogeneous earth information, is still a challenge over the Web.

There are already some Virtual Globe software systems available for the research of geospatial information integration and service. These software systems, though focusing more on intuitive and easy-to-use user experience, have extended the user group of GIS from domain experts to the general public. Example Virtual Globe software systems include the Google Earth provided by Google Inc., the World Wind from the National Aeronautics and Space Administration (NASA), and the Bing Maps from Microsoft Co.

The Google Earth was released in 2005 after the acquisition of Keyhole Inc. It presents a three dimensional model of the Earth by integrating satellite images, aerial images and digital maps using broadband and 3D technologies. In addition to earth browsing, Google Earth provides many interactive functions, including distance/area measurement, place name marking and image uploading. After incorporating the 3D modeling software of SketchUp, Google Earth also supports the import of self-built 3D building models.

The popularity of the Google Earth has a close relationship with its release of Keyhole Markup Language (KML). KML not only describes the content of the data (dot, line, surface and image), but also saves some relevant information for the representation of the data, which can be directly used in Virtual Globe (Wilson 2002). With the support of KML, various users, from the general public to domain experts, can build their own applications on Google Earth. These data can be either used locally by a single user, or can be uploaded directly to the servers of Google Earth and used by global users. Figure 1 is an example of using KML in Google Earth to describe the pathway of a severe tropical storm occurred in Bangladesh in November 2007. Recently, KML has become an international web standard. On the server side, Google Earth utilizes its original Google distributed

Figure 1.   A tropical storm shown on the Google Earth using KML (Bangladesh in November 2007).

file system (i.e. GFS) (Sanjay 2003), Google half-structured concurrent data saving and accessible interfaces (i.e. BigTable) (Fay 2006), and MapReduce (used for the parallel computation of data with a size more than 1TB) (Jeffrey 2004). These distinguished features support servers of Google Earth to be running on a large number of personal computers. As a result, Google Earth can provide high quality web-based geospatial data services to a large number of users, and this is also one of the major reasons that Google Earth becomes so popular.

The World Wind, another Virtual Globe system for 3D visualization, can present images from NASA, U.S. Geological Survey (USGS) and other Web Map Services (WMS) on a 3D Earth model. Users can roam and zoom on the model they observe as they wish, and they can also look up place names and administrative divisions. The most well-known feature of the World Wind is its open architecture, which makes it convenient to extend functions. Figure 2 shows a function provided by its own plug-in, which depicts the seismic information on the Earth in one week. The data comes from USGS. For an earthquake in Figure 2, the colors indicate the time-span passed, while the number of cycles indicates the magnitude.

The World Wind adopts equal latitude/longitude spherical pyramid model. The first level of the pyramid has 10 rows and 20 columns, and the resolution of this level is 18.0 degrees. The World Wind users can download the configuration files from its server, and then render the scenes according to the configuration files. The format of the configuration is XML, and users can edit this XML file conveniently, including setting the order of the overlay and the level of transparency. In addition, the World Wind is an open source software system, which greatly contributes to the wide development of the software platform for multi-source geospatial information integration.

This chapter uses an independently developed Virtual Globe software system, named GeoGlobe, as an example to introduce several key technologies for the sharing and serving of multi-source geospatial information through the technology of Virtual Globe. First, the chapter describes a global multi-source and multi-scale geospatial data model for the seamless organization of global geospatial data. Second, it proposes a distributed multi-servers balanced scheduling method in Wide Area Network (WAN) for multi-user concurrent access to the multi-source geospatial data over the Internet. And finally, it provides a peer-to-peer transmission strategy for the fast transmission of multi-source spatial data over the Internet.

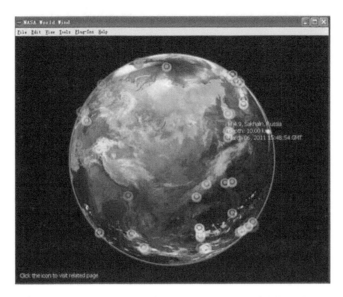

Figure 2.　Global seismic information in one week (09/11/2008) on the World Wind.

2　ORGANIZATION OF MULTI-SOURCE GLOBAL GEOSPATIAL INFORMATION

It is a primary issue to build a global division model for the fast indexing of global geospatial information. Therefore, a global equal latitude/longitude ellipsoidal division model based on a hierarchical quadtree structure (Samet 1990) is proposed and described in this section.

2.1　A global equal latitude/longitude ellipsoidal division model

The equal latitude/longitude ellipsoidal division model means that longitude and latitude interweave evenly on the globe surface to generate a grid by a fixed space (Deng 2003, Zhou 2001). The structure of this division is simple and easy to operate. In addition, when combined with compression algorithms, it shows high efficiency on storage. The equal latitude/longitude grid is particularly suitable for organizing and managing large scale and multi-resolution spatial data. As a matter of fact, it is widely used in the engineering field. This model is multi-scale inherently, where the tile resolution of the same level is the same, while the tile resolution of the neighboring levels is of 2 time relationship. It is convenient to calculate the geographic coordinates of a grid tile according to its row and column numbers, and vice versa.

The following rules are defined: the ranges of longitude and latitude are [−180°, +180°] and [−90°, +90°], respectively, and the values outside the ranges are invalid; the resolution at the k level is 2 times of the resolution at the k + 1 level; at any level, the ratio of the number between the transversal and vertical tiles is 2:1, and the coding order of grid tiles starts from left to right and from bottom to top. The tile number at the 0 level is 2*1, and thus it can be derived that the tile number at the k level is $2^{(k+1)} * 2^k$.

Based on the aforementioned rules, it can be derived in which grid tile a geographic coordinate point is located and the geographic range of a particular grid tile at a certain level. Specifically, given the longitude λ and the latitude φ, its row number and column number at the k level can be calculated using the following formula:

$$RowNo = \lfloor ((\varphi + 90) / (180/2^k)) \rfloor \bmod 2^k \qquad (1)$$
$$ColNo = \lfloor ((\lambda + 180) / (180/2^k)) \rfloor \bmod 2^{(k+1)} \qquad (2)$$

In the formula: $\sqcup$ denotes the downward rounding operator, mod represents the modulus operator, and λ and φ denote longitude and latitude, respectively; k denotes the level number that starts from 0.

Inversely, the geographic range of a grid tile can be computed using the level number of this grid tile k, the row number y, and column number x as follows:

$$west = ((x \bmod 2^{(k+1)}) \times (180/2^k)) - 180 \qquad (3)$$

$$east = west + (180/2^k) \qquad (4)$$

$$south = ((y \bmod 2^k) \times 2^k/180) - 90 \qquad (5)$$

$$north = south + (180/2^k) \qquad (6)$$

Figure 3 illustrates the coding of four-level equal latitude/longitude grid tiles for a globe. The tile resolution at the 0 level is 180° and this level has 1 row and 2 columns. The third level has a tile resolution 22.5° and 8 rows and 16 columns. According to formulas (1)–(6), it can

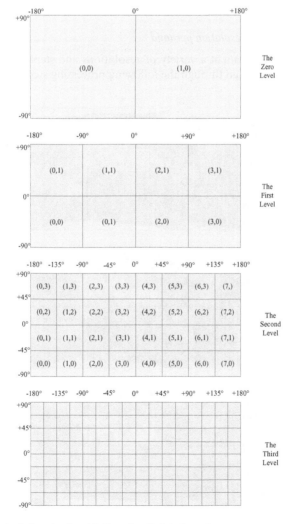

Figure 3.   Equal latitude/longitude grid tile code of global quadtree.

89

be calculated that Beijing (longitude about 116.5°, latitude about 40.0°) is located at the tile with the row number 5 and column number 13 at the third level. The longitude range for this tile is 112.5°–135.0°, and the latitude range for this tile is 22.5°–45.0°.

The pyramid structure formed by equal latitude/longitude grids share much similarity with the quadtree structure: both of them adopt hierarchical model, and the ratio between the neighboring levels is 2. Therefore, equal latitude/longitude grid model often adopts the quadtree structure for the indexing of the tiles. Figure 4 demonstrates the quadtree indexing of the equal latitude/longitude grid. Let k, x, and y denotes the level, column, and row number of a grid tile in the quadtree, respectively. The parent-children and neighboring relationship are presented as follows:

Its left neighboring grid tile is $(k, (x − 1 + 2^{(k+1)}) \mod 2^{(k+1)}, y)$.
Its right neighboring grid tile is $(k, (x + 1 + 2^{(k+1)}) \mod 2^{(k+1)}, y)$.
Its forward neighboring grid tile is $(k, x, (y + 1 + 2^k) \mod 2^k)$.
Its backward neighboring grid tile is $(k, x, (y − 1 + 2^k) \mod 2^k)$.
Its parent grid tile is $(k − 1, \llcorner x/2 \lrcorner, \llcorner y/2 \lrcorner)$, k > 0

Its four children grids are: $(k + 1, 2*x, 2*y)$, $(k + 1, 2*x + 1, 2*y)$, $(k + 1, 2*x, 2*y + 1)$ and $(k + 1, 2*x + 1, 2*y + 1)$s.

2.2 *Organization of multi-resolution pyramid*

In order to provide image data at a variety of resolutions and stepless zooming, the original image data usually must go through the following processing steps including mosaicing,

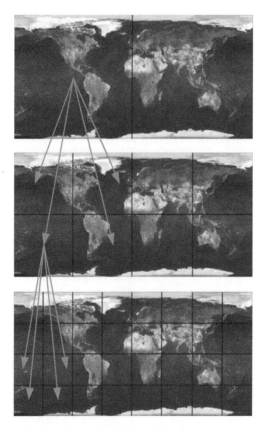

Figure 4.   Quadtree indexing structure of equal latitude/longitude grid.

sampling, tiling, and pyramid building (Li & Wu 2004). Mosaic refers to the process of combining a number of smaller images into a seamless large image. Sampling refers to sample a high resolution image into images at various resolutions by grades according to multiple relationships (it needs to be 2 times for stepless zoom). Tiling means dividing an image into multiple tiles evenly at each resolution level. Pyramid building refers to constructing all tiles into a pyramid, where the neighboring tiles are linked by parent-child pointer.

Based on the pyramid organization, it is applicable to not only provide images at various resolutions, but also support the function of stepless zooming. Firstly, low resolution images are presented, and then high resolution images can be retrieved by the parent-child pointers when users select the region of interest. It is possible to browse and zoom images smoothly, since there is no need to retrieve unneeded tiles for users. It should be noted that tiles are always stored in compressed form, usually in the JPG format, in order to reduce storage cost and provide fast network transmission. Therefore, any two tiles, even at the same level, might have different sizes due to the difference of details.

The implementation of a pyramid usually uses a structure similar to tree structure. Each level of the tree represents a resolution grade of an image; downward from the root node, the nodes respectively represent the zero level tiles, the first level tiles and so on sequentially. Every node not only records the data of the tile, but also stores four pointers linking to tiles at the next level. Such an approach integrates the index and data together in one structure. Search is always started from the root node, according to index information from top to bottom to locate tiles. If the retrieved tile is located at the N level (N is the grade of the pyramid), at least N times I/O operations are needed to retrieve the tile. High cost of the I/O operation is the biggest weakness of this tree-structured pyramid. In addition, the complexity of the structure and saturation of the pyramid tree structure makes the pyramid poor in expansibility.

In order to improve the efficiency of the I/O operation and provide high expansibility, this chapter proposes a pyramid structure, where index and data are discrete (the so-called pyramid hereafter). The core idea is to divest the index information from the structure and organize it independently, i.e. a pyramid structure includes index and data, and they are connected by pointers. In the discrete pyramid structure, every tile is stored in the data section, and it has a record in the index section. The record indicates the level, the code and the size of the tile and the position of data section. Figure 5 gives an example of the discrete pyramid, of which the arrows represent the pointers that point to the data section. In the process of retrieving data, the index section is scanned firstly. The index section is comparatively small (each tile uses only 12 bytes), and it can always be loaded in the memory. Compared

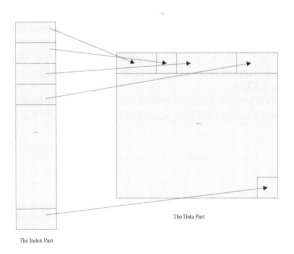

The Data Part

The Index Part

Figure 5. An example of the discrete pyramid.

with the cost of the I/O operation, the cost of scanning memory is relatively small. Secondly, when the tile to be retrieved is located using the index, the tile can be retrieved by one I/O operation.

The discrete pyramid structure we propose here has the following three advantages compared to the traditional pyramid structure: 1) index section and data section are organized independently: the two adjacent tiles in index section are not necessarily adjacent in data section, and vice versa; 2) memory and external storage are consistent. So when system has enough memory, it can load the discrete pyramid structure into the memory without any transformation; and 3) the data structure is simple and easy to be implemented, operated and expanded.

2.2.1 *Data organization based on Morton codes*

The data volume of spatial data from multiple sources can be very large. It is not always applicable to load the whole data into the memory and, therefore, the I/O performance will be a challenge problem. It is obvious that there are rarely any optimizations for single tile accessing, but most of the time the requested tiles are more than one and these tiles are semantically related: either with brotherhood relationship in the horizontal direction, or parent-child relationship in vertical direction.

When two logically adjacent tiles are stored adjacently in a physical disk, the overhead of I/O operation can be reduced as much as possible. This is due to the reason that the seek distance of the magnetic head in the disk will be shortened to a minimum on this condition. Clustered storage and localization can be applied to those tiles with brotherhood or parent-child relationships, thus the I/O accessing performance of the data part can be improved.

For the horizontal access localization, the adjacency of the tiles at the same level should be as high as possible. In two-dimensional space, the space proximity refers to the closeness extent of two-dimensional adjacent locations mapped into one-dimensional codes, which are encoded by a certain encode mode. In many emerged coding modes, Morton code is the most effective one (Michael 1995). Morton codes hash 2D coordinates into one dimension by interleaving the coordinate number bits (Wikipedia 2010b), for example, $M(3, 4) = 26$. The encoding process is presented in Figure 6. Based on Morton codes, another data organization approach (the so-called cascade Morton codes organization below) can be proposed. The data are firstly organized by levels of hierarchy, and at each level, tiles are stored by the descending sequence of Morton codes. For the pyramid in Figure 3, taking the top 2 levels for instance, the data part organized by the cascade Morton codes organization is as follows: <0.(0,0), 0.(1,0), 1.(0,0), 1.(1,0), 1.(0,1), 1.(1,1), 1.(2,0), 1.(3,0), 1.(2,1), 1.(3,1)>. There are two parts of in each item; the first part is the rank of level and the second part is tile number.

For the vertical access localization, the parent tiles should be stored first, and then the children tiles. So the tiles are organized by the depth-first sequence (called Depth-first Organization). For the pyramid in Figure 3, taking the top 2 levels for instance, the data part organized by the Depth-first Organization is as follows: <0.(0,0), 1.(0,0), 1.(1,0), 1.(0,1), 1.(1,1), 0.(1,0), 1.(2,0), 1.(3,0), 1.(2,1), 1.(3,1)>.

Access localizations of horizontal and vertical directions are mutually contradictory. When taking horizontal access localization into account, it will definitely harm that in the vertical

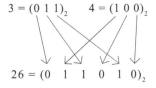

Figure 6. Morton code of coordinate (3, 4).

direction, and vice versa. In practical applications, even with the drilling down operation, all tiles in the geographical range of a vision will be requested from the tile level to which the camera goes down. So compared to horizontal access localization, horizontal access localization is more important, and so cascade Morton codes organization is more suitable than the Depth-first Organization for data organization.

2.2.2 Spatial index organization

Since the index part is relatively small, it can be loaded entirely into the computer memory. Though accessing the memory is much faster than the hard drive, locating the tiles by traversing through the index is still not at high performance. Therefore, a cascade ordered organization for the index part is proposed. That is, at a certain level, the indices of the tiles are arranged by the sequence of tile codes. The start and end index items are pre-recorded in the data dictionary, which allows the use of the binary searching algorithm available for index locating. Thus the time complexity of the search can be reduced from $O(N)$ to $O(\log_2 N)$.

When processing the new request, the internal code is calculated first using the row and column numbers, and then a binary searching is conducted in the index range of the tile level the data belongs to. As aforementioned, the organization of the index part is not dependent on the organization of the data part, which means that the encoding mode of the index part can be even simpler, such as row priority codes (i.e. coding by row sequence) instead of Morton codes. For the pyramid in Figure 3, taking the top 2 levels for instance, the index parts are: <0.(0,0), 0.(1,0), 1.(0,0), 1.(1,0), 1.(0,1), 1.(1,1), 1.(2,0), 1.(3,0), 1.(2,1), 1.(3,1)>.

The data volume of each tile is unlikely to be the same, but the sizes of index items are identical. Based on this fact, a direct locating organization solution is proposed, which makes the overhead of locating tile to the time complexity of $O(1)$.

The storage sequence of tiles is the same in both direct locating organization and cascade ordered organization. But for the direct locating organization, even there are tiles missed, index items can be reserved for each of them. Therefore explicitly recording of the tile codes is not necessary. The location of its index item can be calculated based on the tile's level, row and column number, to retrieve the size of the tile and the location of the data part.

Taking row priority codes as an example, the location of the tiles is calculated as follows:

$$TiledxPos = LevelOffset + (RowNo \times NumCols + ColNo) \times TiledxLen \qquad (7)$$

where TileIdxPos refers to the index location of the requested tile; LevelOffset is the location offset of the index item of the requested tile; RowNo is the row number of the requested tile; NumCols is the columns count of the level that the requested tile belongs to; ColNo is the column number of the requested tile; and TileIdxLen is the length of each index item.

For the index part, the direct locating organization surpasses the cascade ordered organization in terms of the time complexity. However, the space complexity should be considered. Assuming that a pyramid has N levels and the first level has M tiles (including the missing tiles where a tile is considering as missing when it contains no content), the missing rate of tile is S, and the length of the component in an index item is L. There are three components of an index item in the cascade ordered organization: tile code, size and location. The overhead of space is: $3*L*M*(\Sigma^{(i-1)})*S$. On the other hand, an index item consists of two components for direct locating organization: tile size and location, so the overhead of space is $2*L*M*\Sigma^{(i-1)}$.

It can be concluded that when the tiles missing rate is less than 2/3, the direct locating organization is superior to the cascade ordered organization in both time complexity and space complexity. So in most cases, the previous one is more suitable for the index part organization. It is noted that, even the index part is too large to be loaded entirely into the memory, for direct locating organization, only one I/O accessing is good enough to retrieving the index information from hard disk.

3 THE BALANCED SCHEDULING OF DISTRIBUTED MULTI-SERVERS IN WAN

In case of large-scale and multi-user concurrent access, a single global data server is not powerful enough. Distributed multi-server technology is often employed under these circumstances. Different applications may choose different load-balancing framework. Even with the same framework, the adopted load-balancing strategies may differ. Currently, the solutions for load-balancing framework and strategy are mostly available in the computer science domain but cannot support geospatial applications directly.

3.1 *The framework of distributed multi-servers in WAN*

The problem of load-balancing is one of the major research issues in the field of parallel computing and distributed computing. As a classical optimization problem, the best optimal strategy of load-balancing is very difficult to achieve, and even in the simplified case, it is still a NP (nondeterministic polynomial)-complete problem. Generally speaking, load-balancing technologies can be classified into three levels: network level, operating system level, and middleware level.

1. Network level: Currently, the load-balancing technology at this level is most widely used, such as network routers and switches (e.g. IBM's Web Sphere Edge Server, BIG-IP of F5).
2. Operating system (OS) level: OS-level load-balancing technology mainly refers to the technology achieving the load balancing by clustering, process migrating and scheduling in the operating system kernel. Ameous, a famous Amoeba distributed operating system, has used this technology (Kwok & Cheung 2004).
3. Middleware level: The technology at this level is based on the distributed object technology. It is also widely used due to its low cost and flexibility.

In order to deploy service nodes distributed in WAN and solve the rapid response problem caused by the large-scale concurrent access, GeoGlobe system sets up a Catalog Center between servers and clients, as shown in Figure 7. The architecture follows the "publish-find-bind" paradigm of the Web service technology (IBM 2010):

1. Publish: Data server nodes are registered into the Catalog Center. The meta-information of pyramid, such as pyramid identification, lever number, pixel resolution and geospatial bounding box, are registered. For the same pyramid, it is possible to have more than one server node to provide online services at the same time.

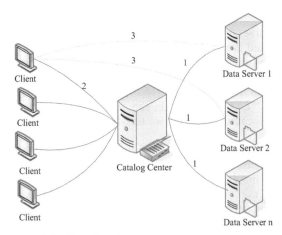

Figure 7.   The architecture of distributed multi-servers.

94

2. Find: Client nodes login into the Catalog Center and discover the pyramids published in the WAN.
3. Bind: Client nodes bind the discovered data node and download pyramid tiles.

Compared with other distributed server architectures, the deployment of this multi-server technology brings the following advantages:

1. Data server nodes can be distributed into WAN instead of the same Local Area Network (LAN);
2. Client nodes can download data tiles from data server nodes directly without passing through the Catalog Center, thus saving the cost of network transmission;
3. A wide range of load-balancing strategies, such as random selection, polling, etc., can be integrated when choosing one among those data servers that can provide the same pyramid tiles; and
4. Data server nodes and the Catalog Center are loosely coupled. Thus more data server nodes can be incorporated easily into the GeoGlobe server.

3.2 *The application related load-balancing technology*

The purpose of designing a load-balancing strategy is to distribute network requests evenly to server nodes as much as possible. Based on the aforementioned distributed server architecture, a simple and straightforward load-balancing strategy is that the Catalog Center randomly selects a server node from nodes providing the same pyramid. Obviously, this method implements the load-balancing only on the probability. The function of random selection can also be assigned to the client side, and this is the load-balancing strategy of client random selection. Compared with the load-balancing strategy of Catalog Center random selection, the latter can support the single-collapse transfer. However, it will increase the communication traffic between the client and the Catalog Center, and also reduces the security.

The above method treats "rational distribution" as "average distribution". In fact, the ability of each server node (CPU, disk and network, etc.) differs greatly. Some nodes may be a large-scale server, while others may be just an ordinary Personal Computer (PC). Therefore, the average distribution load-balancing strategy cannot be truly balanced. In addition, this approach does not take the real-time load conditions of server nodes into consideration. Some nodes may be busy, while others may be idle. As a result, some nodes are overloaded, while others are free.

To solve these problems, Information Technology (IT) experts have already proposed the dynamic load-balancing strategy (Baumgartner & Gammon 1989), which monitors the real-time load conditions of the server nodes, and then selects the server node with the lightest load. In the dynamic load-balancing, the application should select its load factors, specify the weight values for these factors, and design the computation method for the load value based on these factors. There are a variety of load factors. Depending on whether their values change with time, load factors can be divided into dynamic and static categories. Dynamic load factors are usually more important than static factors. Besides the application-specific load factors, some load factors are independent of applications. Static load factors include the processor ability, the memory size, and the speed of disk I/O. Dynamic load factors include the utilization of Central Processing Unit (CPU), available memory, and the status of disk and network.

Comparing the load vector of a single load factor with that of multiple load factors, the latter one is more effective in describing the load status of server nodes. The design of an ideal load vector that can reflect the load conditions of the system should meet the following conditions:

1. The overhead for measuring each load factor is low, which means they can be measured frequently to ensure up-to-date information.
2. It can reflect the load on the competitive resources.
3. Load factors are independent with each other in the measurement and control (Gong 2006).

Many network-based geospatial applications achieve some load balance by using web server systems, such as Apache/Tomcat or Internet Information Service (IIS). However, they cannot integrate load factors specific to geospatial applications. Considering GeoGlobe, the pyramid is the basic unit for access and its main purpose is for visualization. Different pyramids are visited with different probability, owing to their different production date, scale/resolution, and coverage area. Some example situations include:

1. In the same geographical area, if the pyramids have the same scale/resolution, but with different acquisition dates (refers to spatial data production time rather than the pyramid creation time in the rest of this chapter), then the latest pyramids will have greater visiting probability.
2. In the same geographical area, if the pyramids have the same acquisition date, but with different scale resolution, the pyramids with high-scale/resolution will be visited frequently.
3. Compared with the rural and wasteland, cities and scenic areas are more popular to users, therefore those pyramids will be visited with greater probability.

In addition, clients establish connection with the server node according to the pyramid service list returned from the Catalog Center, and then send tile request to the server node based on that connection. A pyramid is requested when it falls within the camera vision. Once a pyramid falls into the camera vision, all visible tiles will be downloaded to the client (except for the local cached tiles). Therefore, the request number and the connection number are two important load factors in GeoGlobe.

1. The connection number (N_{cn}): The connection number between the server and the client is an important factor in measuring the load. The client establishes connection in order to request tiles from the server.
2. The request history frequency (F_{tt}): It is the number of tiles requested during the server's running time. It reflects the load status. The server visited frequently before will also be popular in the future.
3. The current request frequency (F_{ct}): It is the tile number requested recently. It is another important load factor. According to the client's scheduling mechanism, once the client begins to visit some pyramids, the number of the requested tiles may be tens of thousands rather than a few.

In addition, among the load factors that are not specific to applications, the speed of disk I/O is important to GeoGlobe. The reasons are as follows: (1) In GeoGlobe server, disk read operation is used very frequently. (2) Hardware utilization has been reflected by the load factors such as connection number and request number. Factors that affect the network are not considered here since there are too many factors and they are difficult to measure. Thus, the load factors closely related to GeoGlobe are the speed of disk I/O (P_{dk}), the importance of the pyramid (I_{pd}) (decided by the temporal, scale and geographical coverage), the connection number (N_{cn}), the request history frequency (F_{tt}) and the current request frequency (F_{ct}). Among these factors, the first two are static, and the other three are dynamic.

It should be noted that the speed of disk I/O and the importance of the pyramid are specified by the administrator. The faster the speed of disk I/O is, the smaller its load factor value is; and the greater the pyramid importance is, the bigger its load factor value is. Obviously, the importance of a pyramid is decided by all the pyramids the server node possesses.

Based on the design of load factors, a set of weights is introduced, which can be adjusted dynamically to express the importance of each load factor. The final value Load$_{Server}$, reflecting the load status of a server node, can be calculated using the following formula:

$$Load_{Server} = W_1 \times F_{ct} + W_2 \times N_{cn} + W_3 \times F_{tt} + W_4 \times P_{dk} + W_5 \times I_{pd} \qquad (8)$$

The value (W_1, W_2 ... W_5) must be adjusted constantly by repeated experiments according to the actual environment. Table 1 gives a series of values adjusted based on experiments.

Table 1. The load factors of GeoGlobe server and its weight.

Name	Category	Weight
The speed of disk I/O(P_{dk})	static load factor	0.1
The importance of pyramid (I_{pd})	static load factor	0.1
The number of connection (N_{cn})	dynamic load factor	0.3
The frequency of history request (F_{tt})	dynamic load factor	0.1
The frequency of current request (F_{ct})	dynamic load factor	0.4

According to the above formula (8), the Catalog Center collects values of load factors from each server, and calculates the real-time load values of each server when receiving the client's login request. Then the Catalog Center designates the server node for each pyramid from low value to high. There are two ways to collect the load information: cyclical manner and event-driven approach. The former one is to collect the load information periodically according to a fixed time interval. The latter one collects information by trigger-event. Because the GeoGlobe server may encounter a large-scale concurrent access in a short time, the cyclical manner can better reflect the real load situation of the server node.

3.3 *Experiments and analysis*

To test the aforementioned dynamic application-related load-balancing strategy, a series of experiments were conducted. Their performances were compared using the random selection method. In the random selection method, after receiving the connection request, the Catalog Center chooses a server randomly from the candidate ones. Since the connection number and the request frequency have a high weighted impact on the load of a server node, the experiments in this section only choose the load vector composed of the two factors as the load indicator.

All the experiments were arranged in the LAN that contained one client, one Catalog Center, and five data servers. They were connected by 100M/S LAN. The client was a normal PC (Intel(R) Pentium(R) 4 3.0 GHz Duel-core CPU, 512 MB Memory), running the LoadRunner software; the Catalog Center and data server are located on high performance blade machines (Intel(R) Xeon(R) 5110 1.6 GHz Quad-core CPU, 4 GB Memory). All the data servers stored the same datasets: global SRTM DEM (90 m), global BlueMarble images (1000 m), Asia Landsat images (30 m), WUHAN Quick Bird images (0.6 m), NanChang DMC aerial images (0.1 m) and China place names (1:1,000,000).

By recording the browsing history of the GeoGlobe Client to all of the above datasets, the running script of the LoadRunner can be obtained. All of the experiments set the load collection period as 10 seconds. To simulate the real environment, one machine continuously sent data requests to data servers.

3.3.1 *Adjusting the weight factor*

The connection and request number have different impacts on the load of a server node, so their weight coefficients are different. To achieve a good load balance, it was necessary to choose an appropriate load weight coefficient.

In this experiment, the client ran the script recorded before, and the response time of the system with different load vector index was obtained. If the script ran only once, the random factors will affect the system response time. To reduce this effect as much as possible, the experiment repeated the running script for three times and took the average value as the response time. To obtain the optimal load coefficient, eight sets of weight vectors were designed: (10, 1), (5, 1), (3, 1), (2, 1), (1, 1), (1, 2), (1, 0), (0, 1). The experiment results are shown in Figure 8 and summarized as follows.

1. Compared with the load factor of request number, the factor of connection number has a greater impact on the load of the server node.
2. When $Load_{Server} = 2*Connection + Request$, the response time is the shortest, which means the load balance is the best.
3. When choosing (2, 1) as the weight vector, the result is better than a single factor, such as (1, 0) or (0, 1).

3.3.2 Increasing the client number

The response time changes with the client number. To evaluate the exact influence, this experiment sets the client number as 100, 200, 600, 900, and uses the correlation method and random method to evaluate the load balance effect. Similarly, the LoadRunner iterates for three times. The experiment results are shown in Figure 9, from which the following conclusions can be drawn:

1. Whether using the correlation method or random method, the response time increases as the client number increases.
2. The correlation method is better than the random method. With the increase of the client number, the advantage becomes more obvious.

3.3.3 Increasing the iteration number

In the previous two experiments, the run script is iterated for three times to reduce the influence of the random factors. This section describes an experiment that was designed to

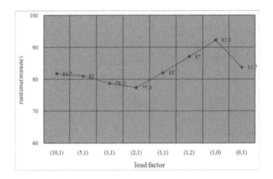

Figure 8. The relation between the load coefficient and response time.

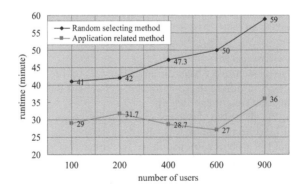

Figure 9. The relation between the client number and response time.

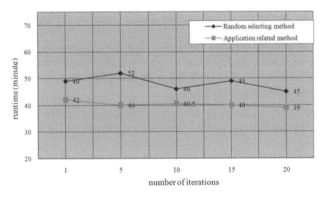

Figure 10. The relationship between iterations and response time.

evaluate the relationship between the response time and iteration. The same methods were used. The experiment results are shown in Figure 10 and reveals:

1. The effect that iterations have on the correlation method can be ignored.
2. The iterations will cause fluctuation of the response time to the random selecting method. This is an inevitable result when the Catalog Center chooses the destination nodes randomly: if the Catalog Center chooses the nodes with lighter load, the response time will be shorter, and inversely the response time will increase. This experiment indicates that three times are good enough to achieve the required precision.

4 GEOSPATIAL DATA TRANSMISSION BASED ON PEER-TO-PEER NETWORK

The progressive transmission of geospatial data, such as image, terrain, 3D model and vector, through the network in real time can be called geospatial data stream transmission. The process of acquiring geospatial data over the Web by Virtual Globe is currently a typical kind of geospatial data stream transmission issue. This type of stream transmission is driven by user's requests from any spatial location. Such interactive geospatial data stream transmission is referred to as Geospatial On Demand Streaming (GeoOD) in this chapter.

4.1 *The geospatial data stream transmission mode for Virtual Globe*

The GeoOD based on a Virtual Globe system is a content acquisition model driven by user's behaviors. It is identified as a nonlinear acquiring method (Mayer-Patel & Gotz 2007, Hu et al. 2008). It has some additional features when compared with the traditional methods.

1. Multi-dimensional data: The geospatial data usually contains spatial information. It means each data block can have two or three dimensions. But the common data streaming usually transmits single-dimensional data. For example, the peer-to-peer file-sharing system, such as BT and EMule (an open source software for network source downloading and sharing, http://www.emule.org), takes the files as one sequence for transmission. The Streaming Media, such as PPLive (a Chinese software for online television watching, http://www.pptv.com/), takes time as the dimension for transmission.
2. Pyramid data organization: Since a Virtual Globe system involves the use of global remote sensing images, the data volume is massive. It is impossible to manage them in a single-file system. So generally speaking, the globe needs to be divided into grids according to geographical area, and a pyramid management mechanism is provided by building spatial index, such as Quad-tree Indexes.

3. User interaction: Because remote sensing data is worldwide, requests are random in GeoOD when compared with traditional streaming media. The interaction of traditional streaming media includes forward, backward, jump, etc. However, the interaction mode of spatial data requests is more dynamic and frequent. Users can switch to any region of the global area. Interaction include many directions, such as upward, downward, left, right, or even any combination of them.

The above characteristics determine the differences between the GeoOD-P2P transmission and traditional P2P transmission. Table 2 presents a summary of these differences.

4.2 *The network transmission architecture of a Virtual Globe system based on peer-to-peer network*

A hybrid structure similar to the BitTorrent system is adopted as the GeoOD-P2P network structure. It shortens the average node distance when compared with the regular grid, and enhances the data transmission speed (Liao et al. 2007). GeoOD-P2P network sets a centralized node as the index server, referred as Super Peer hereafter. Each Peer releases its resources metadata to the Super Peer. If they need to locate resources, they should access the Super Peer to get the metadata. The Super Peer center is a high-performance machine, so it can ensure fast response. From this perspective, GeoOD-P2P network can be classified as a central topology. Though each Peer relies on the Super Peer, the Super Peer does not participate in the specific data transmission and sharing process. It only provides shared resource metadata and helps peer nodes to discover each other. After the discovery stage, the nodes will cooperate to complete data sharing and transmitting process through a series of distributed mechanism. Therefore, the network of GeoOD-P2P is a fully distributed structure at the stage after the node discovering. Based on the previous analysis, GeoOD-P2P system at this stage adopts a fully-distributed and unstructured topology. There is no certain structure between nodes, but they can organize a covering network according to user requests.

GeoOD-P2P network is a hybrid topology designed for the GeoOD transmission. Compared with the existing hybrid structures, such as Napster and BitTorrent, it has some improvements. In the Central Topology part, the Super Peer in the GeoOD-P2P network does not store all the metadata of spatial data, but only records their summary

Table 2. Comparison between GeoOD-P2P, P2Plive and demand.

	P2PLive-streaming media	Demand-streaming media	GeoOD
Data synchronization	Synchronization	Asynchronously	Asynchronously
User behavior	No interaction	Partial interaction	Various and frequent interaction
Driving force	Data itself	Data itself play a leading role, Supplemented by User behavior	User behavior play a leading role, supplemented by data
Acquisition mode	Orderly	Orderly	No fixed order
Predictability	Predictable	Predictable	Unpredictable
Data continuity	Continuous on time dimension	Continuous on time dimension	Continuous on spatial dimension
Resource unit	One streaming media file	One streaming media file	One data block
User goals	One resource unit	One resource unit	Multi resource unit
Data blocks number	Hundred or thousand	Hundred or thousand	Millions or billions
Shareable peers number	more	more	Little
Peers group alternate	Lower frequency	Lower frequency	Higher frequency

information. In the Full Distributed Unstructured Topology Stage, an optimization method is designed for GeoOD-P2P networks to overcome the shortcomings of distributed unstructured network. In some specific applications, when there are some problems in efficiency, scalability, and robustness caused by the Super Peer, GeoOD-P2P systems can easily extend the topological structure into a super node structure, by setting multiple Super Peers and building their communication mechanisms to improve efficiency and robustness.

In GeoOD-P2P networks, the central Super Peer records all information of network nodes and resources. To locate resources, a query request is firstly sent to the Super Peer. Next, the Super Peer will return some nodes that most likely contain resources, and logically define those nodes as the neighbors of the requested nodes. In GeoOD-P2P network, there is a rule that nodes only connect, transmit, and share data with their neighbors. Then the nodes and their neighbors constitute a Neighbor Node Covering Network, which is a part of the GeoOD-P2P's fully distributed topological structure.

In the operational process of GeoOD-P2P network, each node can select a certain number of other nodes to maintain the neighborhood, choose some neighboring nodes to establish transmission relationship dynamically, and build the Transmitting Node Covering Network in the internal of the Neighbor Node Covering Network. Therefore, the main algorithms that influence the GeoOD network topology structure include resources location algorithm, dynamic meshing algorithm, Peer selection algorithm, and redundancy minimum data priority scheduling algorithm.

In GeoOD-P2P network, a neighboring node is determined by peer node's request resource. The connection among nodes is random due to its random requests, but each node and its neighbors stay with relatively fixed relationship once the connection has been determined. Figure 11 outlines the GeoOD-P2P network, in which all Peer nodes (P1, P2 ... P6) connect to the Super Peer in order to request the resource location information.

Let us take P2 in Figure 11 as an example. According to its current request, P2 connects to the Super Peer, determines P1, P3 and P4 as its neighbors, builds connections with them, and forms a Neighbor Node Covering Network. Assuming P4 and P5 provide data for P2, they form a Transmitting Node Covering Network. The Transmitting Node Covering Network certainly belongs to the Neighbor Node Covering Network.

Node joining and exiting is another factor that affects connection. Because there is a centralized Super Peer in GeoOD-P2P networks, once the message of node joining and exiting is reported to the Super Peer, all of the Peer nodes in the whole network will get the information by interacting with the Super Peer. This method is very simple, but its shortcoming is the delay of messages. The information about node joining and exiting recorded by the Super Peer is real-time, but the message on whether the node is online or offline and recorded by the Peer node may be obsolete. It needs to be updated in the next interaction with the Super Peer. One method to update the online/offline message is to use some exploratory auxiliary

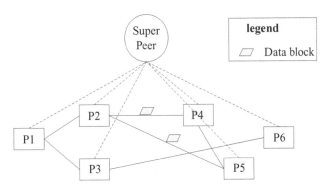

Figure 11.   GeoOD-P2P network.

strategies, such as the heartbeat package strategy. If one Peer node does not receive any Heart Packet in one cycle, it means the node has exited already.

In GeoOD-P2P networks, each node should publish resource metadata information to the Super Peer after the node joining information has been sent to the Super Peer. The Virtual Globe system client runs on the Peer node as entity software. Once the client starts up successfully, the first step is to release resources metadata to the Super Peer, which informs that this Peer node has joined into GeoOD-P2P networks. If the client is closed, it will also send an offline message to the Super Peer, and then the Super Peer will delete this Peer's resources metadata records, causing the Peer node to exit.

4.3 Geospatial metadata publishing and searching based on peer-to-peer network

At the top level of geospatial data pyramid structure, the global data is divided into grids using the Equal Longitude and Latitude Subdivision method. Assuming the corresponding coordinates in degrees of the grid width and height are Grid Degree, the global will be divided into 180/*GridDegree* grids in north-south direction of the globe, and 360/*GridDegree* in east-west direction. The grid identification is:

$$G_{ij} = \left[G_{Ri}, G_{Cj} \right] \tag{9}$$

$$G_{Ri} = \left\lfloor \frac{|-90 - latitude| \% 180}{GridDegree} \right\rfloor \tag{10}$$

$$G_{Cj} = \left\lfloor \frac{|-180 - longitude| \% 360}{GridDegree} \right\rfloor \tag{11}$$

In the geospatial data pyramid structure, a grid's corresponding coordinate range (ROG) is certain. It takes the ROG at the top level as a reference and does not change with the resolution of spatial data. Grid index is created by setting up an 1:N mapping relation between the grids and globe data blocks. Supposing that there is a Peer cache file $F_{i'j'}$, through the following transformation we will get the corresponding grid identification $G_{F_{i'j'}}$.

$$F_{i'j'} = [F_{Ri'}, F_{Cj'}] \tag{12}$$

$$latitude = -90 + (R_{i'} + 1) * TileDegree, \quad R_{i'} = 0, \ldots N_0/2 \tag{13}$$

$$longitude = -180 + (C_{j'} + 1) * TileDegree, \quad C_{j'} = 0, \ldots N_0 \tag{14}$$

$$G_{F_{i'j'}} = \left[G_{Ri'}^F, G_{Cj'}^F \right] \tag{15}$$

$$G_{Ri}^F = \left\lfloor \frac{|-90 - latitude| \% 180}{GridDegree} \right\rfloor \tag{16}$$

$$G_{Cj}^F = \left\lfloor \frac{|-180 - longitude| \% 360}{GridDegree} \right\rfloor \tag{17}$$

If $G_{F_{i'j'}} = G_{ij}$, then $F_{i'j'} \in G_{ij}$; data block $F_{i'j'}$ belongs to grid G_{ij}. It means that grid identification of data block $F_{i'j'}$ is G_{ij}. Therefore, based on the spatial organization and mesh method described in this chapter, building mapping relationship between the grids and data blocks is very simple, using longitude and latitude as medium.

There are three kinds of resources metadata in GeoOD-Peer node: the cache metadata of the node itself, the summary information of the cache metadata, and cache metadata of the neighboring Peer node. As shown in Figure 12, the summary information of cache metadata

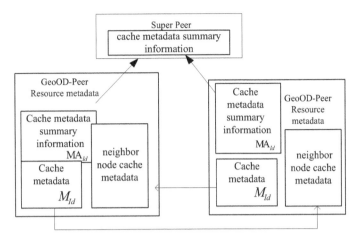

Figure 12.   Resource metadata in GeoOD-P2P network.

in the Super Peer contains the cache metadata of each Peer and the neighbor node cache metadata including all neighbors' cache metadata.

Based on the above peer-to-peer network structure and metadata publishing principles, the process of geospatial resource location in peer-to-peer network is summarized as follows:

1. Confirming the current requested data blocks, and then searching them in its owner cache according to file identification. If those data exist, there is no need to acquire resource location; otherwise, do it.
2. Confirming the grid identification and level identification of the current requested data blocks.
3. Looking up the cache metadata of the neighbor node to judge whether it has this grid's information. If there is no record about this grid, it means the grid only entered recently for the first time. So if there is no resource metadata of this grid, resources location request should be sent to the Super Peer. If the neighbor node's cache metadata has this grid, it means the data of this grid has been requested recently, and the resource metadata in the grid has been recorded. In summary, the cache metadata of the neighbor node should be searched first. If the requested resource is found, we should get the nodes list of it. Otherwise, the request should be sent to data server.
4. After getting the Peer nodes belonging to one grid according to the level identification and grid identification, the nodes list should be returned to the request node. And then these nodes become the neighbor of the request node.
5. Connecting the neighbors one by one.
6. Requesting and getting each neighbor cache metadata about the grid at this level.
7. Updating their neighbor node cache metadata, and then return to Step 3.
8. Obtaining the nodes listing of the requested resource. These nodes are called candidate nodes. Now the resource location process finishes.

4.4   *Experiments and analysis*

Figure 13 shows the network topology and hardware configuration used in experiments. It contains mainly one Main Server using HTTP protocol, one Super Peer Server, and two Peer groups with different network conditions. Peers in group are connected by 100 M/s LAN, and Peer groups are connected through a campus network. The Super Peer server is also deployed on the campus network. However the Main Server, located remotely, is belonging to a different network. The bandwidth between Group A and Group B is 2 M/s, and the bandwidth between the Super Peer and each group is 2 M/s as well.

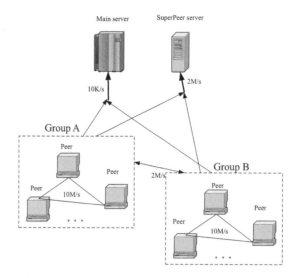

Figure 13.   Network topology and hardware configuration.

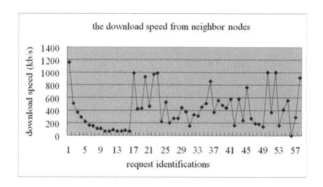

Figure 14.   Efficiency of data acquisition by P2P.

After testing GeoOD-P2P systems, we collect the statistics about the related information of data blocks downloaded by combining the Main Server and P2P network. Figures 14 and 15 represents the downloading time of data blocks.

From Figure 14, we can see that the downloading speed decreases at the beginning. This is because when the P2P starts, a large number of initialization tasks should be completed with the user's viewpoint moving into each Peer grid. The most time-consuming process is the metadata information acquisition step (i.e. Server Peers listing and Server Peer's data distribution). The metadata request priority is higher than data downloading request, so the downloading requests are delayed in the request queue, and the downloading speed is affected. However, once the metadata has been obtained, the user's request will be sent directly to the remote Peer, and the downloading speed becomes faster.

From Figure 15, we can see the average downloading speed is only 4 KB/S. This is because the campus network speed is slow when requesting the remotely located Main Server. Comparing with the 400 KB/S of the P2P network, the downloading speed of the Main Server seems to be too slow. Therefore, using the P2P mode is necessary during the spatial data transmission. It also reveals that the Main server provides 77 responses of request, so the center server is also necessary. This is because there are no neighboring Peer nodes possessing this data. Therefore the method of mixing P2P with center is reasonable.

104

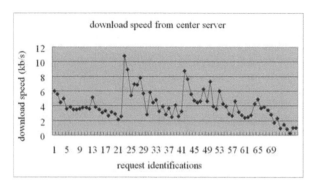

Figure 15.   Efficiency of data acquisition by center server.

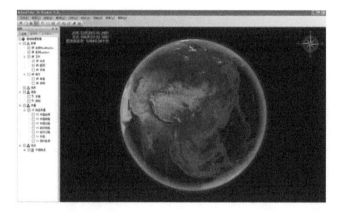

Figure 16.   GUI of the GeoGlobe client.

In summary, the P2P mode not only effectively saves the resource and workload of the server side, but also greatly improves the downloading speed of spatial data flow. Therefore the waiting time of clients decreases.

5   PROTOTYPE IMPLEMENTATION-GEOGLOBE

GeoGlobe is a Virtual Globe software system developed by the State Key Lab of Information Engineering in Surveying, Mapping and Remote Sensing (LIESMARS for short) of Wuhan University in China. Based on the methods proposed in this chapter, GeoGlobe can manage multi-source, multi-scale, and multi-temporal geospatial data efficiently. GeoGlobe users can easily browse (including roaming, zooming, and flying) geospatial data in spherical 3D manner.

GeoGlobe software system consists of three parts: GeoGlobe server, GeoGlobe builder, and GeoGlobe viewer. GeoGlobe server manages all registered geospatial data and provides fast network access. GeoGlobe builder organizes geospatial data into a multi-resolution pyramid, which is the prerequisite for 3D visualization in GeoGlobe. GeoGlobe viewer visualizes location-related data, including image, DEM, vector, place name, 3D building, and service processing result, where the former five items are supplied by GeoGlobe server and the last one is the response from various geospatial web services.

The Graphic User Interface (GUI) of GeoGlobe client is shown in Figure 16. The left is a layer tree, listing all loaded pyramids provided by GeoGlobe server. This tree is manually configured by the server administrator and shared by all GeoGlobe clients.

Figure 17.   3D buildings around the "Window of World" in Shenzhen, China.

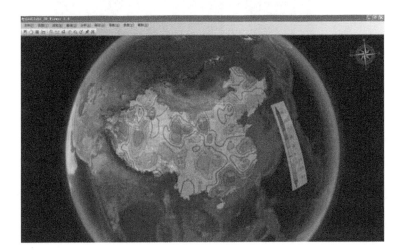

Figure 18.   Monthly precipitation in China.

Figure 17 shows the 3D buildings around the "Window of World" in Shenzhen, China, where the images, collected form aerial photography, have a resolution of 0.6 meter, and the DEM comes from SRTM with a resolution of 90 meters. Besides these geospatial data, other type of data can also be integrated into the GeoGlobe system, such as weather information. In Figure 18, the monthly precipitation of China is shown, where the color indicates the amount of rainfall.

6   CONCLUSION

Virtual Globe has been recognized as an appealing technology to support Digital Earth. Once people have experienced the visually rich and highly compelling nature of data delivered via Virtual Globes with their highly engaging context of 3D, it is hard for them to go back to a flat 2D world. This chapter takes a step further to make Virtual Globes more powerful in information integration and network services. To achieve this goal, several key issues are discussed in detail.

A global seamless multi-source and multi-scale geospatial data model, based on the equal latitude/longitude ellipsoidal division model, is proposed. To implement this logical structure more efficiently, a discrete pyramid structure is introduced, with which any pyramid tile can be located by just a simple computation. This data model has been applied in GeoGlobe system and acts as a key technology to cope with the I/O problem caused by huge data size.

A distributed multi-server architecture, working in WAN and used for meeting the large-scale concurrency access requirement, is proposed. The balanced scheduling method, running on this architecture, is also discussed. To test the efficiency of this design, three set of experiments were conducted and the results manifest its applicability.

Finally, this chapter introduces a GeoOD-P2P transmission network topology and presents the corresponding schedule strategy, used for the fast transmission of multi-source spatial data over the Internet. This method bases on the technology of peer-to-peer transmission, but captures important spatial characteristics. The corresponding experiments have shown that it can greatly improve the downloading speed of geospatial data flow.

Based on above key technologies, we developed a new Virtual Globe system-GeoGlobe. It is not only an earth browser, but also a platform of information integration and service. After its first release in 2006, many applications were developed and deployed, and it is now used by several geographical information centers in China. Recently, GeoGlobe is selected as the core supporting software for the China public service platform of geographical information (see http://www.tianditu.com).

ACKNOWLEDGEMENTS

The work reported has been done within the Innovation Research Group Project, funded by the Natural Science Foundation of China (Grant No. 40721001), and also partly supported by another grant from the Natural Science Foundation of China (Grant No. 41001296). We are grateful to those students and researchers who participated in the development of GeoGlobe system.

REFERENCES

Baumgartner, K.M. & Wah, B.W. (1989) A load balancing strategy in local computer systems with multi-access networks. *IEEE Transactions on Computers*, 38 (8), 1098–1109.

Blower, J., Gemmell, A., Haines, K. et al. (2007) Sharing and visualizing environmental data using Virtual Globes. In: UK e-Science All Hands Meeting, Nottingham, UK.

Boschetti, L., Roy, D.P. & Justice, C.O. (2008) Using NASA's world wind virtual globe for interactive internet visualization of the global MODIS burned area product. *International Journal of Remote Sensing*, 29 (11), 3067–3072.

Declan, B. (2006) Virtual Globes: the web-world World. *Nature*, 439 (7078), 776–768.

Deng, X. (2003) Research of Service Architecture and Algorithms for Grid Spatial Data, Ph.D. Dissertation, University of PLA Information Engineering (in Chinese).

Fay, C., Jeffrey, D., Sanjay, G. et al. (2006) Bigtable: A Distributed Storage System for Structured Data. OSDI, Seattle, USA.

Glaister, D. (2006) *The Google lets homesick President Keep an Eye on the Ranch. The Guardian,* [Online] Available from: http://www.guardian.co.uk/technology/2006/oct/27/news.usnews

Gong, W.H. (2006) Research of Key Technologies in Database Cluster System. Ph.D. Dissertation of Huazhong University, P.R. China (in Chinese).

Goodchild, M.F. (2007) Citizens as voluntary sensors: Spatial data infrastructure in the world of Web 2.0. *International Journal of Spatial Data Infrastructures Research*, 2, 24–32.

Hu, S.Y. & Jiang, J.R. (2008) Plug: Virtual Worlds for Millions of People. *Proc. of P2P-NVE*.

IBM Redbook. (2004) Patterns: Service-Oriented Architecture and Web Services, Retrieved on 22nd December 2010. [Online] Available from: http://www.redbooks.ibm.com/redbooks.nsf/redbooks/

Jeffrey, D. & Sanjay, G. (2004) MapReduce: Simplified Data Processing on Large Clusters. OSDI, San Francisco, USA.

Kwok, Y.K. & Cheung, L.S. (2004) A new fuzzy-decision based load balancing system for distributed object computing. *Journal of Parallel and Distributed Computing*, (64), 238–253.

Li, L. & Wu, F. (2004) The Model and Visualization of Multi-scale Spatial Data (1st edition). Beijing: Science Press (in Chinese).

Liao, W.C., Papadopoulos, F. & Psounis, K. (2007) Performance analysis of BitTorrent-like systems with heterogeneous users. *Performance Evaluation*, 64 (9–12), 876–891.

Mayer-Patel, K. & Gotz, D. (2007) Adaptive streaming for nonlinear media. *IEEE Multimedia Magazine*, 14 (3), 68–83.

Michael, V. & Yannis, M. (1995) Dynamic inverted quadtree: a structure for pictorial databases. *Information System*, 20 (5), 483–500.

Samet, H. (1990) The Design and Analysis of Spatial Data Structures. Addison-Wesley Publishing Company.

Sanjay, G., Howard, G. & Shun-Tak, L. (2003) The Google File System. SOSP, New York, USA.

Schöning, J., Hecht, B. & Raubal, M. et al. (2008) Improving Interaction with Virtual Globes through Spatial Thinking: Helping Users Ask "Whu?". 13th international conference on Intelligent user interfaces, Maspalomas, Gran Canaria, Spain.

Smith, T.M. & Lakshmanan, V. (2010) Real-time rapidly updating severe weather products for virtual globes. *Computer & Geosciences*, available online 28th October 2010.

Spano, S. (2006) Google releases 2006 Google Earth Election Guide. Technical report, [Online] Available from: http://www.google.com/press/pressrel/earth_election_guide.html

Wang, S. (2008) Formalizing computational intensity of spatial analysis. 5th International Conference on Geographical Information Science, Park City, UT, USA.

Wikipedia. (2010a) Virtual Globe. Retrieved on 22nd December 2010 [Online] Available from: http://en.wikipedia.org/wiki/Virtual_globe

Wikipedia. (2010b) Z-Order. Retrieved on 22nd December 2010 [Online] Available from: http://en.wikipedia.org/wiki/ Z-order_(curve)

Wilson, T. (eds). (2002) OGC® KML. Version: 2.2.0. OGC 07-147r2. Open Geospatial Consortium, Inc. 251.

Zhou, Q.M. (2001) The reference model for digital earth. From Image to Digital Earth, pp. 88–95.

Advances in Web-based GIS, Mapping Services and Applications – Li, Dragićević & Veenendaal (eds)
© *2011 Taylor & Francis Group, London, ISBN 978-0-415-80483-7*

Building web services for public sector information and the geospatial web

David Pullar
Geography, Planning and Environmental Management, The University of Queensland, Brisbane, Australia

David Torpie & Tim Barker
Queensland Treasury, Brisbane, Australia

ABSTRACT: The use of commercial geobrowsers, such as Google Maps and Microsoft Virtual Earth, has dominated developments of geospatial web services. This popularity is driven by the capabilities to provide continuous visualization of the Earth's surface for searching and viewing information. This is also an important requirement for disseminating public sector information; but governments have been slow to take advantage of geobrowser technology as they face technical and information ownership issues. We identify requirements to support web mapping of public sector information, these include: i) scaled geography over a variety of levels from local to global scales, and ii) efficient data processing for information retrieval within a client-server computer architecture. Through a prototype application we also explore the use of a mapping paradigm for accessing information and provision of government services. The chapter collects requirements from a state government agency to provide information for gambling licenses and related support services. A geo-enabled web application is developed for these requirements and we discuss issues for integrating services with geobrowsing technology.

Keywords: Spatial infrastructures, spatial web services, mapping, information systems, interoperability

1 INTRODUCTION

Geographical information is quickly evolving to the geospatial web; bringing new usability paradigms and functionality with geobrowsers such as Google Maps and Microsoft Virtual Earth (Scharl 2007). These new developments are underpinned by the growth of consumer mapping web applications and adoption of open standards. A particular interest of our work is to investigate the use of these tools for provision of online government services and the opportunities they provide for more open, responsible and efficient government (GIMO 2009). An advantage of geobrowsers is that they present a continuous visualization of a globe as a reference frame for searching and viewing information (Butler 2006). Google's mission is to deliver information in a meaningful spatial context (Jones 2008), enabling users to search and explore temporal 2D/3D content on the Earth's surface along with indexed web information. Geobrowsing technology builds upon networking communication layers for sending and receiving data via the Internet. This includes support for geographic data formats, such as Keyhole Markup Language (KML) and Geography Markup Language (GML). Within commercial geobrowsers data access is filtered, but it is possible to link published KML and GML as overlays via web services. It is this ability for third parties to create and publish geospatial data as integrated web services that is the focus of this chapter.

The rationale for our work is to explore issues that arise to geo-enable government services by layering them onto geobrowser technology. The aim is to take advantage of existing geospatial web infrastructure (Craglia et al. 2008) while building specialized application services. A project was initiated to test the concepts and solutions by developing a prototype web service for a specific application. The project raised technical and institutional issues for publishing geo-enabled government services. Technical issues related to performance and use of standards, whereas institutional issues related to accountability and control over service delivery. The contribution of this chapter is insights gained from developing geospatial web services for a public sector application. Technical issues need to be addressed to achieve the envisioned goals for the geospatial web, but more importantly governments need to examine the way users interact with the geospatial web for provision of government services.

The next section discusses the geospatial web and the need for specialized representations for spatial data. The section briefly reviews system support in general for geospatial web services, and Section 3 looks more specifically at infrastructure to support public sector information services. Earlier frameworks concentrated on information discovery via metadata and online catalogs, whereas the concepts behind the geospatial web are to use geography to organize information. This presents challenges in rethinking for government services which traditionally provide a listing or directory of the services they offer. A different way of thinking would be to discover these services in the context of a geographic locality. We come back to these challenges at the end of the paper. Section 4 describes a prototype application, its requirements and implementation. It is based upon one application area, but we believe the issues are generic. We explore technical issues on: i) support for scaled geography with level-of-detail data access, ii) query performance, and iii) the software architecture using client-side or server-side processing. Section 5 examines lessons learnt from the technical implementation, and includes a discussion on using the geospatial web for accessing public sector information. The final section relates our work to research needs for spatial data infrastructures and the geospatial web, and the future research we see as necessary.

2   WEB SERVICES AND THE GEOSPATIAL WEB

A web service is an application running on a remote server that responds to requests. Typically web service requests are defined by a URL address along with a query string, and responses are XML-encoded data. The Open Geospatial Consortium (OGC) leads in the development of standards for geospatial web services. They publish specifications for geospatial data models used for data interchange and access. Developing geospatial web services is considered more demanding than general purpose services because of data volumes and the need for spatial query processing (Tu & Abdelguerfi 2006, Chang & Park 2006). Beyond these technical challenges we believe there are other issues related to integrating geospatial content with consumer mapping products like Google Maps and Microsoft Virtual Earth. In particular, support for scaled geography from local to global scales raises a number of processing challenges (Rinner et al. 2008). This requires processing logic and supporting data representations within the application that could be implemented on client-side or server-side software. Hampe & Intas (2006) explore the behavior of Web Feature Services (WFS) to support multiple representations for geospatial data indexed by map scale. They suggest extending existing standards to include metadata on scale ranges to support query filter functions accessing scale dependant feature data. While finding an efficient encoding scheme for multi-scale representations of vector data is a research issue, for raster data it has been technically solved using an image pyramid structure (Richards & Jia 2006). An image pyramid stores raster data in levels corresponding to a scale, where lower levels are divided hierarchically into tiles with a finer spatial resolution. Image pyramids and a tile-based structure were used in early geobrowser developments, such as WorldWind (NASA) and TerraServer (Barclay et al. 2002), to encode satellite imagery and scale-rendered maps. The hierarchical structure for image pyramids can also be used to index vector data which is then rasterized as part of query retrieval. Bradbec & Samet (2008) explored the performance issues for map services to support spatial browsing with raster or vector-rasterized data within different client-server configurations. They found a configuration that supported a hierarchical data structure on the server and caching tiles on the client to be the most efficient. The standards effort is also active in this area; the OGC (2007) is extending WMS to support image tiles. However the challenge for standards is to do more than provide a description of the tile structure and data sources, it also needs to support direct access to tile sets that allow caching on a server. Google Maps and Microsoft Virtual Earth in comparison assume a fixed tiling structure and naming convention, and this allows more efficient direct requests for tile sets. Performance is an important consideration for government web services as they often provide mapping data over large areas. The next section reviews the paradigms used by governments for information access versus the geospatial web.

3   GEO-ENABLED EGOVERNMENT WEB SERVICES

The development of a Spatial Data Infrastructure (SDI) is embraced by governments as a means to access public sector information, actions, decisions, and polices; and is a key focus of international research (Craglia et al. 2008). Governments are largely responsible for establishing a SDI, but have envisioned that partnerships would develop with academia and industry over time (Masser et al. 2008). However the SDI paradigm for organizing information is very different to the geospatial web, it is largely designed as a registrar of datasets and services that organizes data principally by theme and data custodian. This is different to the way the geospatial web has evolved where geography is the principle way used to organize information (Jones 2007, Craglia et al. 2008). Another significant issue is controlled access to information. SDI directs queries to the actual information provider and hence the provider has full control over data access. In contrast, geobrowsers like Google or Microsoft filter data access through their API and potentially can impose licenses or embed advertising with web services. This is a major issue for governments, especially for critical services that they require full control over for mandatory reasons. There are alternatives to commercial geobrowsers using open source products. Notably NASA released a geobrowser named World Wind in 2004. In many respects NASA's World Wind offers more open and advanced programming features over commercial geobrowsers; including an open software development kit and a scientific visualisation viewer able to show animations of global environmental processes (World Wind Central). However it lacks the high resolution global base imagery and consumer information that are popular in geobrowsers. World Wind is more suitable for the scientific community (Butler 2006), whereas commercial geobrowsers have greater market acceptance and a strong focus on the user experience of the wider community. But this marketplace is also a major slice of the target audience for government, and hence the wide adoption of these technologies is a compelling argument for SDI to support applications within geobrowsers. Some governments have already utilised commercial geobrowsers to publish data sources. For example the Western Australian government developed an application called SLIP Enabler that utilises open GIS technology to service data for a variety of management areas; including emergency services, natural resources and land development (Ducksbury 2008). The SLIP Enabler can connect a data source with Google Earth, but this is for single layer access. In the next section we discuss the requirements for an application which we believe reflects generic requirements for public sector information.

4   A PROTOTYPE APPLICATION FOR A GEOSPATIAL WEB SERVICE

Government agencies are increasingly publishing data and services, and are considering how to take advantage of the geospatial web and building services on commercial geobrowsers. While government agencies offer a large class of services we explore the application requirements for a few information services, but we believe these are representative of the types of services provided by government. A prototype web service was developed from the requirements using geobrowser technology.

4.1 *Application requirements*

A key factor in gathering requirements was to differentiate between delivery of core and promotional government services. The distinction is that core services have mandatory requirements whereas promotional services enhance government functions but do not have strict service provision requirements. An example of a core requirement is to publish locations for election polling booths, whereas a promotional service would support finding the nearest booth in a map interface from a user supplied address. Information services with geobrowser technology could provide other data on opening times, accessibility to public transport, and way-finding services.

Our experimental prototype was developed to support a number of needs related to licensing of premises with gambling machines. Gambling licenses are controlled by the government and there is a need to provide client and community support services related to gambling activities. We will elaborate on a few key requirements to highlight developments of web services. Their requirements include provision of:

1. A gambling support service that provides the location of the nearest counseling centers.
2. Support for geographical information that is part of a community impact assessment status (GCIS 2008) that includes:

   – size and type of site (e.g. local tavern, RSL, sporting or community club)
   – patron characteristics
   – size and distribution of membership base (for club sites only)
   – distance (drive time and radial)
   – physical barriers to site access (e.g. major roads, water ways)
   – location of other gaming sites
   – cultural or social factors
   – population density.

The first application required map overlays of gambling sites and support for place-based queries, such as requesting information on the nearest help centre from a given location or address. The geographical distribution of gambling sites and centers varied greatly, so some areas were sparsely mapped and other areas had many overlapping sites. Matching the zooming capabilities of geobrowsers to data content was important for interpreting the data; this required the ability to cluster sites and to present aggregated information as a list. The prototype used concepts of level-of-detail access as supported by geobrowsers to control the generalization of spatial data within the web service.

The second application supported the evaluation process for a community impact assessment which is undertaken as part of the licensing approval process for changes to a site or establishing new gambling venues. A custom geospatial query service was developed to support the information gathering process, particularly for accessing large amounts of geo-demographic data as features or background thematic layers. The demographic information is scale dependant; it is tabulated for neighborhoods, suburbs and districts. We relied upon the zoom display level within the geobrowser to load and render the appropriate scaled geography. This was done efficiently with clustering of graphic features and tiled rasters for the thematic coverages. Features within these coverages were also represented as a point marker at the center of a census polygon. Depending on the display zoom level and their spatial arrangement these were clustered to a single point marker if they overlapped, and at certain ranges they switched to a different scaled geography. We believe providing support for scaled geography in web services is a common requirement for public sector information. The prototype was implemented with both the Google Maps API and Microsoft Virtual Earth API.

The next section discusses a data model developed within the prototype to support scaled geography.

4.2 *Generic requirements*

The general requirements obtained from these applications were for: i) map overlays of information on the geobrowser display, and ii) queries for context relevant information. Map overlays included continuous coverage thematic information such as demographic information, and discrete location information such gambling sites. The information had to be rendered based upon the user zoom scale which required displaying the appropriate level of spatial units, symbolization and legend information. Discrete locations were points which had to be clustered depending on the zoom level.

Almost all queries were spatial and relevant to visible features displayed in the base layers or the application overlays. Users wanted to click on the map to identify the boundaries of spatial units for thematic layers and to obtain a precise value, as opposed to a classified value shown by map symbology and the legend, and other associated information for the region of interest. Likewise by clicking on a discrete location users expected to retrieve relevant information. Additionally users wanted to query nearby associated information, for instance the licensing requirements for clubs included several proximity criteria. Location queries based upon a street address were also required in both applications. There was very little interest in supporting general textual queries; the expectation was that information could be displayed on map overlays and users could retrieve information or query nearby locations by clicking on the map.

The main challenge in these requirements was the contextual nature of queries. Governments hold a vast amount of spatial information; but only a specific subset of information is of interest for a single application. Presenting this information in a scale dependant way on maps and identifying other associated information of interest for a query was difficult.

4.3 *Experimental prototype*

The requirements raised a number of design decisions: i) a data structure to support level-of-detail data access, and ii) building geoprocessing functions to support queries and scaled geography as client-side or server-side processes.

Geobrowsers use an indexing scheme to partition the Earth into a hierarchical system of uniformly sized tiles. Both Google and Microsoft adopt a spatial reference system based upon a spherical Mercator projection and an image-tile pyramid that hierarchically divide the map into quadrants (see Fig. 1). The levels of the pyramid structure define zoom levels where each level corresponds to a certain map scale and level of detail. We developed a simple data model based upon zoom levels for accessing layers (see Fig. 2). An abstract level specification defined the level ranges for displaying a source. The service name is used to identify the specification within a web service catalogue. Specialised classes are implemented for point markers and image tiles. Other specialised classes may be defined for other graphic feature types as needed. Some client-side or server-side geoprocessing is needed to check if the level specification layer is within visible bounds and the zoom level is within the current range. If the later test fails then a query method on the level specification is run to update to the appropriate source data. The source is specified as a URL query string, and in the case of image tiles a URL to a legend image is also returned.

We experimented with implementing the program logic on both client-side and server-side software. Chang & Park (2008) describe the strengths and weaknesses of these approaches for general web services. The data and program logic for running an application can be built within the client browser or on the server. Problems arise in the former case because of the limited support within client-side browsers to run applications and the need to continually access data from the server. Performance problems also arise with a server-centric

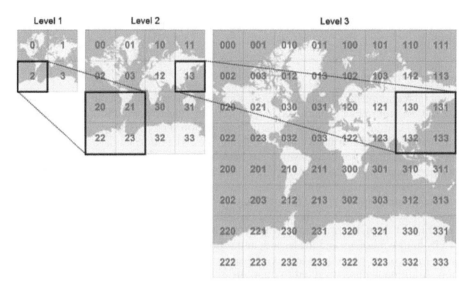

Figure 1. Pyramid structure showing levels for spherical Mercator projection (Source Microsoft Virtual Earth Article http://msdn.microsoft.com/en-us/library/bb259689.aspx).

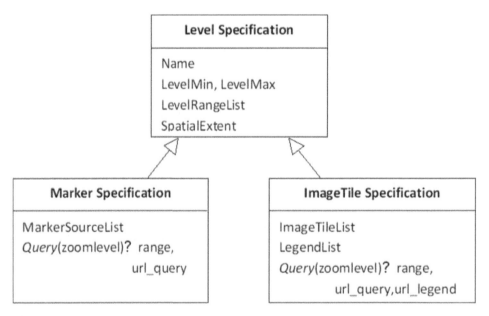

Figure 2. Data model structure used for representing marker and image tile sources for level-of-detail web service access.

application because every user interaction requires a request to the server and for refreshing the geobrowser map display. We developed a server-centric solution as a Microsoft DotNet control that could be added to a web page with object classes for the data model in Figure 2. The server-side objects need to synchronize with proxy (prototype) states and functions implemented on the client-side in JavaScript. This approach is typically favoured by programmers to take advantage of more powerful development frameworks like Microsoft DotNet or Java Enterprise (J2EE). However geobrowser technology uses a mashup architecture where all browser interactions are accessed via the geobrowser's API on the client-side. A mashup

114

architecture is better suited to support client-side scripts that integrate lightweight web services (Liu et al. 2007) and there were definite performance advantages in adopting this approach. This meant that the server-side capabilities had to be well matched to the conventions used by the geobrowser API. In particular; we adopted the same conventions for stream access to raster tiles and their fixed spatial indexing scheme.

There is a debate on the nature of a standard for a tile map service that balances flexibility and scalability (OGC 2007). Standard organizations favor specifications that are more flexible, but this comes with a cost for providing high level support to client applications. Our experience was that you needed high level support to efficiently implement streamed access to raster tiles and for query processing. For instance, a web service to identify spatial objects from a coordinate gives different results depending on the zoom level. It is difficult to implement this query if the web service does not have a shared data model for level-of-detail data access. While there is some differences in naming conventions, the spatial indexing scheme used by Google and Microsoft is the same. We adopted this indexing scheme to support level-of-detail data access for our data model in Figure 2. Just as OGC specifications support map projections, we see the need for standards to support spatial indexing schemes for queries and streamed accessed to raster tile sets.

4.4 *Performance results*

Performance tests were run for accessing points and displaying as a layer in Google Maps and Virtual Earth geobrowsers. The results are shown in Figure 3. Tests were run on a single computer; the results are relative as performance will vary for different computer environments. Two different client-server configurations were implemented (see Fig. 4). In the first configuration the geobrowser is interfaced to a server-based web control which accessed a spatial data server on the host machine via a TCP connection. From figure 3 it can be seen that performance to render point markers in a geobrowser decreases with larger numbers of points; response times from Google Maps decreases rapidly. However, the main difficulty with this configuration is synchronising the application between the geobrowser (which communicates with Google or Microsoft) and server control. The web control may be simplified

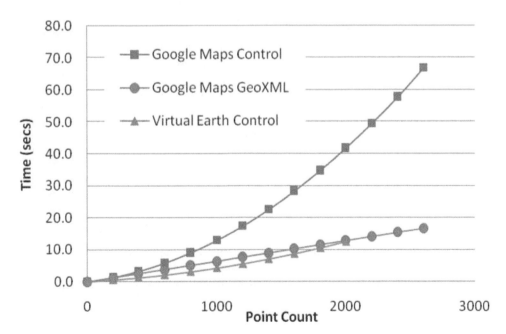

Figure 3.   Display time for overlays in geobrowser.

115

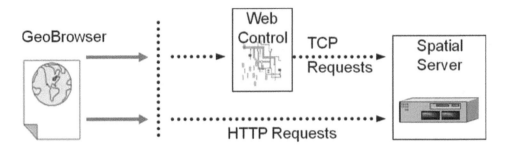

Figure 4. Configurations for geobrowser server access via: i) web control with TCP connection to server, ii) GeoXML with HTTP connection to server.

to just serve spatial data, but the second configuration does this more directly. In the second configuration the geobrowser directly communicates to the spatial server by a HTTP connection. Requests are processed based upon a query string and spatial data is directly streamed to a layer in the geobrowser. Google Maps supports this configuration with its GeoXML interface.

In practice the response time to display external spatial data in geobrowsers becomes unacceptable at counts of above 500 points. Given a geobrowser displaying a map canvas of 500×500 pixels and assuming a uniform distribution of markers clustered to a distance corresponding to 25 pixels, then this produces a maximum of 400 markers. It is rare to have a uniform distribution of points, but this still demonstrates the need to support scale-dependant access and clustering points on the server to be within acceptable performance limits.

5 DISCUSSION

While the public are now familiar with the notion of using the planet to browse and explore information; governments are challenged on how to take full advantage of using geography to organize information. From a technical perspective we explored the potential to use geobrowsers like Google Maps and Virtual Earth for publishing geospatial web services. We demonstrated this was feasible, but it did require adapting data formats to the web for efficient access. All broad coverage datasets were pre-rendered as tiles and all spatial information was structured to suit level-of-detail access. Governments generally prefer to build systems based upon open standards, or at least not to be seen as locked into a particular commercial format. The prototype web service we developed worked efficiently by following a similar software implementation to geobrowser technology where there is well defined data representations to support client-server interactions. We identified the need for scale ranges for spatial access, and to encode this into a query specification rather than handle the complexities of maintaining consistency between geospatial data sources and mapping scale on the client-side machine. While there is some debate on the form for a web tile service specification to keep it flexible (OGC 2007), we found the conventions used by Google Maps and Virtual Earth were flexible. We developed a simple specification for scale ranges based upon level-of-detail index, and used it for accessing raster and feature data efficiently.

While we only focused on one application area with a few web services we developed an appreciation for how public-sector information could be better adapted to the geospatial web. Wide coverage datasets were pre-rendered as image tiles for efficient browsing, but we also supported point queries that returned the background information. Web services queries were developed for feature queries that return associated attribute information, and special services such as identifying the nearest features. For instance, querying what gambling support facilities are near a site and obtaining current information on their open hours and way-finding instructions. We believe that based upon the context for information being displayed, in terms of the business activity, map scale and locality, it is possible to anticipate services of interest to users. This could additionally be constrained by time. For instance, election polling booths may be in operation on a particular day at a particular locality due to special by-elections or local elections. Unfortunately there was only limited testing of these services with a general public audience, but the feedback was that they immediately appreciated the geospatial web provided a different way of viewing public sector information based upon locality.

6 CONCLUSIONS

The paper describes the implementation of a web application to geo-enable government services based on real requirements. Specifically we explore if government information services can be built on top of existing geospatial web infrastructure such as Google Maps or Microsoft Virtual Earth. One aspect of the requirements was to efficiently display continuous map data related to specific themes, for instance population density and geodemographic variables over the whole state of Queensland which covers approximately ¼ of the land area of Australia. These background layers were pre-rendered as raster's and efficiently encoded into a tile structure compatible with Google Maps and Microsoft Virtual Earth. We implemented a web service based upon the GeoXML interface that integrated nicely with these geobrowsers. Population data could be rendered (as an overlay with transparency control) at different map scales for different census geographies (i.e. neighborhood, suburb, district and region) depending on the zoom level of the geobrowser. Another feature of the requirements was support for locational queries, such as clicking on a locality to obtain detailed information. Feature information (currently limited to points) was also supported with web service queries based upon the extent and map scale, results were rendered as markers or a marker cluster depending on the map scale. Future developments will support other graphic objects, such as lines and polygons.

The paper examines the relationship between commercial geobrowser technology with a mass market focus, and SDI which is largely initiated by government. The success of geobrowsers is underpinned by SDI and improved collection of geospatial data by government, but SDI has not taken advantage the different paradigm used for accessing information services by the geospatial web. Craglia et al. (2008) identified a number of research needs to make a better connection between SDI and the geospatial web. The prototype we developed tested some of the technical aspects of these; including integrated web services, tiling schemes, place-based query searches, and IT architectures. We identified that further research could be undertaken to: i) identify associations between activities sensitive to place and time period, and ii) develop better spatial understanding of the sphere of influence of activities based upon scale and distances. This would allow appropriate information to be displayed by a geobrowser relative the user context. Questions will remain as to what public sector information services can be offered using commercial geobrowsers without compromising the integrity of that service (e.g. via the potential for embedded web advertising, licensing fees or the reliability of the service), but the geospatial web is a reality, and the general public will use the geospatial web as their primary means to find information and to make decisions in their lives.

ACKNOWLEDGEMENT

I would like to gratefully acknowledge the support provided by the Queensland State Government of Australia for this project.

REFERENCES

Barclay, T. Gray, J. Strand, E. Ekblad, S. & Richter, J. (2002) *TerraService. NET: An Introduction to Web Services*, Tech. report MSR-TR-2002–53, Microsoft Research, June 2002.

Brabec, F. & Samet, H. (2008). Hierarchical Infrastructure for Internet Mapping Services. In: Sample, J. Shaw, K., Tu, S. & Abdelguerfi, M. (eds.), *Geospatial Services and Applications for the Internet*: 1–30. Springer.

Butler, D. (2006). The web-wide world. *Nature*, 439, 776–778.

Chang, Y.-S. & Park, H.-D. (2006) XML Web Service-based development model for Internet GIS applications. *International Journal of Geographical Information Science*, 20 (4), 371–399.

Craglia, M., Goodchild, M.F., Annoni, A., Camara, G., Gould, M., Kuhn, W., Mark, D., Masser, I., Maguire, D., Liang, S. & Parsons, E. (2008) Next-Generation Digital Earth. *International Journal of Spatial Data Infrastructures*, 3, 146–167.

Ducksbury, M. (2008) Landgate's SLIP Enabler. *Spatial Science Magazine*, Spatial Sciences Institute, Canberra, Australia, Spring'08. pp. 44–45.

Foster, I. (2005) Service-oriented science. *Science*, 308, 814–817.

GCIS. (2008) *Guidelines Community Impact Statement*. Queensland Gaming Commission, Queensland Government, Brisbane, Australia.

GIMO. (2009) *Engage - Getting on with Government 2.0*. Australian Government Information Management Office, Canberra, Australia. [Online] Available from: URL: http://www.finance.gov.au/

Grossner, K., Goodchild, M. & Clarke, K. (2008) Defining a Digital Earth System. *Transactions in GIS*, 12 (1), 145–160.

Jones, M. (2007) Google's geospatial organizing principle. *IEEE Computer Graphics and Applications*, 27 (4), 8–13.

Liu, X., Hui, Y., Sun, W. & Liang, H. (2007) Towards service composition based on mashup. In *IEEE Congress on Services; Proc. Utah USA, 9–13th July 2007*. pp. 332–339.

Masser, I., Rajabifard A. & Williamson, I. (2008) Spatially enabling governments through SDI implementation. *International Journal of Geographical Information Science*, 22 (1), 5–20.

NASA. *World Wind*. [Online] Available from: URL: worldwind.arc.nasa.gov

OGC. *Open Geospatial Consortium, Inc*. [Online] Available from: URL: www.opengeospatial.org

OGC. (2007) *OpenGIS Tiled WMS Discussion Paper v0.3*, OGC Document #07–057r1

Richards, J.A. & Jia, X. (2006) *Remote Sensing Digital Image Analysis*. Springer.

Rinner, C. Keßler, C. & Andrulis, S. (2008) The use of Web 2.0 concepts to support deliberation in spatial decision-making. *Computers, Environment and Urban Systems*, 32, 386–395.

Scharl, A. (2007) Towards the geospatial web: media platforms for managing geotagged knowledge repositories. In: Scharl, A. & Tochtermann, K. (eds.), *The Geospatial Web*. London, Springer. pp. 3–14.

Tu, S. & Abdelguerfi, M. (2006) Web Services for Geographic Information Systems, *IEEE Internet Computing*, 10 (5), 13–15.

World Wind Central. (2008). *Google Earth Comparison*. [Online] Available from: URL: www.worldwindcentral.com/wiki/

Performance

Advances in Web-based GIS, Mapping Services and Applications – Li, Dragićević & Veenendaal (eds)
© *2011 Taylor & Francis Group, London, ISBN 978-0-415-80483-7*

WebGIS performance issues and solutions

Chaowei Yang, Huayi Wu, Qunying Huang, Zhenlong Li, Jing Li, Wenwen Li, Lizhi Miao & Min Sun
Joint Center for Intelligent Spatial Computing and Department of Geography and GeoInformation Sciences, George Mason University, Fairfax, VA, USA

ABSTRACT: WebGIS has become a popular tool for providing data, information, and other services through computer networks including both the Intranet and Internet. Performance emerges as a critical challenge when large numbers of users, complex processing, and large volumes of datasets are involved. The challenge is related to limited computing and networking devices, small bandwidth networks, as well as many other factors. This chapter addresses performance issues in a systematic manner from the aspects of architecture, bottlenecks and performance factors, performance improving techniques, and performance solutions.

Keywords: WebGIS, performance, distributed geographic information processing, geospatial cyberinfrastructure, spatial web portal, cloud computing

1 INTRODUCTION

WebGIS refers to Geographic Information Systems (GIS) operating on a web-based computing platform (Yang et al. 2005). It is a product of the federation between GIS and the web computing platform characterized by different technical protocols, such as HyperText Transport Protocol (HTTP) supported by a web server and web browser. Current WebGISs have extended beyond the traditional HTTP arena. They include all types of GIS clients that can interact with a GIS server through HTTP and web services, such as Web Map Service (WMS) (de La Beaujardiere 2004), Web Feature Service (WFS) (Vretanos 2002), and Catalog Service for the Web (CSW) (Nebert & Whiteside 2005). The technological advancement of HTTP and web services, combined with the growing need for geospatial information has driven WebGIS from the specialized geospatial discipline into the mainstream of information technology (Yang et al. 2011). WebGIS provides a geospatial platform for delivering data, information, and services to a wide variety of users across the planet.

At the global level, commercial WebGIS platforms, such as Microsoft's Bing Maps and Google Earth (Butler 2006), integrate global geospatial data and information to provide popular but relatively simple services, such as viewing geospatial data in 3D as used by realtors and home buyers. At the national level, different national systems, such as the U.S. National Map (USGS 2010) and the Geospatial One Stop (GOS 2010) provide platforms for users to view and download open accessible datasets. Different local governments (e.g. counties) have also deployed WebGIS as a platform to provide public services, such as cadastral querying and evaluation. In addition to the platform functions provided by WebGIS for different services, large quantities of datasets are put online for easy access through open registry, discovery, and integration approach (Nebert & Whites 2005, Li et al. 2010). For example, NASA's Distributed Active Archive Centers (DAACs) provide satellite observation datasets to researchers and global users via the Internet-based WebGIS (NASA 2010). Government agencies, such as EPA (EPA 2010) and U.S. Census Bureau (USCB 2010), also provide their datasets to users online. Many state governments share their geospatial datasets holdings online through WebGIS (e.g. VA 2010). The proliferation of GIS datasets, together with the maturity of WebGIS technologies, and the increasing demand for utilizing geospatial data make WebGISs popular platforms to utilize the web infrastructure and geospatial resources for decision support (Yang et al. 2007). This widespread utilization of WebGIS has also created new challenges. One of the challenges is to design and to develop a WebGIS that is fast enough so that the average Internet user will not be deterred from using the platform. For users, an acceptable waiting time for a web response is about 8 seconds (Corner 2010). This 8 second rule refers to the time between a user request submission and when a response is received. Given the vast amount of datasets, various networks, and the complexity of geospatial information analysis involved, 8 seconds is a very limited time frame. Other issues, such as interoperability, functionality, and reliability, also emerged as WebGIS challenges.

2 PERFORMANCE INDICATORS

Performance becomes one of the most challenging issues of WebGIS when large data volumes and user numbers are involved, especially when a large number of users accesses concurrently (Cao 2008, Zhang & Li 2005). Different from traditional GIS, WebGIS involves computing components consisting of geospatial data servers, analysis servers, web servers, computer networks, and client-side user interactions. Each of these components potentially can create a bottleneck within a WebGIS. Therefore, when performance becomes an issue, computing level strategies need to be deployed to solve bottleneck problems residing on each of the components. This chapter examines the WebGIS performance issue by systematically introducing aspects of architecture and bottlenecks, indicators, improving techniques, and solutions.

WebGIS is developed utilizing the functionality of GIS and web-based computing platforms. Both platforms play significant roles in the architecture. A simplified architecture is depicted in Figure 1, where the system includes a data server, spatial server, graphics (or visualization) server, web server, and web browser. Each of the components and their connections can be a bottleneck for a WebGIS, therefore, when considering performance, each component should be considered.

The performance complexities include four indicators: time, memory, reliability, and interoperability. Different users have different perspectives regarding indicators as detailed in Table 1. There are other indicators, such as security and usability, which might be of concern for some users. Because these indicators are either quite mature via a routine process or systematically addressed in the IT fields, we will elaborate only on the four noted above. We will, however, briefly touch upon security and usability. Security is a concern that is addressed within the server side to make sure that the system, data, and processing functions of a WebGIS are only accessed by users with relevant priority. This is normally achieved through accounts management in the server administration. For example, within ArcGISServer (ESRI 2010) configuration, the map configuration is configured to be only accessible by a map editor, while the service configuration and maintenance can only be accessed by the

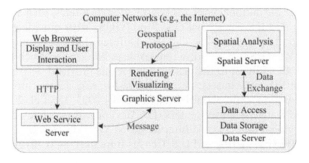

Figure 1.   Architecture of a WebGIS (adopted from Yang 2008).

administrator. Users can only access the published WebGIS portion (ESRI 2010).Usability analysis is typically conducted when developing a WebGIS. Usability analysis often includes a function requirements analysis, graphic user interface tests, and user questionnaires in addition to compilation of access statistics of users to find the most visited functions and the routine habits of general users (Ingensand & Golay 2008).

2.1 Time complexity

Time complexity is the most important factor in performance measurement. It refers to the time spent between the submission of a request by an end user and the response finalized at the client side. Since each component within the WebGIS architecture will spend time accomplishing its own tasks, the time complexity is the total time spent on each component. Therefore, time can be described as Equation 1.

$$T = \sum_{i=0}^{n} Ti \tag{1}$$

Ti refers to the time spent on the ith component. Equation 1 can be simplified as the time spent on data transmission to (send) and from (receive) the server as well as the time spent at the server and client sides for processing as noted in Equation 2. The time spent on sending requests and receiving results depends mostly on the computer networks bandwidth.

Table 1. Different perspectives of WebGIS performance.

	Host and maintainer	Designer and developer	End user
Time	How many resources are needed to reduce time and complexity? Restricted by available budget, network and computing infrastructure.	Strategies, architectures, algorithms and implementations can reduce time complexity. Restricted by available technology, tools, and designer's and developer's knowledge and experience.	How much time is needed to wait for a response? 8-second rule applies. Restricted by the server and network performance and web client computer if client side computing is involved.
Memory	How much memory is needed for a well-performing WebGIS? Restricted by available budget and technology.	What database design, data structure and organization methods can make full use of available memory? Restricted by database and designer's and developer's knowledge and experience.	How much memory will be used by a WebGIS client? Depending on client side function and computing design.
Reliability	How to ensure reliability? Redundancy system and better system configuration.	How to deploy a stable and reliable system? Hardware and software redundancy, exception detect and process.	Fast and smooth access to a WebGIS is desirable.
Interoperability	Transparency? Scalable? Easy to maintain and upgrade?	How many days does it take to develop the system? Can any technologies be reused?	Compatibility with a client hardware and software? Familiar user interface and free and/or open-source clients.

The process time normally depends on the process conducted at the server side for generating a response given the requesting conditions.

$$T = Tsend + Tprocess + Trecieve \tag{2}$$

The time complexities for the process at server and client sides can be described using the traditional time complexities for algorithms. In computational complexity theory, it is denoted by big O notation $O(n)$, where the actual time used is represented by the orders of magnitude in respect to the complexity of the data and process times. The time complexity depends on the network bandwidth. Therefore, when reducing the time complexity of a WebGIS, we need to improve the network throughput according to the network characteristics. We can also try to reduce the process time so the total time needed for a response can be reduced. Normally, we cannot revise the sending and receiving time across the network, which can only be tested and utilized as is. For the server-side processing, we can utilize different techniques to minimize the time for processing through processing algorithm optimization and management strategies. Total processing time can be reduced through these computing techniques and hardware upgrades.

2.2 Space complexity

Another complexity of WebGIS is the computing resource RAM (Random Accessible Memory) needed to conduct processing requests and generating responses. As a valuable resource, RAM is needed by all software and, at the same time, is very limited. Therefore, the less RAM used for the same process and datasets, the better the software is. The memory required within a WebGIS task includes RAM at both the server (S_{server}) and client (S_{client}) side. In a computer network, the memory is observed as the data volume ($V_{network}$) transmitted between the client and the server. Therefore, the space complexity of a WebGIS is a function (Equation 3) of the space used in each of the components of a WebGIS to conduct relevant tasks.

$$S = f(Sserver, Vnetwork, Sclient) \tag{3}$$

The space complexity does not have a linear relationship with time complexity. For example, when the data volume is reduced for transmission, the time can be reduced. When more memory is allocated on the server side, a faster system can be achieved due to the faster server responsiveness. The space complexity of data at the server side and the client side will impact the space complexity for network transmission. Therefore, the space complexity is a complex function of the server, network, and client. According to different specific applications, however, the dominating space complexity can be determined. For example, if a WebGIS function requires significant computing but the data transmitted is limited, then the server or client side computation dominates the space complexity. Moreover, if minimum processing complexity is involved but a large volume of data is involved, then the network space complexity will dominate the space complexity of a WebGIS.

2.3 Reliability

The reliability is another complex performance indicator and refers to the availability and accuracy of a WebGIS service at all times for users at different places. A standard solution to this complexity is to duplicate the components in the system so when one part fails, the duplicated part will be called into operation. The reliability is also related to the accuracy of specific algorithms that can handle different types of errors within the data, user requests, information processing, response interpretation, analysis, and any other WebGIS functions. So the reliability is ensured through the duplication of a critical component as well as through geospatial information processing algorithm. The duplication strategy also benefits performance by leveraging duplicated resources to respond to more concurrent user requests.

2.4 *Interoperability*

Interoperability means that a WebGIS can be reused or shared with other WebGISs. Thus the system can be: a) built once, used many times, b) transparent to other systems, c) easily upgraded and maintained, and d) easily shared or utilized by other systems as a component (Bambacus et al. 2007). Interoperability is normally achieved through the standardization of the functional components' interfaces and the data contents within different WebGISs. For example, WMS and WFS will allow the interoperability among WebGIS to interoperate through same interface. eXtensible Markup Language (XML) (W3C 1996) will allow WebGISs to share data in the same format. Interoperability is addressed within the GIS community through organizations like the International Organization for Standardization (ISO/TC211 2010) and Open Geospatial Consortium (OGC 2010). There are industrial de facto standards, such as Keyhole Markup Language (KML), used by Google Earth.

Interoperability enables WebGIS to be shared and reused, solving one aspect of the performance challenge. At the same time conformity with the standardized interfaces will reduce speed when transforming data content from and into standard formats. For example, the WMS ensures interface interoperability and Geography Markup Language (GML) (Portele 2007) ensures content interoperability. However, the interoperable process impacts the response speed of a system (Yang et al. 2007). The performance indicator will appear complex in an operational WebGIS because the components (Fig. 1) will be geographically distributed and different numbers of components will be involved. There are techniques that can be utilized to improve the performance and the most important and popular ones that can help improve the WebGIS performance (Yang et al. 2005) are described in Section 3. Section 4 describes the solutions based on the techniques for specific applications.

3 PERFORMANCE IMPROVING TECHNIQUES

Besides utilizing more computing resources (such as more processing units and memory), software and geospatial techniques also play different roles in improving WebGIS performance (Yang et al. 2005). According to previous research and experiments (such as Yang et al. 2005 and Yang et al. 2007), the techniques identified in Table 2 can be utilized within different components of a WebGIS to improve the performance by reducing time and space complexity. For example, the first row of Table 2 illustrates that progressive/asynchronous transmission will a) decrease server time and increase server space complexities because data are duplicated at server side, b) increase network time and space complexities because data volume transmitted is increased, and c) decrease client time complexity and increase client space complexity because a faster response is achieved with multiple steps of data transmission.

3.1 *Pyramid and data volume reduction*

Geospatial datasets are voluminous and have multiple dimensions. Client-side user interactions, such as visualization, do not require transmission of a complete dataset. A *pyramid* is one of the techniques used to manage large volume datasets. It is a series of reduced resolution representations of the dataset for both raster and vector datasets. A *pyramid* contains a number of layers with different spatial resolutions, and each layer is re-sampled (a raster pyramid) or simplified (a vector pyramid) at a more generalized level. When a user sends a request, the server will choose the layer within the proper level and quickly identify desired data within the pyramid (see Yang et al. 2005 for a detailed implementation). For this research we used the global land cover dataset of 317 megabytes to test the effectiveness of the pyramid and spatial index. Two web map servers are deployed on the same computer (8 CPU

Table 2. Performance improving techniques and their relevant functional components.

	Server		Network		Client	
	Time	Space	Time	Space	Time	Space
Progressive/Asynchronous transmission	↓	↑	↑	↑	↓	↑
Efficient data organization/Spatial index/Cache	↓	↑	–	–	↓	↑
Parallel computing	↓	↑	–	–	↓	↑
Compression/Encoding	↑	↓	↓	↓	↑	↓

Note:
1. Legend: ↑ – Complexity increased; ↓ – Complexity increased; - – Complexity not significantly affected.
2. For server and client components, space is utilized to trade for time. Usually more space is used to reduce time needed. However, compression technology is used to reduce space usage while an overhead time of encoding and decoding is required. This overhead time is matched by using far less transmission time in network. Please refer to 3.1 and 3.2 for detailed discussion.
3. For network component, more data means more time needed. However, the time increased is paid off by less time consumed at both the server side and the client side or by transmitting only a small part of the whole response data as this first return to improve the user's experience. Please refer to 3.1 for detailed discussion.

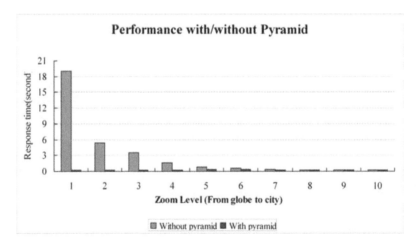

Figure 2. Response time for a GetMap request at different zoom levels.

cores, 2 Ghz, 4G RAM, Linux OS) using MapServer (MapServer 2009). One server has a pyramid data structure and relevant index, the other one does not have a pyramid or index. The response for a single GetMap request during the different zoom levels are recorded for both servers. 10 different detail levels are selected to test.

The result (Fig. 2) shows that the server response time is significantly reduced by utilizing pyramid structure on the server side. The response time for pyramid is almost constant and small compared to the ones without pyramid.

3.2 *Progressive transmission*

For addressing discrepancies in volumes of data between the server (large) and client side requests (small), various approaches were developed to facilitate the transmission between the server and the client. These approaches can be divided into three categories including compressing, partitioning, and simplifying. Compression based transmission reduces the volume of data that is transmitted. The popular algorithms in this category include the

JPEG method (Gonzales & Woods 1993), the wavelet transformation method (Morlet & Grossman 1984), Partitioning and tile transmission refer to sending partial data to the client side and is mainly used for query purposes (Wei et al. 1999). The third category extracts data to form multi-level models as one pyramid. The pyramid is delivered to the client side through progressive transmission (Fig. 3). Here a progressive transmission implementation using multiple-dimension datasets is introduced.

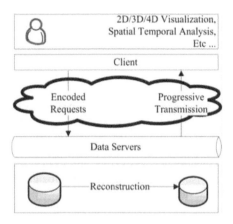

Figure 3.   Progressive transmission in WebGIS.

The most important aspect of progressive transmission is how to construct the pyramid, which varies with datasets. Geospatial data includes vector and raster data. Vector data are described with spatial objects including objects and object relationships (Zlatanova et al. 2004). These are organized in pyramids to provide fast access for editing, queries, analyses and visualization (Buckley 1998). Pyramids can also be integrated in multi-views (Lee et al. 2008). The major issue is how to compress the data in an efficient manner while reserving the original topology and connectivity of the data (Yang et al. 2007). Map generalization is a typical method to assist the generation of multi-level vector data. In addition, the transmission of geometry relies on the construction of progressive meshes using triangulation or polygons (Samet 1990). Section 3.1 provides an example for raster data.

The sequent progressive transmission implementation is paired with the multi-level models of geospatial datasets (Komzak & Slavik 2003). The general idea is to transmit coarser resolution data first and finer resolution data subsequently and gradually. Because low resolution data take less space, the transmission process can first provide an overview of the data in a short time period. Implementations of progressive transmission differ with the specific data characteristics, compression methods, client and server configurations, and visualization purposes (Davis & Nosratinia 1998, Fisher 1995). To illustrate how progressive transmission is applied in multidimensional data, we utilized dust particle density data generated from the NMM-Dust model (Xie et al. 2010). Represented in a series of sequential 3D matrixes, this scalar data is four dimensional and gridded. Each point in the matrixes is defined by latitude, longitude, pressure level and time. All datasets are kept on the server side while the client side requests and processes the data for visualization. The progressive decomposition and transmission of temperature data from the model output are kept on a server with an original data volume of 4.7 MB. Six levels are formed based on index tree composition and the level 6 corresponds to the original dataset. The data volumes at level 6 correspond to that of the original data, thus the transmission time is equal to the original one. The test server has an Internet speed of 23.3 mbps to the server on the campus at the Center for Intelligent Spatial Computing of George Mason University. The test client is located off-campus with a network access of 28 kb bps (wireless) and 79 kb bps (wired).

Table 3. Progressive transmission of 4D temperature data.

| Level | Evaluation | | Transmission speed | |
	Storage	Compression ratio	Wired (ms)	Wireless (ms)
1	3 kb	$6.2*10^{-4}$	37	107
2	4 kb	$8.4*10^{-4}$	50	142
3	10 kb	$2*10^{-3}$	126	357
4	73 kb	$1.5*10^{-2}$	924	2607
5	577 kb	$1.2*10^{-1}$	$7*10^3$	$20*10^3$
6	478 2kb	1	$6*10^4$	$17*10^4$

Table 3 shows the performance of transmissions. Level 6 is equal to the original data and its transmission speed is used as the reference. This table illustrates that progressive transmission has two advantages: 1) the data are transmitted at different resolution levels, which saved the transmission time by offering a series of gradual views; the low resolution with small data volume can be delivered in a short time, and 2) Progressive transmission reduces the rendering complexity by visualizing only a portion of original data. Due to the reduced data volume at low resolution, the time and space complexity of rendering is also reduced. For visualization driven applications, the progressive transmission can be combined with the pyramid, which is expected to further optimize online visualization (Livny et al. 2007). A drawback is that progressive transmission costs more space and preprocessing time on the server side.

3.3 *Spatial index and cache*

A traditional WebGIS usually generates the requested maps dynamically on the server or with client technologies, such as with Java applets or Scalable Vector Graphics (SVG), to publish maps. It is time intensive on the server side since rendering maps from spatial data can be time consuming. *Spatial indexing* utilizes the location of geospatial data information in storage to improve the processing efficiency. In traditional GIS, a spatial index (Samet 1990) is widely used in a geospatial database to optimize the geospatial operations, such as spatial query and spatial analysis. In the open and distributed Internet environment, spatial indexing can also help improve WebGIS performance. Similar to the pyramid and progressive transmission of multiple levels, pyramid data structure and spatial indexing can be utilized to reduce the time and space complexity of a WebGIS (Yang et al. 2005).

When using pyramid structures, a large number of tiles (determined by the data size, tile size and levels of pyramid) will be generated. It is very time consuming for the operating system to locate and read tiles from the hard disk each time a tile is accessed. By caching the frequently requested tiles, the server performance can be enhanced. An experiment was conducted using LandSAT imagery data (~6 gigabytes in GeoTIFF format). A six-level pyramid with the tile size $512 \times *512$ pixels was built. Two tile servers were established to serve the data. One server utilized the embeddable database Oracle Berkeley DB (BDB) (Oracle 2010) to store all the tiles and cache the frequently requested tiles, while the other one did not cache. The average response time for the server with BDB cache was much smaller than the other server (Fig. 4). With the increase of concurrent requests, the average response time for both servers increases. However, the increased speed for a BDB cache server is much slower.

3.4 *Asynchronous transmission*

The increasing applications of WebGIS generate large catalogues and datasets served through the Internet. This brings two more complexities in WebGIS performance: a) some WebGIS

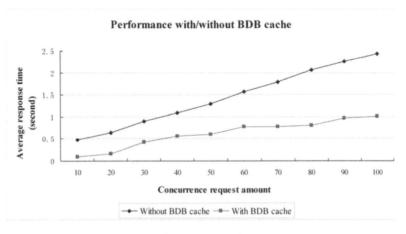

Figure 4.   Average response time using different number of concurrence requests.

operations, such as data searches from large data volumes with large data sets will require minutes before the final results can be returned (Li et al. 2010); b) when multiple servers, such as WMS servers, are accessed, different servers will have different response times ranging from seconds to minutes or hours (Yang et al. 2008). Therefore, it is important to recognize this and issue the request for updating intermediate results in an asynchronous manner, or using asynchronous transmission. As a solution, we can a) request updates every few seconds to get results for large datasets, or b) issue asynchronous requests so that whenever a server responds, the client will receive updates with each asynchronous transmission.

3.5   *Multi-processor and grid computing*

Two or more CPU computing systems can be built as a multiprocessing system to handle multiple concurrent user requests. WebGIS often requires a fast response for large numbers of concurrent users, which can be addressed by a multiprocessing system. However, due to the limitations of scalability of a multiprocessing system, it is normal to have a multiprocessing system with 8–16 CPU at most which is sufficient for some WebGIS with relatively less concurrent user requests and low computational requirements. For applications with massive concurrent users and intensive computing demands, (such as those in Google Earth), a grid computing platform with loosely coupled CPUs can be utilized (Armstrong et al. 2005, Yang et al. 2008). Within a grid platform, each resource is typically a high-performance computer or a cluster of servers. Although most grid computing applications were used to concentrate on the areas of computational science and engineering that have traditionally been major users of high-performance computing, a much wider variety of applications have been explored (Hawick et al. 2003). This technique is demonstrated through a real-time routing application on a grid system. On-line real-time routing often has thousands of concurrent users. Figure 5 shows the system architecture, where data sets include static network characteristics, temporal signal control status, historical link travel times, OD (origin-destination) nodes, and other non-recurring events such as accidents and work zones.

When massive concurrent user requests arrive, the computing power of a grid platform is utilized by dispatching the routing requests to different computing nodes. Each routing task will be running on a single grid node scheduled by a middleware (such as Condor, Thain et al. 2005), and each node will return the optimal path for the requested OD pair. When a routing task is completed, the result will be sent to a web server and passed on to the client.

The real-time routing algorithm (Cao 2007) is used and experiment results of 1 CPU, 4 and 8 CPUs (Fig. 6) show that the response time of more CPUs is less than those with less CPUs.

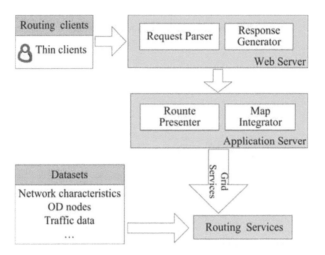

Figure 5. Architecture of real-time routing within a grid platform.

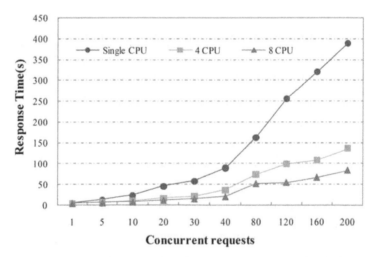

Figure 6. Response time for different number of concurrent requests against 1 CPU, 4 CPUs, and 8 CPUs. (Response time in vertical axis is the total time that the server is used to responding to all concurrent requests. Horizontal axis is concurrent request number.)

4 SOLUTIONS

Considering different WebGIS applications, computing components, software techniques, and data organization approaches, we can design different solutions to solve different performance problems. This section demonstrates the utilization of the techniques to improve performance within three different WebGIS applications including WFS, Water cycle EOS ClearingHouse (WECHO), and Chesapeake bay application.

4.1 WFS performance enhancement

WFS is a standardized interface for serving vector data on the web and is being used in different areas for interoperability issues (Peng and Zhang 2004). WFS uses GML as default encoding method for transmitting vector data from servers to clients (Vretanos 2002). GML is XML (Extensible Markup Language)-based and supports point, polyline, and polygon with

130

a large number of floating numbers. The floating numbers are represented as a string character in GML and interpreted at the client side through decoding and normal GIS operations. This brings performance problems including 1) the encoding from binary numbers to string characters in GML at server side, 2) the parsing and rendering of data from GML at the client side, 3) network transmission of large GML streams becomes a major bottleneck especially when the network bandwidth is limited, such as in a wireless connection. To improve WFS performance, we use data volume reduction (refer to section 3.1) including a binary XML (BXML) strategy for encoding vector data. The datasets are selected from ESRI datasets with different data volumes. Meanwhile, the popular GZIP-based compression algorithm is employed to decrease the data volume in network transmission to improve the efficiency. The test environment we used for the WFS performance test is shown in Figure 7. At the server side, two types of shape file encodings are provided as BXML and XML. Client requests go through the web server to get either binary BXML stream or GML stream. Accordingly, two types of parsers in the client side (BXML parser and GML parser) are used for data parsing and rendering. In addition, the web request includes a flag indicating whether to enable server-side compression or not (no compression as default value).

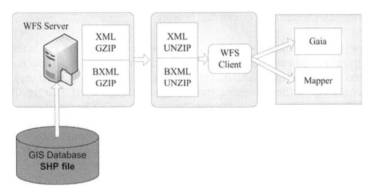

Figure 7.   The overall experimental system architecture.

Figures 8 and 9 illustrate the experimental results under various network connections, encoding methods, and transmission modes. Figure 8 shows the performance of WFS in BXML/XML encoding and with/without server-side compression within wired network.

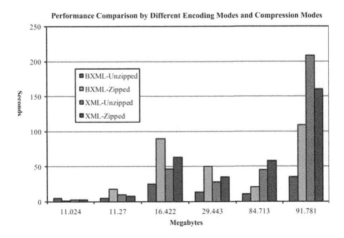

Figure 8.   Performance comparison of WFS in a wired network.

131

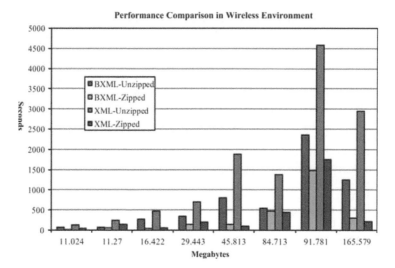

Figure 9.   Performance comparison in a wireless network.

Figure 9 shows the performance within wireless network. The vertical axis is the time spend on processing (zip/unzip, and binary conversion) and transmitting data in seconds.

With a wired network (Fig. 8) data on the server side can be downloaded very fast and the system bottleneck lies in the processing (rendering) of downloaded data. In this situation, although GZIP compression reduces the data amount needed to be transferred via Internet, it also increases the computing requirement significantly. Consequently, for almost all testing cases, we find that the system performance is severely damaged by introducing GZIP compression. Also, without GZIP, BXML is usually 2–6 times faster than its XML format. This demonstrates the higher efficiency of BXML.

The same experiment was performed in a wireless network (Fig. 9). Results show that with wireless links, the network connection becomes a bottleneck. The overall system performance is approximately proportional to the amount of data that needs to be downloaded. Using BXML is no longer the decisive factor to improve system performance. GZIP compression can also significantly reduce the data amount needed to be transferred, and it becomes the solution. The speed performance is a comprehensive result of the processing and transmitting. Therefore, for a wireless network where transmitting time is dominating, the processing can be omitted where it is consistent that zipped BXML will have better performance than others. In a wired connection, the processing time will impact the results, therefore, when the data volume is small, the processed BXML and zipped XML approaches will be more time consuming. When the data volume increases, the transmitting will dominate the results, therefore, unzipped BXML has the best performance. Figures 8 and 9 are used to illustrate the performance among four methods (unzipped XML, zipped XML, zipped BXML, and unzipped BXML) on the wireless and wired networks. The results showed that unzipped BXML has the best performance on wired network (Fig. 8) and that zipped BXML has the best performance (Fig. 9) on wireless network. This is because the zipping and converting processes will take time and will impact the results in wired network when transmitting time is relatively short. Here we pick the popular zip and binary compression within XML and GML communities, additional compression methods may be developed through specific datasets.

4.2   WECHO

WECHO provides an integrated WebGIS solution for scientific Earth observation data cataloguing, searching, visualizing and analyzing. By connecting to ECHO (EOS Clearing HOuse) middleware, a WECHO client is able to effectively search huge amounts of metadata from

ECHO, which stores more than 2,129 data collections composed by 54 million individual data granules and 13 million images. A WECHO client connects to an ECHO database through HTTP/TCP request, then provides functionalities for local data discovery and data holding. WECHO provides three types of search functions, namely simple search, complex search and semantic search (Fig. 11). All three search methods facilitate data discovery for users with different backgrounds and needs. The data found can be in various formats, such as HDF, netCDF, GRIB or binary. WECHO provides a flexible visualization tool to access and display them in both 2D and 3D manners via OPeNDAP protocol. 2D visualization tool is able to dynamically parse and subset the scientific data and visualize them in time-series. Meanwhile, a KML file could also be generated for enabling the 3D animation in Google Earth.

The searching results would very possible include hundreds of records and the time spent could be up to 10 minutes. To achieve good performance with such a big data volume, we adopted multi-thread techniques (3.5) running in the background and asynchronous transmission (3.4). In this way, a flexible and efficient framework is established to improve scientific research and analysis (Fig. 10).

By adopting the multithreading and asynchronous transmission techniques, the results can be updated at a fixed time interval (e.g. every 2 seconds) to show the intermediate results instead of waiting for up to 10 minutes to get final results (Fig. 11).

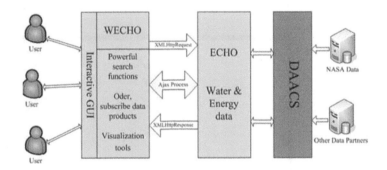

Figure 10.   WECHO architecture (revised from Houser 2005).

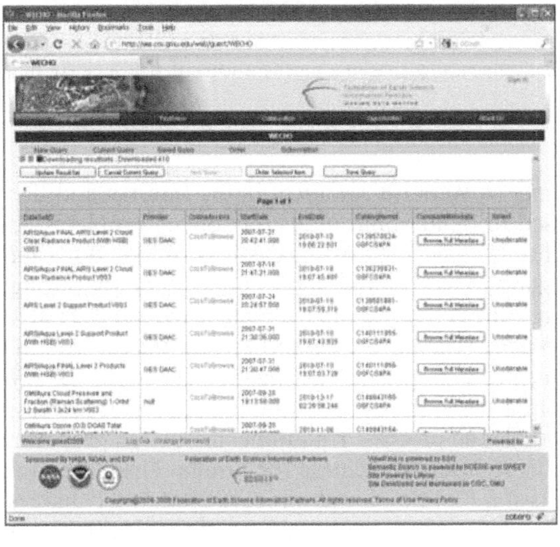

Figure 11.   The WECHO system interface, the progress for querying results can be updated frequently as illustrated by the progress bars and the number of pages at the top of the figure.

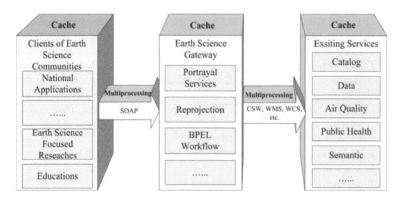

Figure 12.   The ESG interface and architecture.

Figure 13.   Explore the changes of nitrogen concentrations of Chesapeake bay in ESG.

4.3 *Chesapeake bay*

There are large amounts of WebGIS resources in the Internet and they can be dynamically searched and composed to build applications for solving problems. ESG (Bambacus et al. 2007) is a system built to leverage the existing catalogs, WMS, WFS and WCS, and reprojection and other processing tools to build application on the fly through distributed search, access and use of the resources. The example used here dynamically invokes datasets of natural and human phenomena related to Chesapeake bay. Based on a SOA (Service Oriented Architecture) the data, processing, and visualization functions of a WebGIS is composed from different places and there are more than one data server, processing server, and visualization server involved in the process. Traditional method in processing the services one by one would take minutes to hours to achieve the results. Therefore, we adopted cache (3.3), asynchronous transmission (3.4), and multiprocessing (3.5) techniques to improve the performance at the server side, on the gateway, and the client side (Fig. 12).

When an end user issues a request, the client side will first check if the dataset or service requested has been cached before it sends a request to the server(s). The most popular data, such as river and coastal lines, for Chesapeake bay are cached. If a request is sent to the gateway, the gateway will first check the records in the cache. ESG utilizes the cache servers to store the data that are frequently requested. If the requested datasets or services are available, then the ESG will directly return the results to end users (Fig. 13). If the requested data or processing results are not available, a list of asynchronous requests will be sent to data and processing servers. The multiple servers involved will be invoked simultaneously and once a result is returned, the user interface will be updated.

Integrated with all these techniques, ESG provides a fast platform for us to investigate the changes of environment and relevant indicators of the Chesapeake bay under the global climate change in the past decades, for example, nitrogen concentration in Figure 13. Incorporating the results of concentrations of nitrogen and phosphorous and other indicators (e.g. water temperature) in different years, changes of water quality of the Chesapeake bay can be explored. The exploration of the changes of nitrogen concentrations of Chesapeake bay through ESG will invoke multiple servers simultaneously through asynchronous transmission. And cache is widely used in the process to improve performance.

5 DISCUSSION AND CONCLUSIONS

This chapter introduces the performance issues of WebGIS and how improvements can be made. Techniques for improving performance are reviewed, researched, and tested in different network environments. Examples of the utilization of the techniques to improve performance are given to demonstrate how to build systems with good performance.

Different from traditional GIS, WebGIS involves computing components of geospatial data servers, analysis servers, web servers, computer networks, and client-side user interfaces. Each of these components can become a bottleneck within a practical WebGIS (Yang et al. 2008). Techniques for improving the performance include not only adopting more computing resources, such as more processing units and memory, but also software and geospatial techniques, such as dynamic loading, caching, geospatial indexing and geospatial data organization (Yang et al. 2007). Taking into consideration computing components, software techniques, and other techniques, we designed several solutions to solve performance problems in different types of WebGIS applications.

The goal of WebGIS is to serve the general public through a popular web interface that is easy to learn, appealing, and independent of implementation complexities (Egenhoffer et al. 1988). The design of the interface will decide what functions and how much interaction is needed with the server. The interface design will also decide the performance improving techniques and solutions.

The advancement of computing technologies introduces new platforms, such as grid computing and cyber infrastructure (Yang et al. 2010, Huang & Yang 2010), especially cloud computing, which will provide computing as utility (Armbrust et al. 2009). Cloud computing has been defined to have five characteristics including on-demand self-service (for customers as needed automatically), broad network access (for different types of network terminals, e.g. mobile phones, laptops, and PDAs), resource pooling (for consolidating different types of computing resources), rapid elasticity (for rapidly and elastically provisioning, allocating, and releasing computing resources), and measured Service (for automatically controling and billing resource) (Mell and Grance 2009). WebGIS reliability, response time, and memory consumption will be dynamically and automatically provisioned given the characteristics. Further research needs to be conducted in alignment with the three services of Infrastructure as a Service (IaaS), Platform as a Service (PaaS), and Software as a Service (SaaS) to best leverage cloud computing platforms for WebGIS and Interoperability.

Many new initiatives, such as Digital Earth and Global Earth Observation System of Systems (GEOSS) have as their objective to integrate various types of geospatial datasets across the web computing platforms to be able to better respond to global crises, such as water shortages and emergency responses. The implementation of such initiatives and platforms not only requires supporting massive concurrent users, but also requests the integration or mashup of a large amount of distributed geospatial information resources and improvement in real time (Yang & Raskin 2009). Performance will become a tough challenge in such systems and will continue to be an active research topic that requires geospatial professionals 1) to explore new performance problems with domain experts, 2) to develop new performance-improving techniques with computer and network engineers and mathematicians, 3) to identify mature solutions with application domain users, and 4) to share experiences among peers.

ACKNOWLEDGEMENTS

Research and developments reported are supported by FGDC 2008 and 2009 CAP projects (08HQAG0015 and G09AC00103), and NASA project (NNX07AD99G).

REFERENCES

Armbrust, M., Fox, A. & Griffith, R. (2009) Above the Clouds: A Berkeley View of Cloud Computing, University of California, Berkeley, Berkeley, CA, 2009. [Online] Available from: http://www.eecs.berkeley.edu/Pubs/TechRpts/-2009/EECS-2009-28.html [last accessed 7th June 2010].

Armstrong, M.P., Cowles, M.K. & Wang, S. (2005) Using a computational grid for geographic information analysis: A reconnaissance. *The Professional Geographer*, 57 (3), 365–375.

Bambacus, M., Yang, C., Evans, J., Cole, M., Alameh, N. & Marley, S. (2007) ESG: An interoperable portal for prototyping applications. *URISA Journal*, 19 (2), 15–21.

Buckley, D.J. (1998) *The GIS Primer: An Introduction to Geographic Information Systems*. Pacific Meridian Resources.

Butler, D. (2006) The Web-Wide World. *Nature*, 439 (7078), 776–778.

Cao, Y. (2008) Utilizing grid computing to optimize real-time routing. George Mason University Ph.D. Dissertation. p. 107.

Corner, S. (2010) The 8-second rule, [Online] Available from: http://www.submitcorner.com/Guide/Bandwidth/001.shtml [last accessed on 20th April 2010].

Davis, G. & Nosratinia, A. (1998) Wavelet-based image coding: An overview. *Applied and Computational Control, Signals, and Circuits* 1.

De La Beaujardiere, J. (2004) Web Map Service, version 1.3. *OGC Implementation Specification*, [Online] Available from: http://portal.opengis.org/files/?artifact_id_5316, OGC [last date accessed on 05th February 2010].

Egenhofer, M. & Frank, A. (1988) Designing object-oriented query languages for GIS: Human interface aspects. In *Proceedings of the Third International Symposium on Spatial Data Handling*. International Geographical Union Commission on Geographical Data Sensing and Processing, Williamsville, NY. pp. 79–96.

EPA. 2010. [Online] Available from: http://www.epa.gov/tribal/datamaps/ [last accessed on 20th April 2010].

ESRI. (2010) [Online] Available from: http://www.esri.com/software/arcgis/arcgisserver/index.html [last accessed on 14th April 2010].

Fisher, Y. (1995) *Fractal Image Compression: Theory and Application to Digital Images*. Berlin Heidelberg, New York, Springer.

Geospatial One Stop, GOS. (2010) [Online] Available from: http://gos2.geodata.gov/wps/portal/gos [last accessed on 14th April 2010].

Gonzales, R.C. & Woods, R.E. (1993) Digital Image Processing. Addison-Wesley.

Hawick, K.A., Coddington, P.D. & James, H.A. (2003) Distributed frameworks and parallel algorithms for processing large-scale geographic data. *Parallel Computing*, 29, 1297–1333.

Houser, P. (2005) Water and Energy Cycle EOS Clearing House (WECHO). [Online] Available from: http://crew.iges.org/research/WECHO/ [last accessed on 20th April 2010]. Huang, Q. & Yang, C. (2010) Optimizing grid computing configuration and scheduling for geospatial analysis -- An example with interpolating DEM, *Computer & Geosciences*, Available from: DOI: 10.1016/j.cageo.2010.05.015.

Ingensand, J. & Golay, F. (2008) User performance in interaction with Web-GIS: A semi-automated methodology using log-files and streaming-tools. *Lecture Notes in Geoinformation and Cartography,* 433–443.

ISO/TC211. (2010) [Online] Available from: http://www.isotc211.org/ [last accessed on 14th April 2010].

Komzak, J. & Slavik, P. (2003) Scaleable GIS data transmission and visualization. In *Proceeding. of the 7th International Conference on Information Visualization IV03*, IEEE, London.

Lanter, D.P. (1991) User-centered graphical user interface design for GIS. *National Center for Geographic Information and Analysis Report.* pp. 91–96.

Lee, J. & Zlatanova, S. (2008) A 3D data model and topological analyses for emergency response in urban areas. In: Zlatanova & Li (eds.), *Geospatial information technology for emergency response (ISPRS book series)*, London, Taylor & Francis Group. pp. 143–168.

Li. W., Yang. C. & Yang. C.J. (2010) An active crawler for discovering geospatial Web services and their distribution pattern – A case study of OGC Web Map Service. *International Journal of Geographical Information Science,* 24 (8), 1127–1147.

Li. Z., Yang. C., Wu. H., Li. W. & Miao. L. (2010) An optimized framework for seamlessly integrating OGC Web services to support geospatial sciences. *International Journal of Geographical Information, Science,* Available from: DOI: 10.1080/13658816.2010.484811.

Livny, Y., Kogan Z. & El-Sana J. (2007). Seamless patches for GPU-based terrain rendering. In *Proceedings of WSCG 2007.* pp. 201–208.

MapServer. (2009) [Online] Available from: http://mapserver.org/ [last accessed on 14th April 2010].

MapServer. (2009) [Online] Available from: http://www.opengeospatial.org/ [last accessed on 14th April 2010).

Mell, P. & Grance, T. *The NIST Definition of Cloud Computing,* [Online] Available from: http://csrc.nist. gov/groups/SNS/cloud-computing/ [last accessed on 7th June 2010].

Morlet, J. & Grossman, A. (1984) Decomposition of hardy functions into square integrable wavelets of constant shape. *Siam Journal on Mathematical Analysis,* 15 (4), 723–736.

NASA. (2010) [Online] Available from: http://eospso.gsfc.nasa.gov/ [last accessed on 14th April 2010].

Nebert, D. & Whiteside, A. (2005) Catalog Services, Version 2. *OGC Implementation Specification.* [Online] Available from: http://portal.opengis.org/files/?artifact_id_5929 [last date accessed on 5th February 2010].

OGC. (2010) [Online] Available from: http://www.opengeospatial.org/ [last accessed on 14th April 2010].

Oralcle. (2010) [Online] Available from: http://www.oracle.com/technology/products/berkeley-db [last accessed on 14th April 2010].

Peng, Z. & Zhang, C. (2004) The roles of geography markup language (GML), scalable vector graphics (SVG), and Web feature service (WFS) specifications in the development of Internet geographic information systems (GIS), *Journal of Geographical Systems,* 6 (2), 95–116.

Portele, C. (2007) OpenGIS Geography Markup Language (GML) encoding standard. [Online] Available from: http://portal.opengeospatial.org/files/index.php?artifact_id=20509&passcode=p74xj vt23hm328389rks [last accessed on 14th April 2010].

Riedemann, C. & Kuhn, W. *GIS-User interfaces in practice, from specialist to public usage.* [Online] Available from: http://cognition.iig.uni-freiburg.de/events/work&conf/raeumlicheInferenz/web-riede/ text.html [last accessed on 20th April 2010].

Samet, H. (1990) *The Design and Analysis of Spatial Data Structures.* Addison-Wesley, Reading, MA.

USCB. (2010) [Online] Available from: http://factfinder.census.gov/jsp/saff/SAFFInfo.jsp?pageId=gn7 maps [last accessed on 14th April 2010].

USGS. (2010) [Online] Available from: http://nationalmap.gov [last accessed on 14th April 2010].

VA. (2010) [Online] Available from: http://gisdata.virginia.gov/ [last accessed on 20th April 2010].

Vretanos, P.A. (2002) Web Feature Service, Version 1.0. [Online] Available from: http://portal. opengeospatial.org/files/?artifact_id_7176 [last date accessed on 5th February 2010].

W3C. (1996) Extensible Markup Language (XML). [Online] Available from: http://www.w3.org/XML/ [last date accessed on 5th February 2010].

Wei, Z., Oh, Y., Lee, J., Kim J., Park, D., Lee, Y. & Bae, H. (1999) Efficient spatial data transmission in Web-based GIS. In *Proceedings of the 2nd International Workshop on Web Information and Data Management*, Kansas City, Missouri. pp. 38–42.

Xie, J., Yang, C., Zhou B. & Huang, Q. (2010) *High-performa*nce computing for the simulation of dust storms. *Computers, Environment and Urban Systems,* 34 (4), 278–290.

Yang, B.S., Purves, R.S. & Weibel, R. (2007) Efficient transmission of vector data over the internet. *International Journal of Geographical Information Science*, 21 (2), 215–237.

Yang, C., Cao, Y. & Evans J, (2007) WMS Performance and Client design principles. *The Journal of GeoInformation Science and Remote Sensing*, 44 (4), 320–333.

Yang, C., Evans, J., Cole, M., Alameh, N., Marley, S. & Bambacus, M. (2007) The emerging concepts and applications of the spatial web portal. *Photogrammetry and Remote Sensing*, 73 (6), 691–698.

Yang, C., Li, W., Xie, J. & Zhou, B. (2008) Distributed geospatial information processing: sharing earth science information to support Digital Earth. *International Journal of Digital Earth*, 1 (3), 259–278.

Yang, C. & Raskin R. (2009) Introduction to distributed geographic information processing research. *International Journal of Geographical Information Science*, 23 (5), 1–8.

Yang, C., Raskin, R., Goodchild, M.F. & Gahegan, M. (2010) Geospatial cyberinfrastructure: Past, present and future. *Computers, Environment and Urban Systems*, 34 (4), 264–277.

Yang, C., Wong, D., Yang, R., Kafatos, M. & Li Q. (2005) Performance improving techniques in WebGIS. *International Journal of Geographical Information Science*, 19 (3), 319–342.

Zhang, C. & Li, W. (2005) The Roles of Web Feature and Web Map Services in Real-time Geospatial Data Sharing for Time-critical Applications. *Cartography and Geographic Information Science*, 32 (4), 269–283.

Zlatanova, S., Rahman, A.A. & Shi, W. (2004) Topological models and frameworks for 3D spatial objects. *Computers & Geosciences*, 30, 419–428.

*Advances in Web-based GIS, Mapping Services
and Applications – Li, Dragićević & Veenendaal (eds)*
© *2011 Taylor & Francis Group, London, ISBN 978-0-415-80483-7*

Data reduction techniques for web and mobile GIS

Michela Bertolotto & Gavin McArdle
School of Computer Science and Informatics, University College Dublin, Dublin, Ireland

ABSTRACT: Advances in data collection technologies and the widespread use of the
Internet have enabled the availability of incredibly rich data repositories to the wider community. While this presents exciting opportunities, it also causes new challenging problems. One
of them, known as "information overload", occurs when there is too much data available for
users to extract useful and relevant information. Being able to reduce the amount of spatial
data presented to users while increasing relevance with respect to their specific needs is a critical research issue. In Web and mobile environments, data reduction is needed not only to help
users find the information they are looking for, but also to facilitate efficient transmission.
Proposed solutions rely on progressive data transmission. In this chapter we present a survey
of the state-of-the-art research in the field of spatial data reduction which aims to provide
end users with reliable and relevant spatial information in an efficient way. Traditional trends
in Web and mobile GIS for data reduction are reviewed along with newer methods which are
based on personalization techniques. The chapter serves to critically highlight what has been
achieved in the area of spatial data reduction and presents some newer areas where research
is ongoing.

Keywords: Data reduction, GIS, web based, cartography, mapping, real-time, visualization

1 INTRODUCTION

1.1 *Spatial information overload*

The widespread availability of the Internet and the willingness to share data has given rise to
the emergence of exciting new opportunities, not available to many until recently. In the past,
in many domains, access to (electronic) data was the privilege of a few (the so-called "data
keepers"). Nowadays, an incredibly diverse range of data is publicly available. Advances
in data collection technologies have generated rich spatial data repositories which are now
accessible by the wider community.

Although the advantages provided by this new reality are well recognized, the sheer volume of data with which we are often presented highlights how our ability to collect and store
data has far outpaced our capability to retrieve and exploit it. In other words, in many cases
too much data is available for users to be able to find what is really relevant to them and use
it effectively. Consequently, developing effective methods to present users with a subset of
data required by their specific needs is a critical issue. This new challenging problem called
"information overload" is typical of virtually every application domain.

While there is a large research effort in this area, most approaches developed so far are
applicable only to specific application fields such as Web recommender systems (Speretta &
Gauch 2004, Qiu et al. 2009) and relate mainly to non-spatial data.

Information overload is now very evident in the spatial domain where it presents two main
issues: 1) too many datasets are available, and 2) too much information is packed within a
given dataset. In the first case, providing better tools to assist users in identifying the datasets they need can resolve the problem (Flewelling & Egenhofer 1999). Proposed solutions

include the use of improved spatial metadata or spatial summaries. In some cases spatial metadata (in the form of textual descriptions), if effectively structured, can greatly help users locate their datasets of interest. To overcome the second issue, spatial summaries in the form of "subsets" of data that can be used instead of the fully detailed version are required. These permit a preliminary analysis, to assist users to select only the fully detailed portions of data they require. A simple example in the case of raster data repositories is provided by the use of thumbnail images that might be sufficient to convey the content of high resolution images. The second issue also involves being able to effectively reduce the amount of spatial data presented to users, and, at the same time, providing the most relevant information. This is of critical importance for Web and mobile GIS applications.

Although not discussed in detail in this chapter, the semantic geospatial Web, which aims to develop a new framework for retrieving geospatial information based on the semantics of spatial ontologies, is becoming relevant when discussing spatial data reduction methods. The semantic geospatial Web will provide users with the opportunity to obtain data which is more relevant to their requirements than existing text-based search methods as the semantic information associated with the underlying spatial data is considered (Egenhofer 2002, Scharl & Tochtermann 2007). Additionally, the semantic geospatial Web can be used to support the personalization techniques presented later in this chapter.

1.2 *Data reduction approaches in Web and mobile GIS environments*

In Web and mobile environments, when the dataset of interest has been identified, users need to download and use it. However, the resolution and level of detail of currently available spatial datasets has become so high that they occupy very large files often containing a lot of irrelevant information. This causes impediments to both the use and transmission of such datasets.

Therefore, data reduction is needed not only to help users find the information they are looking for, but also to facilitate its use and transmission. Much research is currently being dedicated to the development of methods for reducing the level of detail of a specific dataset (while maintaining relevant information) as well as towards techniques to speed up data exchange.

A viable solution relies on the progressive transmission of data at increasing levels of detail: a subset of the data is sent first and then incrementally refined in subsequent stages. The advantages of progressive transmission include: efficient and more reliable data transmission (as smaller files are sent), quick response and, possibly, transmission of only the most relevant detail. This is especially important for mobile users trying to download datasets of interest while in the field. Even if the fully detailed version is needed, they can start working with a coarser version while more detail is being added progressively. Besides being able to perform preliminary operations on temporary versions of the data, users can realize that the detail of the currently displayed representation is sufficient for their purpose. They can then interrupt the downloading of more detailed representations (stored in larger files) and save both time and disk space.

Progressive transmission can be used for delivering both raster and vector spatial datasets (Zhu & Yang 2008). While for raster images techniques are well established, more challenging issues arise for vector data as reducing the detail can cause topological inconsistencies. Depending on the intended use for the dataset, preservation of the topological structure can be a strict requirement. Several implementations have been developed for the special case of triangular meshes (for terrain modeling and 3D visualization); however, the investigation and development of progressive transmission techniques for more general vector spatial data still present many challenges. This chapter discusses these issues and surveys recent techniques developed for progressive spatial data transmission. These techniques focus on reducing the size of the dataset while preserving consistency. They do not however take into consideration user preferences; this is instead the aim of personalization techniques. Matching user preferences enhances the relevance of the data presented to users.

1.3 *Chapter content*

This chapter presents a review of existing techniques to improve performance of spatial information systems by examining traditional methods for efficient data reduction and transmission, as well as newer approaches which rely on data personalization. In all instances, the issues of information overload are addressed. The review clearly identifies which areas have been explored while also highlighting future research directions in this field.

The chapter considers the new context in which there has been a shift away from the requirement of highly accurate datasets for professional applications, to the use of maps for a diverse range of uses where high accuracy is not a critical issue. These two uses largely differ in their requirements. For the first case, sound cartographic principles are needed. For the latter case, when visualization is the only purpose, raster data can be utilized. Even when queries are issued, if a visual answer is sufficient, the calculation can be done on the vector representation on the server and the result displayed in the form of a raster map generated on the fly.

Many freely available software packages such as Google Earth provide advanced capabilities and visual interfaces that allow non-professional users to access the information they need. While these packages can be useful, they present information in a static way without adapting to users' requirements. The only way to personalize the content of the default maps is by switching layers on/off to add/remove detail. However these systems do not provide any other mechanism for personalizing the content of the map. Goodchild (2008) provides a critical review of the shortcomings and advantages of such geo-browsers. In this chapter we discuss new approaches for data reduction based on personalization techniques that can improve the relevance of spatial datasets presented to users as well as transmission techniques to ensure efficient data exchange.

An important field of research emerging recently is that of Distributed Geographic Information Processing (DGIP). Yang and Raskin (2009) have identified six key research themes within this research area, namely: *architecture, spatial computing, models, interoperability, intelligence and applications*. Following this classification, progressive transmission cuts across the architecture and spatial computing thematic trends. On the other hand, newer data reduction techniques such as personalization fall within the scope of the intelligent area of geographic information processing. Standards such as WMS (Web Map Service) and compliance with the Open Geospatial Consortium (OGC) standards, which fall into the interoperability research theme, are becoming increasingly important issues across all areas of this domain (Yang et al. 2007a).

The remainder of this chapter is organized as follows. Section 2 reviews successful implementations for data reduction and progressive transmissions of raster data over the Internet. The problem of compressing and transmitting spatial data in vector format is analyzed in Section 3. Three main types of techniques are reviewed: mesh-based approaches, cartographic approaches, and more recent approaches based on alternative methods including clustering, streaming, tiling and pre-fetching. Section 4 describes data reduction based on map personalization techniques. Finally, in Section 5 we discuss open issues in the field of spatial data reduction for Web and mobile GIS, while in Section 6 we outline some concluding remarks.

2   DATA REDUCTION AND TRANSMISSION OF RASTER IMAGES

Several mechanisms have been developed for compression and transmission of raster images in Web environments. Effective compression techniques for such data are available and provide good compression ratios while causing low information loss. Furthermore, they are efficient and relatively easy to implement. Consequently progressive transmission systems that rely on these techniques have been successfully implemented. Such approaches generally allow the reconstruction of the full resolution image on the user's machine by gradually adding detail to coarser versions.

These longstanding mechanisms are also applied in the spatial domain to exchange raster geo-spatial datasets (including high-resolution satellite images, aerial photos and scanned maps) progressively from a server to a client. The first successful implementations of progressive raster transmission rely on interleaving techniques. The simplest approach consists of randomly extracting sub-sets of pixels from the image and incrementally completing it by adding pixels. If the implicit row/column ordering of images is exploited, pixels can be sampled in a uniform fashion alternatively; hierarchical structures, such as quadtrees can be used to extract more pixels from parts of the image with higher density of detail (Lindstrom et al. 1996).

More sophisticated compression techniques have generated very effective progressive raster transmission methods (Rauschenbach & Shumann 1999, Srinivas et al. 1999). The most commonly used image compression method is the JPEG (Joint Photographic Experts Group) format. Alternative techniques are based on wavelet decompositions (Morlet & Grossman 1984). Wavelet methods produce a robust digital representation of a picture, which maintains its natural look even at high compression ratios (see Davis & Nosratinia (1998) for an overview). Wavelets are well suited for progressive transmission because, besides providing efficient compression, they naturally represent image data as a hierarchy of resolution features (Rauschenbach & Shumann 1999).

Other work on image compression relies on fractal theory (Fisher 1995). The challenges in fractal-based compression include finding a small number of affine transformations to generate the image as well as subparts of the input image that have self-similarity properties. Hybrid compression methods, combining fractals and wavelets, have also been defined (Davis 1998).

When these transform-based methods are used to compress an image, during transmission over the Internet, instead of the image, the function coefficients are transmitted, and the image is subsequently reconstructed by inverting the transformation. Both transmission and reconstruction are very efficient, as they can be performed simply by means of a single pass through each color band of the image.

Much work has been dedicated to the combination of different techniques to improve the performance of WebGIS systems in exchanging raster data. An example is provided by Yang et al. (2005b) where pyramid (Tu et al. 2001) and hash indexing methods are applied for fast access of large raster images. Cluster and multithreading (Chen et al. 2000) as well as dynamic caching are also employed in Yang et al. (2005b). Experiments conducted show the advantage of organizing large datasets using a hierarchical data structure with a super-imposed hash index.

The purpose of raster data exchange over the Internet is very often only for visualization. In this case, techniques for progressive raster transmission are usually suitable and efficient. Even when measurements need be performed on the images (e.g. for photogrammetric purposes), common compression mechanisms (such as JPEG) are effective (Kern & Carswell 1994). However, some applications involve direct object handling and manipulation for which a vector representation is required. Data reduction in this instance is more complex. This topic is discussed in the following section.

3   DATA REDUCTION AND TRANSMISSION OF VECTOR DATASETS

A raster version of the data might not be adequate in certain applications, including contexts in which actual object manipulation is involved. In these cases a vector representation of the data is required. Vector spatial datasets consist of collections of spatial entities in the form of points, polylines and polygons that are connected through spatial relations (e.g. topological, metric, and direction relations (Egenhofer & Franzosa 1991)). Examples include: thematic maps, road network maps, city maps and mesh-based digital terrain models.

In the past much attention has been devoted to the development of data reduction techniques for vector data. Depending on the type of data used, different techniques must be applied. Several approaches based on effective compression mechanisms have been defined and successfully implemented for the particular case of triangular meshes (described in Section 3.1).

More challenging is the compression of other types of datasets such as map data. Proposed approaches are described in Section 3.2, while Section 3.3 reviews additional recent techniques for vector data reduction and transmission.

3.1 *Mesh-based approaches*

Part of the spatial data in vector format is represented by means of triangular meshes. For example, triangulations are used for digital terrain modeling and for real objects surface reconstruction and rendering.

Several compression methods for triangular meshes are based on optimal point decimation techniques and exploit combinatorial properties of triangulations for efficient encoding. Among them, the progressive mesh scheme proposed by Hoppe (1996) represents a fundamental milestone for progressive transmission. However, this method does not have a satisfactory performance as, like other traditional methods (e.g. Zorin et al. 1997), it is topology preserving. This is important for data consistency but it limits the level of simplification. Other methods achieve higher compression by slightly modifying the topology of the input mesh.

Junger & Snoeyink (1998) developed a parallel point decimation technique to simplify a Triangulated Irregular Network (TIN) for progressive transmission and rendering. McArthur et al. (2000) have applied wavelet transforms to generate hierarchical polygonal databases from raw elevation data for application to view-dependent mesh display in WebGIS. Similarly, Zhang et al. (2007) employed wavelet-based compression for interactive visualization of 3D meshes. Yang et al. (2005a) combine TIN generation methods with a Grid approach for multi-resolution terrain model construction and visualization over the Internet. Although effective, these particular techniques can only be applied to mesh data.

3.2 *Cartographic approaches*

The investigation and development of progressive transmission techniques for more general vector spatial data still present many challenges. This is the case for map data (e.g. city street maps, road networks, hydrography networks and administrative boundary maps) whose compression is constrained by several factors. When decreasing the level of detail of a map, a critical issue relates to the preservation of topology to guarantee reliable query results. For example, random selection of objects (to be eliminated or simplified) does not usually generate a topologically consistent representation at coarser detail. Consistency is an essential property for data usability: the answer to a query on the coarser map must be consistent with the answer obtained when the same query is applied to the original (fully detailed) map.

The level of detail of a vector map can be decreased by applying the so-called map generalization process. Such a process relies on cartographic principles whose complete formalization is still an open problem (Li 2007). Although a few automated solutions and algorithms have been proposed for specific cases, they involve complex and time-consuming calculations and therefore cannot generally be applied for real time Web mapping. Given the complexity of the process, two complementary approaches can be adopted for progressive map data transmission (Cecconi & Galanda 2002): "on-the-fly" map generation versus "pre-computation" of multiple map representations. The first approach aims at generating new generalized maps upon user request. The second approach relies on the pre-computation of a fixed sequence of map representations that are stored and transmitted progressively in the order of increasing detail.

Several cartography-based approaches for data reduction and transmission in WebGIS rely on the application of line simplification algorithms. The classical Douglas-Peucker algorithm (Douglas & Peucker 1973) is the most commonly used line simplification algorithm in both GIS research and commercial applications. Despite its popularity, such an algorithm presents a major drawback as it does not guarantee preservation of topology. Improvements that overcome this issue have been proposed (Saalfeld 1999). Saalfeld's modification has been applied by other researchers for the development of progressive transmission of simplified maps (Boujou et al. 2005, Zhou & Bertolotto 2005). These methods are mostly applicable to linear datasets and rely on the pre-computation of a discrete and fixed set of maps at increasing levels of detail.

Another method based on the simplification algorithm has been described in Yang (2005) and applied for progressive transmission over the Internet. Such a method, unlike those previously mentioned, extracts continuous levels of vector data with increasing detail. However it applies only to polygonal datasets. Also applicable only to this type of data is the approach proposed in Ai et al. (2004). The work developed in Yang (2005) has been subsequently completed and extended to the case of datasets including different types of features (specifically both lines and polygons) by Yang et al. (2007b). An implementation has been developed and results show a significant improvement in the transmission efficiency.

A drawback of the previously mentioned methods is the fact that they apply only line simplification for generalising the map while other (topological) operators such as region thinning, region collapse and region merge cannot be utilized. A hierarchical data structure for efficiently storing and transmitting a sequence of multiple map representations generated by applying a set of topological generalization operators has been defined in Han and Bertolotto (2004) based on the formalization of Bertolotto and Egenhofer (2001). In such a data structure, only the coarser map representation is completely stored while intermediate levels just store increments. Vertical links are explicitly kept to facilitate full reconstruction of intermediate levels. The work developed in Bertolotto & Egenhofer (2001) relies on topology only and preserves topological consistency at the conceptual level. However, in order to apply the corresponding data structure for map generalization of real datasets, the integration of geometric as well as semantic criteria for generalization is needed. Thus far, this remains an open problem.

3.3 *Additional approaches*

In the previous two sections several methods for data reduction used for efficient transmission of spatial datasets, in vector format, over the Internet were discussed. These were subdivided into two main categories: 1) methods applied to mesh data, and 2) cartography-based methods. Although some of these methods are efficient for particular datasets, none of them can be applied for generic sets of data or operations.

Additional approaches for data reduction of vector datasets that are orthogonal to the methods presented here have been developed. These are described below:

- *Pre-fetching and caching techniques* attempt to improve response time by prefetching and storing, in a cache database, tiles which the user is likely to request in the near future. Kang et al. (2001) and Tu et al. (2001) have demonstrated such techniques.
- *Clustering techniques* aim to define meaningful groupings within the input data, and then select a represnetative for each group; an example is developed in Shekhar et al. (2002).
- *Out-of-core methods* utilise external memory for handling and progressive visualization of GIS and terrain data efficiently. Lindstrom & Pascucci (2002) and Cignoni et al. (2003) among others have used this approcah successfully.
- *Streaming techniques* utilise view-dependent refinement based on hierarchical level of detail data structures and have been employed by Guthe & Klein (2004).
- *Indexing-based techniques* can be used to speed up access to vector data on both the client and server by utilising clustering and multithreading. This approach is demonstrated by (Yang et al. 2005b, Zhang et al. 2007).
- *Tiled vector techniques* split the dataset into a smaller pieces and transmit these packets to the client where they are reassembled based on predicted user behaviour using a tiled caching mechanism (Antoniou et al. 2009).

Several different techniques can be combined to improve the performance of Web and mobile GIS systems. For example, while Koshgozaran et al. (2006) utilize Huffman encoding combined with line aggregation for fast vector data transmission over the Internet, the technique has only been applied to road network databases.

Other methods relate to the area of data-intensive computing (Raicu et al. 2009). This domain involves clusters of distributed computers being used to process complex data. Within this area, Grid computing has been identified as a key research topic in GIS (Yang & Raskin 2009). It can be utilized to handle the large quantities of spatial data which are now avaialble and to resolve similar problems which data reduction addresses. One such example is a National Secience Fundation (NSF) funded TeraGrid project which focuses on developing a Web-based GIS to provide access to the geospatial services on the grid (Wang & Liu 2009). In this area new frameworks have been proposed to exploit the power of grid computing. For example, one approach provides an additional tier to the traditional 3-tier architecture. This 4th tier supports the other tiers with distributed Web services (Zhang & Tsou 2009). These approaches have been successfully used to resolve a number of GIS related issues (Wang et al. 2009).

More recently techniques for reducing vector data content have been based on a new approach that tries to increase relevance by taking into account user specific needs. Such an approach relies on personalization and is discussed in detail in the next section.

4 MAP PERSONALIZATION

Personalization is employed by recommender systems on the Web to improve the accuracy of client recommendations. Different techniques used when providing personalization include user modeling (Fischer 2001), implicit profiling (Kelly & Belkin 2001), and data mining (Eirinaki & Vazirgiannis 2003).

User modeling enables application developers to model specific preferences of users based on explicit or implicit user input. Implicit profiling is used to ascertain user interests by unobtrusively monitoring user behavior as they browse Web pages. Examples of implicit behavior include accessing Web pages, scrolling within Web pages, and selecting hyperlinks and bookmarking. Data mining techniques allow implicit analysis of the captured data, and assists in making assumptions about user interactions and use of information.

4.1 *Initial map personalization applications*

Several existing GIS applications provide personalized services to their clients in an attempt to address spatial information overload (Dunlop et al. 2004, Schmidt-Belz et al. 2002). Many such applications focus on personalizing non-spatial content as opposed to spatial map content. Examples include presenting users with the opening times and menus of recommended restaurants (van Setten et al. 2004) or informing the user that a nearby attraction is closing for lunch (Cheverst et al. 2000).

Personalizing map feature content in GIS remains an open issue. Some applications rely on explicit user input to help retrieve more focused map content (Burigat & Chittaro 2005). However, this approach places additional interaction demands on the user by soliciting them for input pertaining to their interests. Such an overhead can break the user's focus and significantly increase interaction length. This can be arduous, especially when operating on a mobile device.

A significant number of GIS mapping systems typically generate either tourist maps (Schwinger et al. 2009) or route maps (Agrawala & Stolte 2001). With tourist maps, personalization is usually restricted to recommending points of interest within a geographical region (Schwinger et al. 2006) or to providing tourist services (Poslad et al. 2001). Recommended services are typically linked to non-spatial information, e.g. booking accommodation online or locating restaurants specializing in a specific type of food.

With route maps, personalization is typically restricted to the display of recommended routes or to the content of the actual routes themselves (Liang et al. 2008). In general there is lack of actual map feature content personalization in both tourist maps and route maps.

4.2 *More recent work*

Personalizing map content provides a means of drawing a user's focus to the feature content which is most relevant to their current task. This type of map personalization contributes to addressing the issue of spatial information overload by reducing the amount of feature content presented to users. This can be achieved by omitting irrelevant feature content when requests for maps are made.

An approach based on this principle is described in Weakliam et al. (2005) where maps are personalized by generating profiles for users based on their interactions with spatial data. Examples of interactions include operations such as adding or removing a specific layer of features of a map (implicitly signifying interest or disinterest). All user interactions are logged and analyzed and a dynamic user profile is generated and used to produce a map that matches the ascertained preferences. Spatial content personalization can occur at two different levels: at the spatial layer level (only layers deemed interesting are included) and at the spatial feature level (only a subset of features within a spatial layer are included). An interesting characteristic of this approach is that it does not require any explicit input from users. Instead, their interactions with the system are implicitly monitored and analyzed. The work in MacAoidh & Bertolotto (2007) goes one step further and focuses on not only standard interaction operations (e.g. pan/zoom) but also mouse movements and mouse clicks. Experimental evaluation has reported some interesting results linking user's mouse movements and interests when interacting with spatial data.

Other work focuses on the development of context-sensitive models, where context includes the location and the computational context of a specific user, as well as implicitly/explicitly calculated user profiles. Context-based techniques for map adaptation are being investigated by Petit et al. (2008a, b). A drawback of some of the context-based applications that have been proposed thus far is that they tend to identify user preferences at design time rather than runtime. Ballatore et al. (2010) try to overcome this issue by combining context with mouse interactions as indicators of interest in map features.

Although map personalization and adaptation techniques have attracted a lot of attention, it is not surprising that they have mostly targeted recreational applications. Indeed, the aim is data reduction to help users read/interpret the maps. As a result, techniques proposed so far are quite simple and do not rely on sound cartographic principles. This is a problem when professional use of the data is required.

Research in this area is still in its infancy. Further investigation is needed in this promising direction. Once well established map personalization and adaptation techniques have been implemented, they could be effectively used as a data reduction technique and combined with progressive transmission to improve data exchange and relevance in Web and mobile GIS systems.

5 DISCUSSION

One of the biggest problems of the digital era in which we live is information overload: too much information regarding a given topic or task is available. Information overload is now particularly evident in the spatial domain where users find it difficult to locate information which is of interest on a map. This can be a case of users issuing queries to locate specific features using Web-based GIS tools. For example, it is impossible to quickly read or interpret an overloaded and cluttered map such as the one reported in Figure 1, where a user has requested to view hotels in a relatively small area of New York City.

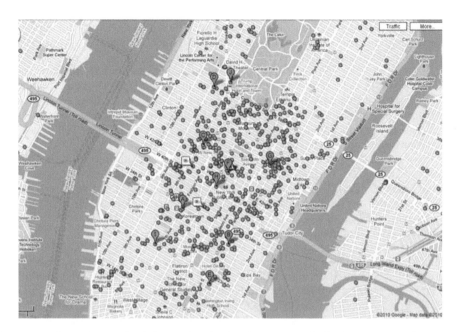

Figure 1. Information overload: A map showing hotels in New York City (Google Maps 2010).

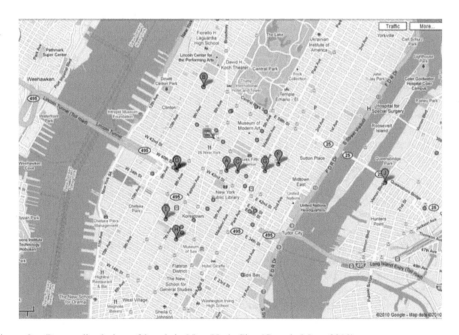

Figure 2. Personalized view of hotels in New York City (Google Maps 2010).

Although in some cases users can de-clutter maps by removing specific layers, in many instances the information packed into maps is often too detailed and not required by most users. A far more advantageous view of the same data is provided in Figure 2, where the information displayed has been personalized to match the preferences of a specific user.

With highly detailed datasets stored in very large files, in Web and mobile environments, such data reduction is needed not just to alleviate confusion for users and help them complete

147

their tasks, but also to speed up transmission. The issue is therefore two-fold; firstly data reduction is required in order to speedily deliver data which is manageable and secondly the data must be relevant to the user given their current context.

It is important to keep in mind the fact that many spatial data users are non professional and non expert GIS users. This impacts on the level of accuracy that is required. For such users it is important to access the information they need efficiently and to have it displayed as clearly as possible in order to interpret and use it effectively. In these cases, clear, readable maps and simple, well documented associated information are important while high accuracy (e.g. in terms of measure of distances) is not necessarily an issue.

For these kinds of applications (which will quickly, if not already, represent the majority of spatial data use over the Internet), developers should concentrate their efforts on data reduction based on effective map personalization by removing unwanted detail and increasing the relevance of the data presented. This form of data reduction will instantly speed up download and rendering time without relying on more complex and time-consuming cartographic methods of simplification.

When adopting this approach, there is a tradeoff. Cartography-based and other well established approaches for data reduction can generate accurate and highly consistent datasets which can reliably be used and manipulated. These are generally essential for professional users. However, such techniques do not provide personalization (i.e. they are not based on specific user preferences). The content of the produced datasets is based on geometric, visual, semantic, and other cartographic knowledge criteria which, in the majority of cases are independent of users' tasks and preferences. This is in contrast to map personalization techniques which do not necessarily generate fully consistent or accurate datasets.

These tradeoffs represent a fundamental challenge in determining which approach to take for data reduction. Should future efforts be devoted to personalization and the benefits which it brings for map presentation and delivery or towards sophisticated cartography data reduction techniques? Perhaps the ideal solution is to combine both approaches in order to produce an application that can be effectively used by expert and non expert GIS users. For example, user profiling and personalization can be used to guide data reduction when selecting the layers which are deemed important during the map generalization process for specific users. This approach should improve both the relevance and performance of Web and mobile GIS. Research in this area is still open, and to the best of our knowledge there are no concrete examples of projects dedicated to examining this issue, as a result further work is required to determine the benefits of such an approach.

6   CONCLUSIONS

In this chapter, we reviewed and surveyed existing approaches for data reduction and efficient transmission of spatial data with applications to Web and mobile GIS. Many traditional and well established approaches are limited to the transmission of raster images and triangular meshes. Although recently some new similar techniques have been developed for vector data, several challenges and open issues still remain. Bertolotto (2007) did a complete overview and discussion of such challenges within this area.

This chapter has critically surveyed and analyzed approaches based on traditional, sound cartographic principles. More recent methods, based on novel techniques, have also been described with a view to combine different approaches within the same system. These include: clustering, pre-fetching and caching, grid computing, indexing and map personalization. Among these techniques, personalization seems very promising in addressing specific user needs and therefore represents a worthwhile research direction to further study.

The new generation of on-line spatial information systems should be able to employ different data reduction techniques in an automatic manner depending on the application for which the data is required. For example, both raster-based and vector-based methods should be available.

Indeed, the ideal solution would combine many different approaches into an intelligent, adaptive system that will select the most suitable method to generate the best dataset according to user task and preferences. For example, an intelligent system could first provide an initial personalized visual representation of the data while continuing the downloading of further more accurate detail in the background. With the increasingly wider ranges of GIS users, further investigation is also needed for the development of effective and adaptive GIS interfaces within this area (Neis & Zipf 2008).

ACKNOWLEDGEMENTS

Research presented in this chapter was funded by a Strategic Research Cluster grant (07/SRC/I1168) by Science Foundation Ireland under the National Development Plan. We gratefully acknowledge this support.

REFERENCES

Agrawala, M. & Stolte, L. (2001) Rendering effective route maps: improving usability through generalization, *Proc. 28th International Conference on Computer Graphics and Interactive Techniques* (SIGGRAPH2001), pp. 241–250.

Ai, T., Li, Z. & Liu, Y. (2004) Progressive transmission of vector data based on changes accumulation model, *Proc. 11th International Symposium on Spatial Data Handling*, pp. 85–96.

Antoniou, V., Morley, J. & Haklay, M. (2009) Tiled vectors: a method for vector transmission over the web, *Proc. W2GIS'09*, Maynooth, Ireland, Lecture Notes in Computer Science, Springer-Verlag. pp. 56–71.

Ballatore, A., McArdle, G., Kelly, C. & Bertolotto, M. (2010) RecoMap: An interactive and adaptive map-based recommender, *Proc. 25th ACM Symposium on Applied Computing (SAC)*, pp. 888–892.

Bertolotto, M. (2007) Progressive techniques for efficient vector map data transmission: an overview (book chapter), In: Belussi, A., Catania, B., Clementini, E. & Ferrari, E. (eds.), *Spatial Data on the Web: Modeling and Management*, Springer-Verlag. pp. 65–84.

Bertolotto, M. & Egenhofer, M.J. (2001) Progressive transmission of vector map data over the world wide web, *GeoInformatica*, 5 (4), 345–373.

Boujou, A., Follin, J.M., Bertrand, F. & Stockus, A. (2005) An increment based model for multiresolution geodata management in a mobile system, *Proc. W2GIS'05*, Lausanne, Switzerland, Lecture Notes in Computer Science, Springer-Verlag. pp. 42–53.

Burigat, S. & Chittaro, L. (2005) Visualizing the results of interactive queries for geographic data on mobile devices, *Proc. 13th ACM International Symposium on Advances in Geographic Information Systems* (ACMGIS 2005), pp. 277–284.

Cecconi, A. & Galanda, M. (2002) Adaptive zooming in web cartography, *Proc. SVG Open 2002*, Zurich, Switzerland.

Chen, S., Wang, X., Rishe, N. & Weiss, M.A. (2000) A high-performance web-based system design for spatial data access, *Proc. 8th ACM International Symposium on Advances in Geographic Information Systems*, pp. 33–38.

Cheverst, K., Davies, N., Mitchell, K. & Friday, A. (2001) Experiences of developing and deploying a context-aware tourist guide: the GUIDE Project, *Proc. Conference on Mobile Computing and Networking* (MobiCom), ACM Press. pp. 20–31.

Cignoni, P., Ganovelli, G., Gobetti, E., Marton, F., Ponchio, F. & Scopigno, R. (2003) Planet-sized batched dynamic adaptive meshes, *Proc. IEEE Visualization'2003*, pp. 147–155.

Davis, G. (1998) A wavelet-based analysis of fractal image compression, *IEEE Transactions on Image Processing*, 7 (2), 141–154.

Davis, G. & Nosratinia, A. (1998) *Wavelet-based image coding: an overview, Applied and Computational Control, Signals, and Circuits*, 1 (1).

Douglas, D.H. & Peucker, T.K. (1973) Algorithms for the reduction of the number of points required to represent a digitized line or its caricature, *The Canadian Cartographer*, 10 (2), 112–122.

Dunlop, M., Morrison, A., McCallum, S., Ptaskinski, P., Risbey, C. & Stewart, F. (2004) Focused palmtop information access through starfield displays and profile matching, *Proc. Workshop on Mobile and Ubiquitous Information Access 2004*, pp. 79–89.

Egenhofer, M.J. (2002) Toward the semantic geospatial web, *Proc. 10th ACM International Symposium on Advances in Geographic Information Systems*, pp. 1–4.

Egenhofer, M. & Franzosa, R. (1991) Point-set topological spatial relations, *International Journal of Geographic Information Systems*, 5 (2), 161–174.

Eirinaki, M. & Vazirgiannis, M. (2003) Web mining for web personalization. *Transactions on Internet Technology (TOIT)*, 3 (1), 1–27.

Fischer, G. (2001) User modeling in human-computer interaction. *User Modeling and User-Adapted Interaction*, 11 (1–2), 65–86.

Fisher, Y. (ed.) (1995) *Fractal Image Compression: Theory and Application to Digital Images*, Springer-Verlag, New York.

Flewelling, D. & Egenhofer, M. (1999) Using digital spatial archives effectively, *International Journal of Geographical Information Science*, 13 (8), 1–8.

Goodchild, M.F. (2008). The use cases of digital earth. *International Journal of Digital Earth*, 1 (1), 31–42.

Guthe, M. & Klein, R. (2004) Streaming HLODs: an out-of-core viewer for network visualization of huge polygon models, *Computer & Graphics*, 28 (1), 43–50.

Han, Q. & Bertolotto, M. (2004) A multi-level data structure for vector maps, *Proc. ACMGIS'04*, Washington, DC, USA. pp. 214–221.

Hoppe, H. (1996). Progressive meshes, *Proc. SIGGRAPH'96*. pp. 99–108.

Junger, B. & Snoeyink, J. (1998) Importance measures for TIN simplification by parallel decimation, *Proc. Spatial Data Handling '98*, Vancouver, Canada. pp. 637–646.

Kang, Y.K., Kim, K.C. & Kim, Y.S. (2001) Probability-based tile pre-fetching and cache replacement algorithms for web geographical information systems, *Proc. ADBIS'2001*, ACM Press. pp. 127–140.

Kelly, D. & Belkin, N.J. (2001) Reading time, scrolling and interaction: exploring implicit sources of user preferences for relevance feedback during interactive information retrieval, *Proc. 24th Annual International Conference on Research and Development in Information Retrieval* (SIGIR 2001), pp. 408–409.

Kern, P. & Carswell, J.D. (1994) An investigation into the use of JPEG image compression for digital photogrammetry: does the compression of images affect measurement accuracy, *Proc. EGIS'94*, Paris, France.

Koshgozaran, A., Khodaei, A., Sharifzadeh, M. & Shahabi, C. (2006) A mulit-resolution compression scheme for efficient window queries over road network databases, *Proc. 6th IEEE International Conference on Data Mining* (ICDMW'06).

Li, Z. (2007) Digital map generalization at the age of enlightenment: a review of the first forty years, *The Cartographic Journal*, 44 (1), 80–93.

Liang, Z., Poslad, S. & Meng, D. (2008) Adaptive sharable personalised spatial-aware map services for mobile users, *Proc. GI-Days 2008 Conference*, Munster, Germany. pp. 267–273.

Lindstrom, P. & Pascucci, V. (2002) Terrain simplification simplified: a general framework for view-dependent out-of-core visualization, *IEEE Transaction on Visualization and Computer Graphics*, 8 (3), 239–254.

Lindstrom, P., Koller, D., Ribarsky, W., Hodges, L., Faust, N. & Turner, G.A. (1996) Real-time continuous level of detail rendering of height fields, *Proc. ACM SIGGRAPH'96*. pp. 109–118.

Mac Aoidh, E. & Bertolotto, M. (2007) Improving spatial data usability by capturing user interactions, in the European Information Society - Leading the Way with Geo-Information, *Proc. AGILE 2007* (10th AGILE International Conference on Geographic Information Science), Aalborg, Denmark, Lecture Notes in Geo-Information and Cartography, Springer-Verlag. pp. 389–403.

McArthur, D.E., Fuentes, R. & Devarajan, V. (2000) Generation of hierarchical multiresolution terrain databases using wavelet filtering, *Photogrammetric Engineering and Remote Sensing*, 66 (3), 287–295.

Morlet, J. & Grossman, A. (1984) Decomposition of Hardy Functions into square integrable wavelets of constant shape, *Siam Journal on Mathematical Analysis*, 15 (4) 723–736.

Neis, P. & Zipf, A. (2008) Extending the OGC OpenLS Route Service to 3D for an interoperable realisation of 3D focus maps with landmarks, *Journal of Location Based Services*, 2 (2), 153–174.

Petit, M., Claramunt, C., Ray, C. & Calvary, G. (2008a) A design process for the development of an interactive and adaptive GIS, *Proc. W2GIS'08*, Shanghai, China, Lecture Notes in Computer Science, Springer-Verlag. pp. 100–111.

Petit, M., Ray, C. & Claramunt, C. (2008b) A user context approach for adaptive and distributed GIS, *Proc. 10th International Conference on Geographic Information Science: AGILE*, Lecture Notes in Geoinformation and Cartography, Springer-Verlag. pp. 121–133.

Poslad, S., Laamanen, H., Malaka, R., Nick, A., Buckle, P. & Zipf, A. (2001) CRUMPET: creation of user-friendly mobile services personalized for tourism, *Proc. Conference on 3G Mobile Communication Technologies*, pp. 26–29.

Qiu, J., Liao, L. & Li, P. (2009). News recommender system based on topic detection and tracking, *Proc. 4th International Conference on Rough Sets and Knowledge Technology*, Lecture Notes in Computer Science, Springer-Verlag. pp. 690–697.

Raicu, I., Foster, I.T., Zhao, Y., Little, P., Moretti, C.M., Chaudhary, A. & Thain, D. (2009) The quest for scalable support of data-intensive workloads in distributed systems, *Proc. 18th ACM International Symposium on High Performance Distributed Computing*, pp. 207–216.

Rauschenbach, U. & Schumann, H. (1999) Demand-driven image transmission with levels of detail and regions of interest, *Computers and Graphics*, 23 (6), 857–866.

Saalfeld, A. (1999) Topologically consistent line simplification with the Douglas-Peucker algorithm, *Cartography and Geographic Information Science*, 26 (1), 7–18.

Scharl, A. & Tochtermann, K. (2007) The geospatial web, how geobrowsers, social software and the web 2.0 are shaping the network society, *Advanced Information and Knowledge Processing Series*, Springer.

Schmidt-Belz, B., Poslad, S., Nick, A. & Zipf, A. (2002) Personalized and location-based mobile tourism services, *Proc. Workshop on "Mobile Tourism Support Systems" in conjunction with Mobile HCI '02*.

Schwinger, W., Grün, C., Pröll, B., Retschitzegger, W. & Werthner, H. (2006) Pinpointing tourism information onto mobile maps –a light-weight approach, *Proc. 17th International Conference on Information Technology and Travel & Tourism* (ENTER 2006), Lecture Notes in Computer Science, Springer-Verlag.

Schwinger, W., Grün, C., Pröll, B., Retschitzegger, W. & Schauerhuber, A. (2009) Context-awareness in mobile tourism guides, Handbook of Research on Mobile Multimedia (2nd ed), Information Science Reference.

Shekhar, S., Huang, Y., Djugash, J. & Zhou, C. (2002) Vector map compression: a clustering approach, *Proc. 10th ACM International Symposium on Advances in Geographic Information Systems*, pp. 74–80.

Speretta, M. & Gauch, S. (2004) Personalising search based on user search histories, *Proc. 13th International ACM Conference on Information and Knowledge Management* (CIKM).

Srinivas, B.S., Ladner, R., Azizoglu, M. & Riskin. E.A. (1999). Progressive transmission of images using MAP detection over channels with memory, *IEEE Transactions on Image Processing*, 8 (4), 462–75.

Tu, S.R., Xe, X., Li, X. & Ratcliff, J.J. (2001) A systematic approach to reduction of user-perceived response time for GIS web services, *Proc. 9th ACM International Symposium on Advances in Geographic Information Systems*, pp. 47–52.

van Setten, M., Pokraev, S. & Koolwaaij, K. (2004) Context-aware recommendations in the mobile tourist application COMPASS, *Proc. Adaptive Hypermedia 2004*, pp. 235–244.

Wang, S. & Liu, Y. (2009) TeraGrid GIScience gateway: bridging cyberinfrastructure and GIScience, *International Journal of Geographical Information Science*, 23 (5), 631–656.

Wang, Y., Song, H., Hong, L. & Chen, X. (2009) GUPTDSS: Grid based urban public transport decision support system, *Proc. 6th International Symposium on Neural Networks on Advances in Neural Networks*, Lecture Notes in Computer Science, Springer-Verlag. pp. 1171–1180.

Weakliam, J., Bertolotto, M. & Wilson, D. (2005) Implicit interaction profiling for recommending spatial content, *Proc. 13th ACM International Symposium on Advances in Geographic Information Systems*, Bremen, Germany, ACM Press. pp. 285–294.

Yang, B.S. (2005) A multi-resolution model of vector map data for rapid transmission over the Internet, *Computers & Geosciences*, 31 (5), 569–578.

Yang, B.S., Zshi, W. & Li, Q. (2005a) An integrated TIN and Grid method for constructing multi-resolution digital terrain models, *International Journal of Geographical Information Science*, 19 (10) 1019–1038.

Yang, B., Purves, R. & Weibel, R. (2007b) Efficient transmission of vector data over the Internet, *International Journal of Geographical Information Science*, 21 (2), 215–237.

Yang, C., Cao, Y. & Evans, J. (2007a) WMS performance and client design principles, *The Journal of GeoInformation Science and Remote Sensing*, 44 (4), 320–333.

Yang, C., Wong, D.W., Yang, R., Kafatos, M. & Li, Q. (2005b) Performance-improving techniques in web-based GIS, *International Journal of Geographical Information Science*, 19 (3), 319–342.

Yang, C. & Raskin, R. (2009) Introduction to distributed geographic information processing, *International Journal of Geographical Information Science,* 23 (5), 553–560.

Zhang, L., Yang, C., Tong, X. & Rui, X. (2007) Visualization of large spatial data in networking environments, *Computers & Geosciences*, 33 (9), 1130–1139.

Zhang, T. & Tsou, M.H. (2009) Developing a grid-enabled spatial web portal for Internet GIServices and geospatial cyberinfrastructure, *International Journal of Geographical Information Science*, 23 (5), 605–630.

Zhou, H. & Bertolotto, M. (2005) Efficiently generating multiple representations for web mapping, *Proc. W2GIS'05*, Lecture Notes in Computer Science, Springer-Verlag. pp. 54–65.

Zhu, H. & Yang, C. (2008) Data Compression in Network GIS, *Encyclopedia of GIS*, Springer-Verlag. pp. 209–213.

Zorin, D., Shroder, P. & Sweldens, W. (1997) Interactive multiresolution mesh editing, *Proc. SIGGRAPH '97*, pp. 259–268.

*Advances in Web-based GIS, Mapping Services
and Applications – Li, Dragićević & Veenendaal (eds)*
© *2011 Taylor & Francis Group, London, ISBN 978-0-415-80483-7*

A load balancing method to support spatial analysis in XML/GML/SVG-based WebGIS

Haosheng Huang
Institute of Geoinformation and Cartography, Vienna University of Technology, Vienna, Austria

Yan Li
Spatial Information Research Center, South China Normal University, Guangzhou, China

Georg Gartner
Institute of Geoinformation and Cartography, Vienna University of Technology, Vienna, Austria

ABSTRACT: This chapter aims at introducing methods for load balancing spatial analysis into XML/GML/SVG-based WebGIS. We propose that the decision on where to execute spatial query operations (server side or browser side) should be based on the network communication cost versus the computational cost. This chapter mainly focuses on the network communication cost. After analyzing the workflow of spatial analysis, we address the following issues: GML (server side)/SVG (browser side) –based spatial information representation, spatial query language, and load balancing middlewares for dispensing spatial queries to either server side or browser side. Finally, we design some case studies to evaluate the proposed methods. The results prove that our methods are feasible and operable to support spatial analysis in the Web environment, and have a better performance compared to the server side solution and the browser side solution. The methods enable users to access spatial analysis functions simply with a web browser with SVG support.

Keywords: XML/GML/SVG, WebGIS, spatial analysis, load balancing, network communication cost, spatial data modeling, SVG-based Spatial Extended SQL

1 INTRODUCTION

Recent years have witnessed rapid advances in web-based geographic information systems (WebGIS), which aim at providing GIS functionality and services (such as web mapping and spatial analysis) to users through a common web browser, such as Internet Explorer and Firefox. The ability to support spatial analysis is viewed as one of the key characteristics which distinguish GIS from other information systems. In order to meet the increasing need of spatial information applications in the Web environment, spatial analysis should be introduced into WebGIS.

Scalable Vector Graphics (SVG) was proposed by the World Wide Web consortium (W3C) as a XML-based standard for describing two-dimensional vector graphics and graphical applications in 2001 (W3C 2003). Since then, SVG, together with XML and GML (Geography Markup Language, proposed by the Open Geospatial Consortium) (OGC 2003), have been increasingly considered as a promising solution for solving the problems of spatial data integration and sharing from the syntactic (data format) level. In this solution, GML is used as a coding, storing and transmitting standard of multiple spatial data on the server side, while SVG is considered as a visualization tool for displaying and interacting with vector spatial data on the browser side.

In the literature, some researchers used SVG's graphic elements (such as line and path) and graphic styles (such as fill, stroke and opacity) for spatial information visualization and developed some prototype systems (Neumann & Winter 2010, Peng et al. 2004, Köbben 2007). Furthermore, there is also some research focusing on transformation of GML to SVG (Guo et al. 2003, Tennakoon 2003, Herdy et al. 2008), SVG for LBS (Location Based Services) applications (Jeong et al. 2006), etc.

However, most of the WebGIS applications, especially XML/GML/SVG-based WebGIS applications, have been designed for visualization (web mapping) only, and "avoid the access to spatial analysis functions" such as spatial topological queries, map overlay, and buffer analysis that are vital to spatial information applications (Lin & Huang 2001). When performing spatial analysis tasks, we often install corresponding GIS software (e.g. ESRI's ArcGIS) on our computer, and carry out the tasks in a stand-alone or Local Area Network (LAN) environment. In the meantime, these applications only support their own spatial data formats. There are also some WebGIS applications which carry out all the spatial analysis tasks on the server side and then send the results to the browser side for visualization. These server-side solutions, sometimes, become impractical, as the server cannot handle a large volume of concurrent users. In order to solve this "bottleneck" problem, some of them introduce load balancing technology to the server side, which dispenses users' spatial queries and analysis requests to different servers in a server cluster (Supermap 2010, Qin & Li 2007, Luo et al. 2003, Wang et al. 2004). However, spatial analysis is a complex process; users often have to try different queries before they are satisfied with the results. Since every query may result in a large volume of data (such as intermediate results which users may not need), there will be a high transmission overload between the server side and the web browser side when using this server-side solution. As such, in order to improve the performance, not all the spatial operations should be implemented on the server side. For example, as the "buffer" operation often results in more data output than input, it may be implemented on the browser side in order to reduce the network transmission load.

Recognizing these limitations, this chapter proposes that *in order to meet the increasing need of spatial information application in the web environment, spatial analysis based on load balancing technology should be introduced into XML/GML/SVG-based WebGIS.* Load balancing spatial analysis carries out spatial queries and analyses *either on the server side or the browser side* depending on the *network communication cost* versus the *computational cost.* However, little work has been done on this aspect.

The objective of this chapter is to introduce methods for load balancing spatial analysis into XML/GML/SVG-based WebGIS. After analyzing the workflow of spatial analysis (i.e., defining the goal and evaluation criteria, representing the needed geospatial data, carrying out spatial query and analysis with GIS tools, and result appraisal and explanation), we focus on the following three key issues: *GML (server side)/SVG (browser side) –based spatial information representation, spatial operators and spatial query language* for both server side (for GML) and browser side (for SVG), and *load balancing middlewares* for dispensing spatial operations to the server side or the browser side. With these suggested methods, users can easily carry out spatial analysis tasks in the Web environment.

Having provided a background to load balancing spatial analysis, the rest of this chapter is structured as follows. In Section 2, we analyze the workflow of spatial analysis, and identify the key issues of introducing spatial analysis into XML/GML/SVG-based WebGIS. Section 3 uses the theory of spatial data modeling to design GML/SVG-based spatial information representation models. Based on the representation models, we design some spatial operators and a spatial extended SQL (SSESQL) to support spatial query and analysis directly on GML/SVG in Section 4. Section 5 describes the load balancing middlewares. Some case studies are implemented to evaluate our suggested methods in Section 6. Finally, Section 7 draws some conclusions and presents the future work.

2 WORKFLOW OF SPATIAL ANALYSIS

Spatial analysis is "a set of methods whose results change when the locations of the objects being analyzed change" (Longley et al. 2005, p. 217). It plays a key role in GIS. The ability to analyze spatial data using the techniques of spatial analysis is the key function of GIS. With spatial analysis, value-added products from existing datasets can be provided for users (Goodchild 1998). According to Wu (2002), the workflow of spatial analysis includes four steps: 1) defining the goal and evaluation criteria; 2) representing the needed spatial dataset; 3) carrying out spatial query and analysis with GIS tools; and 4) result appraisal and explanation. Steps 1 and 4 require domain knowledge and are mainly carried out by domain experts. For Step 2 and Step 3, GIS tools are needed to support and assist human-computer interaction. Figure 1 depicts the workflow of spatial analysis.

The chapter mainly focuses on designing methods to support and assist Step 2 and Step 3. For Step 2, we design GML/SVG-based spatial information representation models which can be used to represent the needed spatial datasets in GML and SVG (Section 3). For Step 3, we design and implement some spatial operators and integrate them into an SVG-based Spatial Extended SQL (SSESQL) to support spatial query and analysis on spatial datasets represented in GML and SVG (Section 4). In order to improve the performance, we also introduce load balancing middlewares to reduce the network transmission load (Section 5). However, the load balancing middlewares are completely transparent to the end users.

3 SPATIAL DATA MODELING FOR SVG/GML-BASED SPATIAL INFORMATION REPRESENTATION

In this section, we employ the theory of spatial data modeling to design SVG/GML-based spatial information representation models for representing spatial data on both server side and browser side.

3.1 SVG-based spatial information representation model

In Huang et al. (2008), a theoretical framework is set up for SVG-based spatial information representation based on the theory of spatial data modeling, which includes three steps: 1) choosing a conceptual model which can abstract the real world most appropriately,

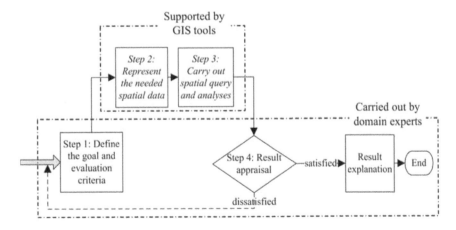

Figure 1. Workflow of spatial analysis.

2) choosing a suitable data structure to represent the conceptual model, and 3) designing a file format, or an appropriate method to record or store the data structure in Step 2.

Based on that theoretical framework, Huang et al. (2008) adapted the OGC's Geometry Object Model (GOM) (OGC 1999), and developed an SVG-based spatial information representation model (Fig. 2).

In this model, we use `<svg>` element to represent *Map* (the dataset), and use *viewBox* attribute to represent its bounded range. *Layer* is represented as `<g>` element. *Point, Curve, Surface* are represented as `<circle>`, `<path>`, and `<path>`, respectively. `<g>` element is used to represent the *Multipoint, Multicurve, Multisurface* and *Multigeometry*. Both spatial and non-spatial attributes of spatial objects are represented as corresponding SVG element's attributes. In the model, if B is PART-OF A (i.e. composition/aggregation relationship), B is represented as a child element of A. For example, Layer is PART-OF Map, so `<g>` element which represents Layer is a child element of `<svg>` element which represents Map. With this model, users can use SVG to represent the needed spatial information. Figure 6 illustrates an example of such representation.

3.2 *GML-based spatial information representation model*

Similar to our SVG-based representation model, we developed a GML-based spatial information representation model (Fig. 3) (Huang & Li 2009).

In this model, we use GML *"FeatureCollection"* element to represent the *Map* (dataset), and use *"boundedBy"* and *"srsName"* to represent the bounded area and spatial coordinate system of the *Map*. *Layer* is represented as element which is inherited from *"gml:_Feature"*. If B is PART-OF A (composition relationship in Fig. 3), B is represented as a child element of B. Both spatial attributes and non-spatial attributes are represented as the child elements of the corresponding GML elements.

3.3 *Transformation from GML to SVG*

By using the models depicted in Figures 2–3, users can easily use GML/SVG to represent the needed spatial data on both the server side and the browser side. Because they are based on the same conceptual model and spatial data structure (logical model), GML-based spatial

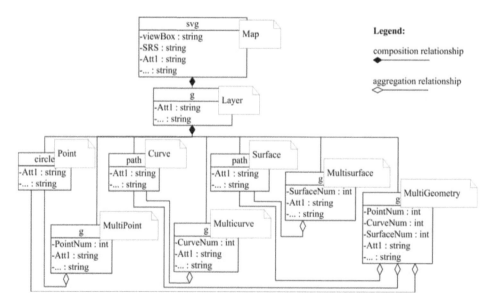

Figure 2.   SVG-based spatial information representation model.

156

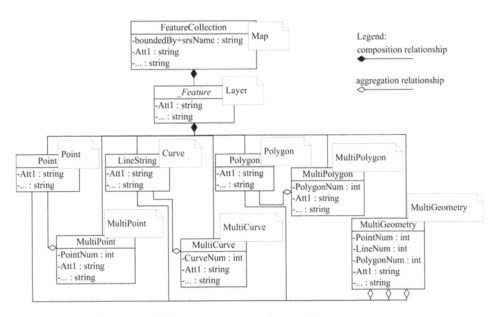

Figure 3. GML-based spatial information representation model.

Table 1. Comparison of GML and SVG-based spatial information representation models.

Description	GML elements	SVG elements/ attributes
Map	FeatureCollection	svg
Bounded area of the map	boundedBy/ Envelope	viewBox
Point object	Point	circle
Curve object	LineString	path
Polygon object	Polygon	path (end with "Z")
Compound object	MultiPoint, MultiLine, MultiPolygon, MultiGeometry	g

datasets on the server side can be easily and losslessly converted to SVG-based datasets for the browser side. Table 1 compares the different elements/attributes in the GML and SVG-based spatial information representation models.

When transforming GML datasets to SVG datasets for visualization, we should also make some transformations to the coordinate systems in GML. Figure 4 compares the differences between the coordinate systems used in GML and SVG. Therefore, if we do not make any transformations to the coordinate systems, the map represented in SVG may become headstand.

We use SVG's "translate" attribute to implement this transformation. The transformation parameter is set as "translate (0, 2*min_y + height), scale(1,-1)", where (min_x, min_y, width, height) represents the bounded area of the map, and can be found in the "viewBox" attribute of the root "SVG" element. Because all the layers have to be translated to avoid headstand, we use <g> element to group all the layers together.

Figure 5 shows a map of Guangdong province (China) represented in SVG. We use the Internet Explorer with SVG plug-in to visualize the map. The map includes two layers: gd_river_line and gd_city_polygon. In Figure 6, we compare the GML and SVG codes which represent the map using our suggested models (Figs. 2–3). The code "<g transform ="translate(0, 4940959), scale(1,-1)">" in Figure 6 also gives an example about the transformation of the coordinate systems.

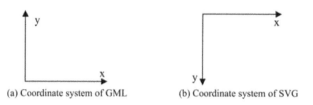

(a) Coordinate system of GML   (b) Coordinate system of SVG

Figure 4.   Comparison of the coordinate systems of GML and SVG.

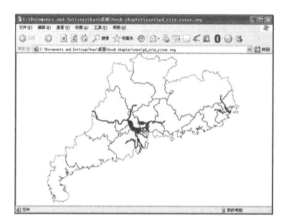

Figure 5.   An SVG map of Guangdong province with two layers (gd_river_line and gd_city_polygon).

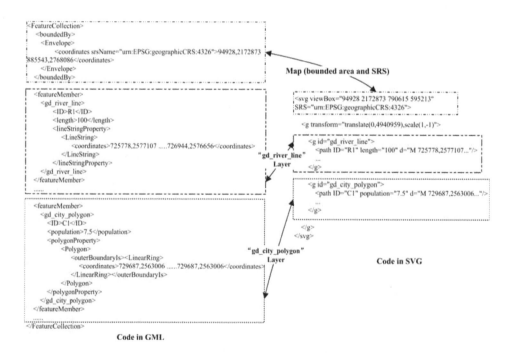

Figure 6.   Comparison of codes in GML and SVG.

158

4 SPATIAL OPERATORS AND SSESQL

This section focuses on how to carry out spatial data query on GML and SVG. It is important to note that XQuery is often employed to query XML-based data, and is also suitable for querying GML/SVG data. However, XQuery has a very special and complicated syntax, and has not become familiar to many people. In contrast, SQL (Structured Query Language) has a relative simple and intuitive syntax, and has been familiar to many technical users. More importantly, SQL is more powerful in querying data. Thus, in consideration of end users' acceptance and the processing capabilities of SQL, an extended SQL is employed for spatial data query on GML and SVG.

Several efforts have been made, attempting to make spatial extensions to SQL (Lin & Huang 2001, Egenhofer 1994). These extended SQL introduce spatial data types (e.g. point, line and polygon) and spatial operators, allowing users to inquire spatial features, primarily in terms of spatial relationships and metric constraints (Lin & Huang 2001). It is widely acknowledged that these spatial operators and SQL like languages can be used for spatial analysis (Frank 1984).

According to Section 3, spatial information is organized as "map - layer - spatial object" structure. A map represented by SVG/GML can be viewed as a database, the layers as tables of the database, the attributes (spatial and non-spatial) of spatial objects in a layer as columns of the corresponding table, and spatial objects as records of the corresponding table. Thus SQL like languages can be used for spatial query and analysis on SVG and GML.

In this section, we design some spatial operators and integrate them into our SVG-based Spatial Extended SQL (SSESQL). The SSESQL uses the basic spatial data types discussed in Section 3: Point, Curve, Surface, Multipoint, Multicurve, Multisurface and Multigeometry. It can be used on both server side and browser side for spatial query and analysis on GML and SVG.

4.1 *Spatial operators*

Spatial operators are mainly designed to access spatial attributes, calculate spatial relationships, and perform geometrical operations. We introduce five types of operators: *attribute access operators* (GeometryType, Centroid, Length, Area, and Envelope), *spatial topological operators* (Disjoint, Touch, Crosses, Within, Overlap, and Contain), *spatial order operators* (East, East_South, South, West_South, West, West_North, North, and East_North), *spatial metric operators* (Max_Dist, Min_Dist, and Mean_Dist), and *geometrical operators* (Intersection, Union, Difference, and Buffer).

These five types of operators can meet the basic needs of spatial analysis. For network analysis, we can use Touch operator and Length operator to find out the touched spatial object (e.g. road) and the distance. We can use Buffer operator and topological operators to carry out buffer analysis, such as "which cities are around 100 km of the Danube River". Also, we can use Intersection, Union and Difference operators to carry out overlay analysis.

It is important to note that, the above spatial operators can be used as APIs (Application Programming Interface) and integrated into other programs. However, we integrate them into an SVG-based Spatial Extended SQL (SSESQL) to support end users' interaction.

More detail of the spatial operators can be found in Huang et al. (2008).

4.2 *SSESQL and some query examples*

As SSESQL is designed for spatial query, there is no need to consider data insert, update and delete. As a result, we integrate the above spatial operators to the original SELECT clause of SQL. The EBNF (Extended Backus-Naur Form) of SELECT clause of SSESQL can be found in Huang et al. (2008).

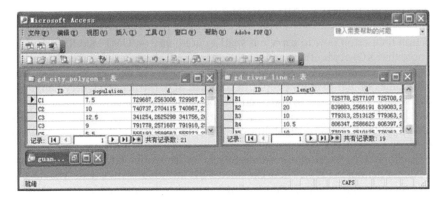

Figure 7. Tables of gd_city_polygon and gd_river_line shown in Microsoft Access.

The following are some query examples. The two layers represented in the Figures 5–6 can be viewed as the following tables: `gd_city_polygon` (ID, population, d) and `gd_river_line` (ID, length, d) in Figure 7.

1. Query example 1: Lists the neighbor cities of city `"C1"`.
```
SELECT city1.ID AS "Neighbors_of_C1"
FROM gd_city_polygon city1, gd_city_polygon city2
WHERE Touch (city1.d, city2.d) = True AND city2.ID = 'C1'
```

2. Query example 2: Lists the cities which are crossed by the river `"R1"`.
```
SELECT c.id AS cid
FROM gd_river_line r, gd_city_polygon c
WHERE r.id = "R1" AND Crosses(r.d, c.d) = True;
```

3. Query example 3: There is some toxic contamination throughout the River `"R1"`. This toxic contamination affects the cities, which are 100 km around the river. Please list all affected cities.
```
SELECT c.id
FROM gd_river_line r, gd_city_polygon c
WHERE Overlap(c.d, Buffer(r.d, 100000)) = True AND r.id = "R1";
```

Since our GML-based spatial information representation model uses the same conceptual model and spatial data structure (logical model) as the SVG-based spatial information representation model, we can also apply the SSESQL to carry out spatial queries and analyses on GML-based spatial data on the server side. The only difference in applying SSESQL for spatial analysis on SVG and GML is the implementation of the SSESQL compiler.

5 LOAD BALANCING MIDDLEWARES

In this section, we focus on designing the load balancing middlewares to dispense a spatial query to either the server side or the browser side based on the cost of that spatial query.

5.1 *General principle*

The general principle of our load balancing algorithm is to compare the costs of server side solution (C_{server}) and browser side solution ($C_{browser}$). If C_{server} is less than $C_{browser}$, execute the spatial query on the server side, and send the result data to the browser. Otherwise, send the data (input data) to the browser, and carry out the spatial query on the browser side.

The cost of a spatial query includes the computational cost (execution of the spatial query) and network transmission cost (input data and output data of spatial operations) and both of them are often measured by delay. As a normal PC's processor performance has been drastically improved, the difference between the server side and the browser side's computational cost becomes minimal or less significant for most of the spatial queries. Therefore, the difference between C_{server} and $C_{browser}$ mainly depends on the network transmission cost. Our discussions below will focus on comparing the network transmission cost.

5.2 Architecture

In order to compare the network transmission costs, we design middlewares for both server side and browser side. Figure 8 shows the architecture.

5.2.1 Browser side's middleware

The browser side's middleware provides an *SSESQL compiler for SVG*, a `cache struc-ture`, and a *decision-maker*. The SSESQL compiler carries out syntax, sentence, and semantic analysis for users' SSESQL sentences. The cache structure records which spatial data have been delivered to the browser side. Currently, we choose a very coarse granularity for organizing and transmitting the spatial data: layer as a unit of organizing and transmitting spatial data. The following is the data structure of the cache structure.

```
Class cache_structure
{
    String objectID; // the ID of the layer
    Bool bExecuted; //True or False, whether the layer is created
                    by browser side' execution,
                // bExecuted = True means that the layer can
                    only be found on the
                //browser side.
    Bool bTransmitted; //True or False, whether the layer has
                        been delivered to the browser
}
ArrayList< cache_structure > cache; // record the transmitted data
```

The decision-maker figures out the needed spatial data (layer) from the **FROM** clause of the SSESQL sentence (spatial query) by invoking the SSESQL compiler, and checks whether all the needed spatial data have been delivered to the browser side (by checking the cache structure). If necessary, the decision-maker sends the request to the server side's middleware.

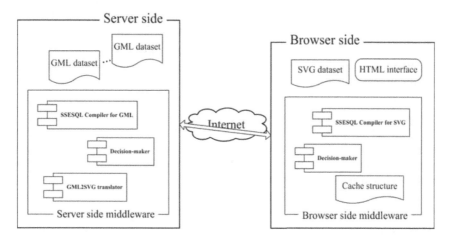

Figure 8.   Architecture of the load balancing middlewares.

161

5.2.2 *Server side's middleware*

The server side's middleware includes an *SSESQL compiler for GML*, a *GML2SVG transla-tor*, and a *decision-maker*. The SSESQL compiler carries out syntax, sentence, and semantic analysis for users' SSESQL sentences to query GML data on the server side. The GML2SVG translator is employed to translate GML data into SVG data. The decision-maker is respon-sible for receiving the requests from the browser side's middleware, and employing some inference rules to dispense a spatial query to either the server side or the browser side based on the network transmission cost of that spatial query.

5.3 *Inference rules in the decision-makers*

Every time users submit a spatial query (by SSESQL) through their browser (e.g. Internet Explorer), the browser side's decision-maker will figure out the needed input data (layer) from the FROM clause in SSESQL sentences, and check whether the needed spatial data have been delivered to the browser side. If all the needed input data are available on the browser side, this spatial query will be executed by the SSESQL compiler on the browser side. Otherwise, the spatial query sentence, the names (IDs) of the needed input data (only those which have not been delivered to the browser side, i.e. bTransmitted = False), and the data size $S_{browser}$ of the other input data whose bExecuted is True (the data can only be found on the browser side) will be sent to the server side.

The server side's decision-maker receives the information, and employs some inference rules for the dispensation. In order to identify these inference rules, we analyze the spatial operations provided for GML and SVG in Section 4. We find that only the "Buffer" and "Union" spatial operators often result in more data output than input. As a result, we design the following inference rules:

1. All the needed input data are available on the server side: if the needed output data do not involve with "Buffer" and "Union" spatial operators (it can be figured out from the SELECT clause in SSESQL sentences), the spatial query will be carried out by the SSESQL compiler on the server side, and the result (output) data will be sent to the browser side. Otherwise, the needed input data will be translated into SVG, and sent to the browser side, the spatial query will be executed by the SSESQL compiler on the browser side.
2. The needed input data are located on different sides (browser side and server side): if the data size S_{server} of the needed input data (only those which have not been delivered to the browser side) is smaller than $S_{browser}$ (the data size of the needed input data which can only be found on the browser side), the needed input data will be sent to the browser side, and the spatial query will be executed by the SSESQL compiler on the browser side. Otherwise $(S_{server} > S_{browser})$, those needed input data which can only be found on the browser side will be sent to the server side, and the spatial query will be executed by the SSESQL compiler on the server side; after that, the result (output) data will be sent to the browser side.

For all the above cases, the cache structure on the browser side's middleware will be updated. It is important to note that the above inference rules are still very simple, and need to be improved further.

6 IMPLEMENTATION, CASE STUDIES AND DISCUSSIONS

In this section, we discuss how to implement the proposed method. In order to evaluate the method, we design three case studies, and compare different solutions (server side, browser side, and load balancing) for accomplishing these tasks.

6.1 *Implementation*

Most of the spatial operators in Section 4 involve geometrical computation. The field *compu-tational geometry* provides rich collection of algorithms for solving pure geometrical problems.

As well, Java already provides some basic computational geometry APIs (Java 2D API). We implement the spatial operators with algorithms of computational geometry and Java 2D API. For the SSESQL, a compiler is needed to carry out syntax, sentence, and semantic analysis for SSESQL sentences. Levine et al. (1992) discussed how to implement an SQL compiler and provided some c++ codes. As our SSESQL is based on the original SQL, we adapt their c++ codes, and implement our own SSESQL compiler by combining the spatial operators. Currently, we do not introduce *query optimization* into our compiler. However, as our SSESQL only introduces some spatial operators into the original SQL, all the technologies for query optimization can be used to improve the performance of our SSESQL compiler.

For the server side's middleware, spatial operators, SSESQL compiler, GML2SVG, and decision-maker are implemented as Java Servlets. For the browser side's middleware, spatial operators, SSESQL compiler, decision-maker and the cache structure are developed as Java Applets, and embedded in HTML. A user interface is also embedded in HTML for inputting SSESQL query sentences (Fig. 9). We use JavaScript to access SVG document and invoke the Java applets, which can help to receive users' input and execute the spatial query on the browser side. In order to facilitate the interaction between server side and browser side's middlewares, we use AJAX (Asynchronous JavaScript and XML) technology (mainly XML-HttpRequest Object). Users can access spatial analysis functions simply with a web browser (such as Internet Explorer) which has an SVG plug-in or SVG viewer (such as Adobe SVG Viewer). In the last several years, more and more web browsers (Firefox, Opera, Google Chrome, etc.) start to provide native SVG support (Wikipedia 2010).

6.2 Case studies

We design several case studies to evaluate the proposed methods. It is important to note that a good design of spatial analysis procedure is vital to a spatial analysis task. This means that

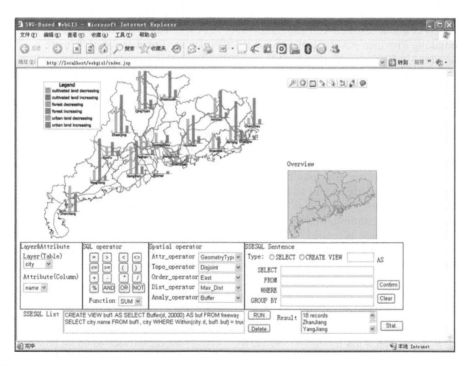

Figure 9. The land uses change along the freeway (Legend for bar graphs: for the first bar, light color for cultivated land decreasing, dark color for increasing; for the second bar, light color for forest decreasing, dark color for increasing; for the third bar, light color for urban land decreasing, dark color for increasing).

users (domain experts) do have to plan the analysis task carefully, even if they have the best spatial analysis tool.

The cultivated lands are very important in a highly populated area like Guangdong Province in China. Our case studies focus on the problem of whether the land uses change with the growth of transportation networks.

Our first case study tries to find out *how the land uses along the freeway change in Guangdong Province between 1987 and 1996*. In order to investigate this issue, we have to define a buffer for the freeway, and find out which cities are located in this buffer, and then analyze the changes of land uses for these cities. We carry out this task based on the workflow described in Figure 1. First, we identify the needed data and the evaluation criteria by carefully analyzing this case study. And then based on the suggested model in Figure 3, we use GML to represent the needed spatial data (freeway and city layers) on the server side. We also represent the district boundary layer in SVG and deliver it to the browser side as the initial User Interface (UI). We then submit SSESQL sentences on the browser side to carry out the spatial queries by the following steps: 1) Calculate a buffer of freeway (using Buffer operator). In this case, we use 20 km as the buffer size. 2) Find out which city centers are located in this buffer (using Within operator). 3) Use the statistics function to generate the bar graphs of changes of land uses for every identified city. While working completely transparent to the end user, the load balancing middlewares dispense the SSESQL sentences (requests) to either the server side or the browser side based on the cost of those requests automatically.

Figure 9 shows the user interface and the result of this case study. Functions of zoom in/ out, pan, layer control, query, and statistics are also added to this interface. The names of cities involved in the calculation are listed in the box shown at the right-bottom corner.

As can be seen from the bar graphs, there is a decreasing trend of cultivated lands and an increasing trend of urban land during 1987 to 1996 in all related cities. These trends are coincident with the situation of Guangdong Province between 1987 and 1996. At that time, with the growth of its transportation networks, especially freeway, all the cities were greatly expanded by converting cultivated lands into urban lands. However, there is an increasing trend of the forest between 1987 and 1996. An explanation for this would be that the Guangdong Province government announced a policy for forest in 1985: "Create a Green Guangdong in 10 Years" (Guangdong Forest 2010). From this case study, we can understand that land uses change with the growth of transportation networks. In order to draw some quantitative conclusions, more case studies should be done by using our proposed method.

In order to illustrate the benefit of introducing load balancing technology, we compare different solutions (server-side solution, client-side solution, and load balancing solution) for carrying out this task. We mainly compare the data amount of the network transmission (excluding the request/response sentences) between the server side and the browser side. Table 2 depicts the result. As can be seen from the table, our load balancing solution has a smaller network transmission load between server side and browser side.

Our second case study tries to address *the relationship between changes of land use and road (railway, freeway, and province level road) density in Guangdong Province*. Similar to the first case study, at the beginning, we only deliver the district boundary layer to the browser side as the initial UI. The task is carried out in the following steps: 1) Calculate the total road length for every city (using Interaction and Length operators). 2) Calculate the area for every city (using the Area operator). 3) Calculate the road density for every city and color every city

Table 2. Comparison of the first case study (data amount is measured by Byte).

	Server-side solution	Client-side solution	Load balancing solution (our solution)
Step1: Buffer	256,347 B	92,122 B	92,122 B (executed on the browser)
Step2: Within	741 B	2,043 B	2,043 B (executed on the browser)
Step3: Stat.	2,168 B	11,839 B	2,168 B (executed on the server)
Total	259,256 B	106,004 B	94,293 B

accordingly (using SQL's "/" operator). 4) Use the statistics function to generate bar graphs of changes of land use for every city. The load balancing middlewares dispense the SSESQL sentences to either the server side or the browser side based on the cost of those requests automatically.

Figure 10 shows the result from this case study. The result lists the name of cities and their road density in the right-bottom box. Every city is marked with different color according to its road density, and with a statistic bar graph showing changes of land use. The legend of the bar graph is the same as that in Figure 9. As a result, we can make the following qualitative conclusion: the changes of land uses differ with different road densities.

We also compare the amount of transmitted data between different solutions for accomplishing this task. Table 3 depicts the result.

We also use our proposed method to calculate *the river density (double line river) in Guangdong Province*. Figure 11 shows the result. In this case study, the city with the highest river density is Foshan city, located in the Pearl River Delta which is the most economically dynamic region in China, whose river density is 16.12 km^2/100 km^2. Other cities (Guangzhou, Dongguan, and Zhuhai) in the Pearl River Delta also have very high river density. Table 4 depicts the similar comparison among different solutions for accomplishing this task.

The implementation of the above case studies shows that our suggested method for SVG-based spatial analysis is feasible and operable. Although we use some very simple and static inference rules for dispensing a spatial query to either the server side or the browser side, and a quite coarse granularity (by layer) for organizing and transmitting spatial data, our load balancing spatial analysis solution decreases the amount of data transmitted over the network between the server side and the browser side, and accordingly improves the overall performance. Our method enables users to access spatial analysis functions simply with a web browser (such as Internet Explorer and Firefox) with SVG support. This greatly improves the functions of current WebGIS applications, most of which have been only employed for web mapping.

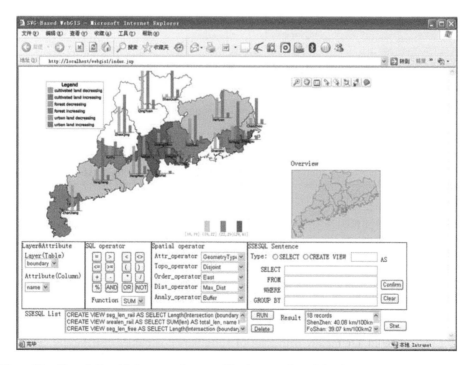

Figure 10. The relationship between changes of land uses and road density.

165

Table 3. Comparison of the second case study (data amount is measured by Byte).

	Server-side solution	Client-side solution	Load balancing solution (our solution)
Step1: Length	1,336 B	1,125,073 B	4,381 B
Step2: Area	1,345 B	0 B	0 B (executed on the browser)
Step3: Density	1,774 B	0 B	0 B (executed on the browser)
Step4: Statistics	2,168 B	11,839 B	2,168 B (executed on the server)
Total	6,623 B	1,136,912 B	6,549 B

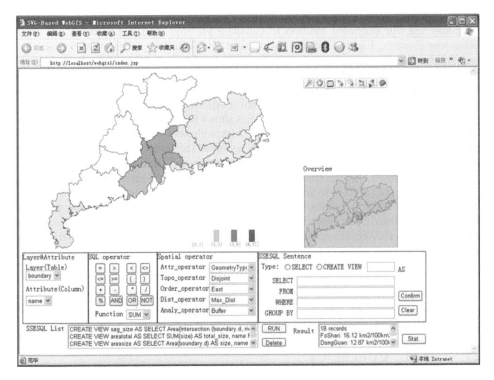

Figure 11. The river density (double line river) in Guangdong Province.

Table 4. Comparison of the third case study (data amount is measured by Byte).

	Server-side solution	Client-side solution	Load balancing solution (our solution)
Step1: River area	1,333 B	439,454 B	2,206 B
Step2: City area	1,345 B	0 B	0 B (executed on the browser)
Step3: Density	1,771 B	0 B	0 B (executed on the browser)
Total	3,449 B	439,454 B	2,206 B

To summarize, our solution has a good support for cross-platform (e.g. Microsoft Windows and Linux) and cross-browser (e.g. Internet Explorer, FireFox and Opera). As a result, our proposed methods are especially suitable for providing spatial analysis functions to heterogeneous Web clients, such as to the public users who may use different platforms and different web browsers and to a company with a lot of subsidiary companies distributing on different places. At this moment, we are cooperating with some municipal Land and Resource Administration Bureaus in China which are interested in sharing spatial data and spatial analysis functions with their branches. The method can also be extended to support spatial analysis on mobile devices (e.g. PDA and smart phones), which would be very useful for field work.

7 CONCLUSIONS AND FUTURE WORK

Currently, most WebGIS applications in general and SVG-based spatial applications in particular only offer web mapping function, and do not provide spatial analysis functions that are vital to spatial information applications. This chapter focuses on introducing load balancing spatial analysis into XML/GML/SVG-based WebGIS. We proposed that the decision on where to execute spatial query operations (server side or browser side) should be based on the network communication cost versus the computational cost, mainly focusing on the network communication cost.

The contributions of this chapter are: 1) identification of the key issues of load balancing spatial analysis in WebGIS, 2) a lossless GML2SVG translator according to the GML/ SVG based spatial information representation model, 3) design of load balancing middlewares to dispense spatial operations to the server side or the browser side, which reduce the network transmission load, 4) improved functionality of current WebGIS applications, especially XML/GML/SVG based WebGIS which have been mostly employed for web mapping only.

Our next step is to make more investigation on the granularity of organizing and transmitting spatial data, and to develop more flexible and precise inference rules for dispensing spatial queries to the server side or the browser side. We are also interested in introducing query optimization to improve the performance of our SSESQL compiler. Additionally, more complex case studies will be carried out to evaluate our suggested method.

ACKNOWLEDGMENT

This work has been supported by the National Natural Science Foundation of China (project no. 60842007), and the Science and Technology Department of Guangdong Province (project no. 2002B32101, 2004B32501001, 2005B30801006). Haosheng Huang and Georg Gartner also thank the support from the project UCPNavi funded by the Austrian FWF. We are also grateful to our anonymous reviewers and editors for their truly helpful comments.

REFERENCES

Chen, S., Lu, X. & Zhou, C. (2001) *Introduction of GIS* (in Chinese). Beijing: Science Publish.
Egenhofer, M. (1994) Spatial SQL: A Query and Presentation Language. *IEEE Transactions on Knowledge and Data Engineering (TKDE)*, 6 (1), 86–95.
Frank, A. (1982). Mapquery-database query languages for retrieval of geometric data and its graphical representation. *ACM Computer Graphics*, 16 (3), 199–207.
Guangdong Forest (2010) Guangdong Forest's 30 years. [Online] Available from: http://www.gdf.gov.cn/ index.php?controller=front&action=view&id=10005572 (In Chinese), [Accessed in Mar. 2010].
Guo, Z., Zhou, S., Xu, Z. & Zhou, A. (2003) G2ST: A novel method to transform GML to SVG. *Proc. the 11th ACM GIS*.
Herdy, K., Burggraf, D. & Cameron, R. (2008) High Performance GML to SVG Transformation for the Visual Presentation of Geographic Data in Web-Based Mapping Systems. *Proc. the 6th International Conference on Scalable Vector Graphics*.
Huang, H. & Li, Y. (2009) Load balancing spatial analysis in XML/GML/SVG based WebGIS. *Proc. ESIAT 2009, Wuhan, 4–5th July 2009*, Los Alamitos, IEEE Computer Society.
Huang, H., Li, Y. & Gartner, G. (2008) SVG-based spatial information representation and analysis. *Proc. W2GIS 2008, Shanghai, 11–12th December 2008*, Berlin Heidelberg, Springer.

Jeong, C., Chung, Y., Joo, S. & Lee, J. (2006) Tourism Guided Information System for Location-Based Services. In: Shen, H.T. et al. (eds.), *APWeb Workshops 2006*, Harbin, 16–18th January 2006. Berlin Heidelberg, Springer.

Köbben, B. (2007) RIMapperWMS: a Web Map Service providing SVG maps with a built-in client. In Fabrikant, S.I. & Wachowicz, M. (eds.), *The European Information Society: Leading the Way with Geo-information*, Berlin Heidelberg, Springer.

Levine, J., Mason, T. & Brown, D. (1992) *Lex & Yacc (2nd)*. O'Reilly & Associates.

Lin, H. & Huang, B. (2001) SQL/SDA: A query language for supporting spatial data analysis and its web-based implementation. *IEEE Transactions on Knowledge and Data Engineering (TKDE)*, 13 (4), 671–682.

Longley, P.A., Goodchild, M.F., Maguire, D.J. & Rhind, D.W. (2005) *Geographic information systems and science (second edition)*. Chichester, John Wiley.

Luo, Y., Wang, X. & Xu, Z. (2003) A dynamic load balancing policy for agent-based DGIS. *Proc. Asia GIS Conference 2003*, Wuhan, 16–18th October 2003.

Neumann, A. & Winter, A. (2010) *Cartographers on the net*. [Online] Available from: http://www.carto.net, [Accessed on Mar. 2010].

OGC. (1999) *OpenGIS simple features specification for SQL (Revision 1.1)*. [Online] Available from: http://www.opengeospatial.org/standards/sfs, [Accessed in Mar. 2010].

OGC. (2003) *Geography Markup Language*, [Online] Available from: http://www.opengeospatial.org/standards/gml, [Accessed in Mar. 2010].

Peng, Z. & Zhang, C. (2004) The roles of geography markup language (GML), scalable vector graphics (SVG), and Web feature service (WFS) specifications in the development of Internet geographic information systems (GIS). *Journal of Geographical Systems*, 6 (2), 95–116.

Qin, G. & Li, Q. (2007) Dynamic resource dispatch strategy for webgis cluster services. *Proc. CDVE 2007*, Shanghai, 16–20th September 2007. Berlin Heidelberg, Springer.

Shekhar, S., Coyle, M., Goyal, B., Liu, D. & Sarkar, S. (1997) Data models in geographic information systems. *Communications of the ACM*, 40 (4), 103–111

SuperMap. (2010) *SuperMap IS.NET 2008*. [Online] Available from: http://www.supermap.com.cn/gb/products/fwskf.htm, [Accessed in Mar. 2010].

Tennakoon, W.T.M.S.B. (2003) Visualization of GML data using XSLT. Master thesis of International Institute for Geo-Information Science and Earth Observation.

W3C. (2003) *Scalable Vector Graphics (SVG) 1.1 Specification*, [Online] Available from: http://www.w3.org/TR/SVG11/, [Accessed in Mar. 2010].

Wang, P., Yang, C., Yu, Z. & Ren, Y. (2004) A load balance algorithm for WMS. *Proc. Geoscience and Remote Sensing Symposium 2004 (IGARSS'04)*, Anchorage, 20–24th September 2004. Piscataway: IEEE.

Wikipedia. (2010) *Scalable Vector Graphics*. [Online] Available from: http://en.wikipedia.org/wiki/Scalable_Vector_Graphics, [Accessed in Mar. 2010].

Wu, X. (2002) *Principles and methods of GIS* (in Chinese), Beijing: Publishing House of Electronics Industry.

Augmentation and location-based services

*Advances in Web-based GIS, Mapping Services
and Applications – Li, Dragićević & Veenendaal (eds)*
© 2011 Taylor & Francis Group, London, ISBN 978-0-415-80483-7

Geolocating for web based geospatial applications

Bert Veenendaal & Jacob Delfos
Department of Spatial Sciences, Curtin University, Perth, Western Australia, Australia

Tele Tan
Department of Computing, Curtin University, Perth, Western Australia, Australia

ABSTRACT: The geospatial location of stationary and mobile objects and phenomena are increasingly important to geospatial applications. The availability and ease of access to geospatial information and digital earth technologies via the Internet are helping to drive a greater demand for geolocating fixed and mobile features, users, assets, services and other phenomena. A range of methods are used to determine spatial locations of fixed and mobile objects and properties including GPS, WiFi, RFID, geocoding, sensors and IP locating. Current research is investigating methods to identify user locations on the internet. The internet forms an attractive platform for Location Based Services (LBS) since it provides a nearly ubiquitous medium and web browsers provide a highly standardised application environment. However, security mechanisms in browsers complicate communication with devices used for positioning. IP positioning can provide a viable alternative, particularly since the computer network hierarchy is related to spatial hierarchies. This chapter introduces location based services and the importance of location. A range of methods for obtaining geospatial location are introduced and the contexts in which they are used are described. Finally, some current research is outlined where a methodology known as VRILS was developed to enable the use of IP addresses for geolocating in a web based environment.

Keywords: Geolocating, spatial location, positioning, location based services, IP positioning, positioning techniques, web GIS

1 INTRODUCTION

Spatial location is the building block of geographic information and is inherent to geospatial applications and location based services. Web-based geospatial systems are rapidly evolving, aiming to offer improved services and more advanced features. An important component of such systems is the ability to access and obtain the position of fixed and mobile geographic features and users.

The most commonly used form of positioning utilizes Global Positioning Systems (GPS). These are increasingly being embedded in applications such as recreational navigation devices, transportation navigation systems, mobile phones and Personal Data Assistants (PDAs). However, the practical limitations to GPS devices mean that they cannot be used indoors or under canopy cover and are also less effective in highly built-up areas due to limited line-of-sight (Kim et al. 2008). A further limitation is that web-based applications do not readily talk to GPS devices. The net effect of this is that web-based geospatial applications are struggling to obtain the information that they most need: spatial location. What are needed are additional positioning methods, such as WiFi and IP-locating, to augment existing positioning systems.

This chapter outlines the different methodologies typically used for geolocating in geospatial applications. It discusses how positioning methodologies may be adapted for web-based use and which new opportunities may exist specifically for applications on the Internet. The constraints of different methodologies are highlighted in the context of different uses, and potential remedies to these constraints are investigated.

2   LOCATION AWARENESS

Geospatial location is a necessary and important part of any geospatial application. The position of objects and processes in space can be used to identify context, associations, relationships, proximity and other characteristics related to geography and events occurring at a location on the earth's surface. While many locations are fixed over time, others will vary as features and objects may be mobile and change position over time. This is where the concept of positioning comes in, to ensure that locations are updated and retrieved at any required point in time. Applications that utilize mobile locations include car navigation systems, traffic monitoring and management, fleet and logistics management, asset capture and management and object tracking including that of persons, animals, features and dynamic processes.

2.1   *Locations and location based services*

Location refers to a specific place on the earth's surface. A place can refer to a position of some simple or complex object such as a road, town, user or service, at a particular location. A position can be represented by anything from a simple geometric coordinate to a complex structure represented by multiple coordinates and multidimensional in nature. Geospatial locations have coordinates that are based on known coordinate reference systems and can be used as a basis for building relationships with other features and phenomena with known positions (Kupper 2005).

Location Based Services (LBS) utilize the location of a user to improve the service(s) provided (Delfos et al. 2008, Kupper 2005). Figure 1 illustrates the fact that LBS involve multiple actors, including data producers such as data and location providers, as well as data consumers such as users and applications. Location providers may be cell phones, PDAs, wireless network adaptors, sensors, GPS receivers, etc. whereas data providers may be database servers providing location, location context and/or other data for geographic features, users or services. Users and applications use the location and value added data to view and integrate into applications such as navigational systems, sensor networks, geocoding services, etc. Specific

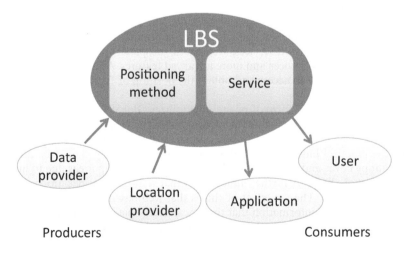

Figure 1.   Actors in a location based system.

actors may take on multiple roles that include both producing and consuming activities. For example, a person with a mobile device such as a cell phone may use their own current location to identify restaurants in their vicinity, and have the addresses and phone numbers of the restaurants displayed on their mobile device (Spaccapietra et al. 2005).

Producers as well as consumers may be fixed or mobile. Continuing with the same example, the person inquiring about a restaurant is a mobile location provider, sending the location of their mobile device at the current location. A database server may be the fixed data provider extracting the contact details and locations of restaurants in geographic proximity to the mobile location. The person making the enquiry then may also take on the role of the user, displaying the restaurant locations in a map view and obtaining menus and contact details for the restaurants.

2.2 *Positioning and context*

Positioning of a user device can be obtained either by the device itself which uses the signals it receives to determine its location, or remotely by measuring the signals to and from multiple receivers and using the geometric relationships to calculate the position (Zeimspekis et al. 2002, Bill et al. 2004). Global navigation satellite systems (GNSS) are an example of self positioning techniques whereas remote positioning techniques include wireless, RFID and sensor network positioning methods.

Manesis & Avouris (2005) discuss how the role of location can be used in applications to determine a user's geographic position and register their movements over space. User's location provides context information for the user and can be used to identify the relationship to features and other persons in geographic proximity. This relationship can be used in human-space interaction applications where users' motions and habits can be observed and monitored, for example, in libraries, classrooms, museums and other public spaces.

What these and similar applications reveal is that location is used in a context. The geographic locality as well as the user preferences, actions and intentions are part of a context in which location based services are provided. Location awareness is an important part of the context that is used in geospatial applications. As the awareness of location and its importance in a diversity of applications increases, so too will the need to provide fixed and real-time location (Kupper 2005). Recent developments have seen an integration of mobile positioning with the publication of messages and status updates by users, for example in twitter (Dorsey 2006) and Ushahidi (Ushahidi 2011). By allowing users to associate a location with information they provide, it becomes possible to create spatial datasets by means of public participation. This so-called "crowd-sourced" data can then be presented visually in context with other spatial information in a variety of systems. The popularity of social networking applications in recent years (eUniverse 2003, Zuckerberg 2004, Boyd & Ellison 2007) has created vast opportunities for integration with positioning methodologies, as the location and proximity of friends and contacts is a very practical piece of information. At present, there are no major social networking applications in existence that have mature support for providing locations for users.

3 METHODS FOR OBTAINING GEOSPATIAL LOCATION

Within an application, geospatial location needs to be captured, represented and accessed. A location is usually represented as a geometric coordinate within a database. However, end-users may not readily identify with a coordinate location, and hence other means of specifying location may be utilized. Examples include street addresses, textual descriptions describing features at that location or in relative proximity, directions relative to landmarks and other features, and vernacular terms used commonly or locally to describe locations (Adeva 2008). No matter how location may be referred to by users, it is generally represented within an application as a coordinate value.

Within an application, how is geospatial location to be obtained? Obviously, for fixed location features, their coordinates can be represented in a spatial database together with the feature definitions as points, lines, polygons or cells. For mobile features or devices, there needs to be some way of capturing their locations at particular times, over a period of time or in real time. The coordinate position or a series of positions obtained may represent the locations of features at one or more particular points in time, or the extent of a multidimensional feature. For example, the former may represent the tracking of a green turtle over a period of time whereas the latter may represent the extent of a road captured as a series of points. Various devices and methods may be used to capture such locations and these are considered in the following sections.

3.1 *Global Positioning Systems*

Global Positioning Systems (GPS) provide a means of obtaining coordinate locations on mobile receiving devices utilizing satellite signals in real-time. The broader term defining these systems is Global and Navigation Satellite Systems (GNSS) of which a number are in operation or being developed, including the United States' NAVSTAR Global Positioning System, Russia's GLONASS, European Union's Galileo, Indian Regional Navigational Satellite System, etc. However, the handheld devices readily available and used by the public are referred to as GPS devices, as they make use of the NAVSTAR GPS System (ARINC Engineering Services 2004) which is currently the only GNSS that is fully operational. Hence, we will use the more familiar GPS term for referring to GNSS.

GPS receivers can be embedded into handheld devices solely designed for the purpose of obtaining and recording the location of that device, or they can be embedded into broader purpose devices such as cell phones, in-car navigation systems, marine navigation systems, mobile mapping systems, etc. GPS or GPS-enabled devices are becoming more commonly available and provide an important means of obtaining spatial location in real-time. For example, GPS-enabled 3G cell phones are able to geolocate the cell phone and display its location within a mapping application on the phone device utilizing streamed image data from the Internet.

Once a position "fix" is obtained in real time, an identifier or descriptor can be associated with it to relate it to a feature or phenomenon at that location. This data can then be downloaded or transmitted to be integrated and used within some geospatial applications.

The limitation of positioning using GPS is that the device or user must be in line-of-sight of the satellites. Hence, only outdoor positions without overhead obstructions such as forest canopies or building structures can be determined (Kim et al. 2008). This constrains the ability to use GPS to obtain positions for indoor and obstructed outdoor locations. For such situations, the GPS system can be augmented by additional receivers of known locations together with mobile wireless devices to obtain location (Zeimspekis et al. 2002).

High precision GPS positioning down to centimeter resolution can be obtained using temporary base stations or permanent GPS Continuously Operating Reference Systems (CORS) networks that cover an entire nation (Gordini et al. 2006). The use of geolocations at such high precision is important for applications such as detecting movements in plate tectonics and deformation monitoring (Taylor & Blewitt 2006), and mobile uses such as automated machine guidance for earth moving equipment in road-building or mining, and precision farming. As many parts of the world receive sufficient coverage, the usage and applications of high precision positioning will continue to increase. The Internet provides an underlying infrastructure for obtaining, integrating and embedding such locations into applications.

3.2 *WiFi and cellular positioning*

The most common form of communication between mobile devices uses wireless network protocols (WiFi) and other radio frequency signals. Wireless devices can be positioned using one or more access points (also referred to as base stations) that are within range of the device and receive/transmit radio frequency signals from/to the device (Bahl & Padmanabhan 2000, Chen et al. 2007, Koppen et al. 2006, Lee & Chen 2008). The access points can be located such that the cells identifying their geographic vicinity of reach overlap with each other. These access points provide the interface from the wireless environment to the wired network. Triangulation algorithms use receiver signal strengths to triangulate from the known locations of base stations to determine the location of the mobile device (Cheng et al. 2005). Signal strength is used as an indication of distance to the access point since attenuation causes signal strength to decrease gradually with distance.

The quality of the position is variable and depends largely on how much interference the signal receives as a result of obstacles between the access point and the receiver, and on whether the locations used for the access points are accurate. Interference causes a signal to be weaker, which would make the distance to the access point appear larger than it really is, potentially biasing a triangulation in a particular direction. For WiFi positioning, locations that have been triangulated are found to be accurate to about 13–20 metres in built-up areas using access point locations determined by means of a GPS (Cheng et al. 2005, LaMarca et al. 2005). In the case where access point locations are exactly known, it is possible to achieve higher precisions of around 5 metres (Köppen et al. 2006).

These techniques can be utilized at wide area cellular telephone networks and receiving stations, or at a local-area sites using wireless receivers linked via a local area network to the Internet (Karl 2004, Lee & Chen 2008). The limitation of positioning within cellular networks is the poor precision obtained which ranges from 100 meters to many kilometers depending on the cell size involved.

3.3 *IP locating*

Every computer on the Internet can be uniquely identified via an Internet Protocol (IP) address. These addresses are assigned hierarchically and can typically be related back to geographic regions at different scales. Because of the relationship of the IP address scheme to geography, it can be used in linking each IP address to a geographic location. A number of IP lookup applications exist that are able to provide the country or even city for a particular IP address. An example is HostIP (Gornall 2005), a freely available IP locating database based on community-contributed data, and GeoIP (MaxMind 2010), a commercial application for determining locations and connection details of IP addresses. These applications are based on IP address look-ups that store locations for broad IP address ranges. The ranges are normally based on *classes* which are the naming convention for groups of IP addresses that have a size of 1 to 4 bytes where a 4-byte group constitutes an individual and complete IP-address. Most IP-lookup applications store locations at a level of 3 bytes, for example 192.168.0.*. This obviously limits the maximum granularity possible, which relates to geographic granularity and therefore limits the ability to identify a specific location.

An alternative approach to populating a look-up table was proposed by Padmanabhan & Subramanian (2001). Their GeoCluster method is based on assigning locations to clusters of computers if enough location consensus exists for the computers within that cluster. The clusters do not use *classes*, as is the case with normal look-up approaches, but are defined using bits rather than bytes. This method to define clusters, called Classless Inter-Domain Routing (CIDR), provides 32 levels of granularity to define a cluster, as opposed to only 4. The number of bits used to define a cluster is called the CIDR-value. When an IP-address in a cluster has a location that conflicts with the location of the cluster, the cluster is broken up into multiple smaller clusters with their own individual location. As these sub-clusters are smaller, their CIDR value is larger, because more bits are required to define this smaller, more precise cluster.

With mobile computers and devices, their position may of course vary over time. Wireless positioning techniques may need to be utilized in conjunction with IP locating methods linked to multiple wireless receiver locations to obtain a current position for a mobile user at a particular point in time.

3.4 *Geocoding*

The geocoding process involves obtaining a geospatial position from a textual address such as a street address or feature/building/site name. Information relating to localities, street names and address ranges are used to determine locations of given addresses or features. These locations are either calculated in the geocoding process or they are pre-determined and stored in an address reference file which can then be utilized in the geocoding process.

For example, a hungry user may have the name of a restaurant that they are trying to navigate to. The name of the restaurant, or alternatively its street address, can be geocoded to provide an actual location which can be displayed on a map display. The reverse process can obtain the name or other details of a feature given the location, and is referred to as reverse geocoding. For example, the user may query the closest feature from a given (current) location, and reverse geocode the location to identify the street address, name or other details of that feature.

Where the textual address information supplied is correct and complete, the positions can readily be determined. However, where there is missing or incomplete data, or the terms describing the address involve aliases or locally known features, the determination of a position may be difficult or impossible. Current research is being undertaken to investigate spatial reasoning and agent based systems to build intelligence into the geocoding process (Liu et al. 2008, Hutchinson & Veenendaal 2005). Agents that represent properties, streets, localities and postcodes, interact with each other and utilize known and contextual address and feature information to determine the best choice(s) for geocodes that represent the locations of the addresses/features.

Geocoding can also be utilized in the context of organizing and searching information and resources on the Internet. Geography can provide a means of organizing information. Angel et al. (2008) describe a methodology for automatically geocoding web pages based on textual addresses, phone numbers, place names and other geographic cues that may relate the page to a spatial location.

3.5 *RFID and sensors*

Radio frequency identification (RFID) technology is based on radio signals being exchanged between readers and tags (Kupper 2005). The tags can be attached or mounted to mobile objects such as persons, animals, assets or other features so that they can be monitored, recorded or tracked. Active tags have their own power source and generally have more intelligence and a greater communications range, but can be expensive to produce. Passive tags extract energy from the radio signals emitted by the readers in order to power the chip and broadcast a weak signal. Although they have poor communications range, production costs for these tags can be very low. An example of passive tags are the stickers used in shops to detect theft; they have a spiral-shaped coil that functions both as an antenna and as an instrument to turn an electro-magnetic field into electricity (Lee 1998).

Wireless geosensor networks comprise a large number of inexpensive, low-power communication devices that interact with each other and may be deployed over a geographic region to measure and monitor features and properties (Culler et al. 2004). Applications include, for example, habitat and environmental monitoring where properties such as temperature, rainfall, wind speed, air pollutants, etc. are captured and delivered to a central server (Duckham et al. 2005). Issues with sensor networks arise with low power supplies and poor communication links resulting in the need to minimize communication frequencies and volumes.

Skibniewski & Jang (2009) describe the design of a sensor network for managing and tracking materials on a construction site. More traditional methods based on GPS and RFID are limited because of inaccuracies in positioning, obstruction of buildings and multipath signal interference in urban construction environments and the high cost of RFID receivers. Skibniewski & Jang (2009) proposed a method that would combine radio frequency and ultrasound signals in sensors attached to construction assets to achieve enhanced accuracy performances in location determination.

RFID and its allied technologies have a key role to play in the healthcare industry. Besides using RFID to manage the supply chain in hospitals, there is an increasing trend to leverage RFID as a platform to deliver greater situation awareness of patients from the admission stage right down to post recovery (Cangialosi & Monaly 2007). The introduction of physiological monitoring sensor technologies (e.g. ECG, temperature, blood pressure, glucose level, etc.) onto RFID has opened a window of opportunity for them to be used to track and monitor patients with greater reliability. This development together with the ongoing research to improve location precisions and accuracies (Cheng et al. 2009) will see a much greater healthcare related deployment of RFID beyond the current 10% adoption rate (Cangialosi & Monaly 2007).

4 CONSTRAINTS AND ISSUES WITH CURRENT POSITIONING TECHNIQUES

Although a range of positioning technologies and techniques are available, they have different characteristics that vary in their capabilities and constraints. Further, even though they could be used to complement each other, limitations currently exist regarding their interoperability and integration.

One of the most significant differences between the positioning techniques is the precision at which a geospatial location can be obtained. Figure 2 illustrates the range from less than 1 meter to many kilometers. The cellular based approaches and current global IP lookup methods provide poor resolution results whereas the GPS, RFID and WiFi approaches can provide much better spatial precision to meters or even less than one meter, depending on the technique utilised. Geocoding techniques depend on the reference address database precisions adopted and relate to the primary sources for capturing point locations of features, buildings and properties.

Although providing better location precision, the GPS techniques are constrained by the ability to receive signals under canopies or interference. The RFID and WiFi techniques are limited in their range to receive signals and require a large number of receivers in close geographic proximity. It would be beneficial if the various techniques could be integrated so as to take advantage of the relative benefits of each. However, their interoperability and ability to integrate, particularly via the web and web-based applications, remains currently

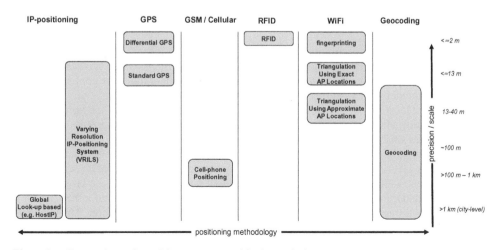

Figure 2. Comparison of precisions among positioning techniques.

constrained due to standards and security limitations. For example, web browsers have security mechanisms that are designed to prevent unauthorized access to computer resources and this makes it difficult to link external devices to provide a geospatial location value directly within a web browser.

The spatial resolution of geolocations using IP positioning techniques is dependent on how well IP addresses can be linked to a geographic location on the earth. IP addresses are associated with computer locations, which themselves are related to positions of users, features and other phenomena in proximity of the computers. Because IP addresses bear a relationship to locations of users, features and services, it is potentially possible to use IP addresses as a fairly good means of geolocating. The following section describes some research involving a technique to use the IP addresses and their structure along with the potential relationship to spatial location to perform geolocating.

5   VARYING RESOLUTION IP LOCATION SYSTEM

The research on positioning and systems that require positioning, although often focused on mobile users (Bill et al. 2004, Völkel & Weber 2005), is also relevant for static users who remain in the same location. In fact, although usage of mobile devices is growing, it is realistic to expect that in the short to medium term, the majority of web-users will continue to have a fixed location. As static users tend to be indoors, GPS-based methods are not available to them. Desktop computers typically also do not have wireless adapters, making WiFi positioning unavailable to them. There is therefore a need for a high-availability positioning method that accommodates this large group of users and this has motivated the development of the Varying Resolution IP Locating System (VRILS).

VRILS was designed to provide an IP-positioning capability for static web users that exceeds the spatial precision of look-up based IP-positioning methods, without requiring the location of every IP-address on a network as input. It is capable of doing so by closely relating the network hierarchy to the spatial hierarchy of devices at varying geographic scales. The IP address of a web user is the only information that can be counted on to exist, making VRILS a relatively dependable positioning methodology compared to methods dependent on particular hardware devices which indoor users may not have. Although VRILS aims to provide positioning for static users, it is also capable of providing locations for mobile users who use a static WiFi access point, as access points themselves have IP-addresses.

5.1   *Relating spatial and network hierarchy*

Computer networks rely on unique addresses that are used to identify devices. As networks become larger, pools of IP addresses are assigned to larger subnets, which typically coincide with administrative units of some sort, such as buildings, cities, states, or countries. If the hierarchy of the network is sufficiently well described, it becomes possible to make assumptions about the locations of individual devices within subnets, if the spatial location of at least one other device within that subnet is known. GeoCluster (Padmanabhan & Subramanian 2001) also uses this principle, but it differs from VRILS because it only assigns locations to the smallest subnets of its data structure, which are the leaf-nodes of the network at a relatively fine geographic scale. VRILS differs by assigning locations of larger scales to the smaller subnets, while also maintaining broader, small scale locations for the larger subnets they constitute. Doing so vastly increases the likelihood of being able to find a location for a particular IP-address, even if this location may not be of maximum precision. The principle behind VRILS is that determining a low-precision, small scale location is far more preferable than not being able to obtain a location at all.

What makes it possible to obtain geospatial locations from IP addresses is the relationship of geographic space to the structure of IP addresses. Figure 3 illustrates how the spatial

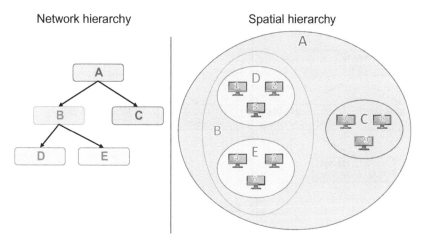

Figure 3. Spatial hierarchy and computer network hierarchy.

hierarchy of computer locations relates to the computer network hierarchy. This relationship is used to assign locations to each subnet in VRILS.

Figure 3 shows how 9 computers have been organized into different subnets. Subnets D and E are in different locations, but both are part of the larger subnet B, which in itself has a location that encompasses the locations of both subnets D and E. Subnet C is in a different location from subnet B. But both B and C are part of an even larger subnet A, which itself has a location assigned to it that contains both locations B and C. In this particular scenario, the locations for each of the subnets could have been determined by knowing the location of one computer in each subnet. If the locations for computers 2, 6 and 8 were provided, then the locations for every other computer could have been determined.

VRILS can be applied to any addressing scheme that can be expressed numerically, but for the purpose of consistency, all examples given here use the 32 bit TCP/IP addressing scheme on which most networks, including the Internet, are based.

5.2 *Adding new entries*

If the location of an IP address is inserted into the VRILS data structure, this location propagates to the subnet and all parent subnets that the IP-address is part of. Because VRILS uses bits to define subnet clusters, this means that if 32 bit addressing is used, each IP address has 31 parent subnets of increasing size. The size of a subnet equals $2^{(32-CIDR)}$, where the CIDR value is the precision in bits used to define the network. The highest precision subnet is specified by 32 bits, and has a size of 1 IP-address. The largest subnet is defined at a precision of 0 bits, and contains all IP-addresses. Unless there is information to the contrary, all parent subnets are assumed to be in the same location as that of the IP-address for which a location is supplied. Locations at different spatial resolutions are assigned as a result of the conflict resolution algorithm that VRILS uses, which lowers the spatial precision of parent clusters when sub-clusters are inserted that have a different location. How this happens is illustrated in Figure 4.

Figure 4 shows how the data structure changes when a new cluster is inserted. This example uses two levels of spatial resolution: country and city. In this case, an IP-address 160.0.12.200 is added which is located in Chicago, USA. This location is propagated to the parent clusters, such that eventually a new cluster is added below the existing cluster of 160.0.12.128-255, which can also be written as 160.0.12.200/25, as it has a 25 bit precision. The cluster 160.0.12.200/25 currently has a location of "USA, Los Angeles" which was propagated from its other sub-cluster, the sibling of the newly inserted cluster. Because the cluster

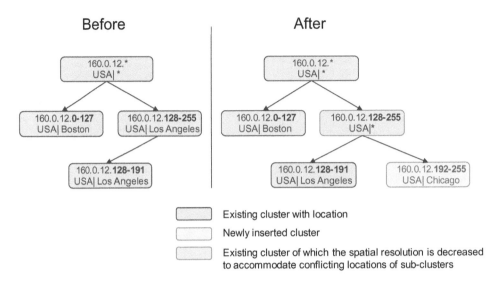

Figure 4.   Scenarios before and after inserting a new IP address into the VRILS data structure.

160.0.12.200/25 is now a parent to two subnets in different cities, but in the same country, its spatial resolution has to be reduced to country-level precision.

5.3 *Finding a location*

Geolocating a particular IP address is a matter of trying to find the subnet of highest spatial resolution that this IP address belongs to. Traversing the subnets sequentially is inefficient because each IP address is stored at 32 levels of precision, each representing a subnet that the IP address could potentially be part of. VRILS therefore uses a binary search algorithm to poll subnets at different CIDR values. If a subnet is not of maximum spatial resolution, it looks for a smaller subnet. If a subnet is not defined at a certain CIDR value, it looks for a larger subnet. The search process is illustrated in Figure 5. This example uses 4 levels of spatial resolution to store locations. The levels used are Country, State, Local Government Area (LGA), and City.

Figure 5 shows how a location is sought for IP address 184.65.20.88. The binary search algorithm starts the search in the middle, at a CIDR value of 16. At this level in the network hierarchy, the subnet it belongs to is 184.65.*.*, which has a location at Country level. Because this is not the maximum spatial resolution possible, the binary search continues at a CIDR value of 24. This yields a value at State level, which is more precise, but also not of maximum resolution. The next subnets visited are at CIDR levels 28 and 26, for which no entries exist. The last subnet searched is at a CIDR value of 25 bits, which is the smallest spatial resolution subnet that can be identified for this IP address. This subnet has a location at LGA level, which in this case is the county of Los Angeles.

5.4 *Evaluation of performance*

VRILS was implemented as part of a prototype LBS called Positioning Mapping Information and Communication System (POMICOS) to provide a positioning capability for non-mobile users (Delfos et al. 2008). POMICOS was implemented on a university campus and serves as a framework to test proposed enhancements to LBS systems, including modern positioning methods such as WiFi and IP positioning. This implementation of VRILS is web-based using the PHP programming language and linked in to a PostgreSQL database. It was configured to use 4 levels of spatial resolution, namely: campus, building, floor-level,

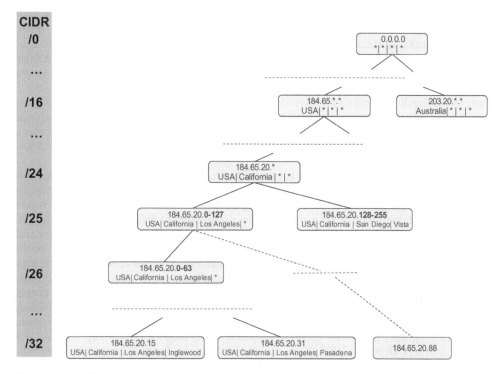

Figure 5.   Graphical representation of the search process for geolocating an IP address.

and room. Volunteers across different campuses were requested by email to visit a website where they could provide their location, which was then stored along with their IP address. In total, 461 positions of IP addresses were collected representing 2.3% of the total size of the university network. These addresses were spread across 42 buildings on 7 campuses.

An analysis was performed where VRILS was populated with increasing amounts of data, to test how performance related to available input data. For the largest quantity of input data, VRILS was populated with 431 entries, where the remaining 30 were used to test accuracy. The result was that at campus-scale (the lowest spatial resolution) 100% of the addresses were correctly positioned. At building-scale (medium resolution) an accuracy of 70% was obtained. At more precise resolutions accuracies were significantly poorer. Using a very small amount of base data (50–100 records) caused VRILS to return high precision locations with reduced accuracy. By populating VRILS with additional data, the precision of the returned locations would initially drop due to the positional conflicts caused by the newly introduced subnets, but accuracy increased with fewer incorrect locations being returned at low spatial resolutions. From 250 entries and beyond the accuracy of higher spatial resolutions would increase again as a result of the overall data structure becoming more detailed.

In general, it could be seen that accuracies were consistently worse for some buildings than for others. This was the case where multiple buildings were managed by the same IT group, and shared a common pool of IP addresses. As VRILS relies on spatial clustering of network clusters, performance degrades significantly where spatial clustering is limited. A more detailed description of the performance analysis can be found in Delfos et al. (2008).

6   CONCLUSIONS AND FUTURE DIRECTIONS

The lack of web-based systems that are able to determine a user's location show that there is much opportunity for research in the area of positioning. For web-based solutions, positioning is particularly difficult as most positioning methods rely on interfacing with additional hardware or devices of some sort. Each of the techniques involving GPS, WiFi, cellular networks, IP locating, geocoding and RFID, has particular characteristics and constraints that reflect their capabilities. These characteristics and constraints relate to their ability to receive signals because of geographic conditions and distances, and relate to the precision of the location that can be obtained. Integrating with web-based applications is currently difficult to implement due to browser security mechanisms that were designed to prevent unauthorised access to computer resources. However, advanced client-side applications using modern technologies may form a solution to this problem, as they permit user's to be prompted to allow access to these devices, thereby forming the link between location and the web.

The IP address positioning method presented here offers increased potential over the simple look-up based methods traditionally used (Gornall 2005). VRILS is a universally applicable methodology for any numeric addressing scheme that provides locations by closely matching spatial hierarchy to network hierarchy using different levels of spatial resolution. As VRILS does not require any information other than an IP address, it is relatively easy and cheap to implement, and has a high reliability for being able to provide a location. However, performance of VRILS is dependent on a network exhibiting spatial clustering, which may not always be the case. VRILS requires a relatively small amount of data, as it tries to position clusters of IP-addresses, rather than individual addresses. However, if insufficient data is used to populate the data structure, accuracy and spatial resolution will be less than optimal, and provided locations should be treated with caution. VRILS is highly suitable for static users, who do not normally have access to devices typically used for positioning, such as GPS or WiFi adapters. However, VRILS can also provide locations for mobile users, by determining the location of the WiFi access point that mobile users make use of.

REFERENCES

Adeva, J.J.G. (2008) Translating Vernacular Terms into Geographical Locations. In: John, Sample, T. Kevin Shaw, Shengru Tu, & Mahdi Abdelguerfi, (eds.), *Geospatial Services and Applications for the Internet:* Science+Business Media, New York, Springer. pp. 139–157,

Angel, A., Lontou, C., Pfoser, D. & Efentakis, A. (2008) Qualitative geocoding of persistent web pages. *Proc. the 16th ACM SIGSPATIAL international conference on Advances in geographic information systems*, 5–7th November 2008, Irvine, New York, ACM.

ARINC Engineering Services. (2004) *Navstar GPS Space Segment/Navigation User Interfaces*, IS-GPS-200. Navstar GPS Joint Program Office (SMC/GP), El Segundo, USA.

Bahl, P. & Padmanabhan, V. (2000) RADAR: an in-building RF-based user location and tracking system. *Proc. the 19th Annual Joint Conference of the IEEE Computer and Communications Societies, 26–30th March 2000, Tel Aviv.*

Bill, R., Cap, C., Kofahl, M. & Mundt, T. (2004) Indoor and outdoor positioning in mobile environments – a review and some investigations on WLAN-positioning. *Geographic Information Sciences*, 10 (2), 91–8.

Boyd, D.M. & Ellison, N.B. (2007) Social network sites: Definition, history, and scholarship. *Journal of Computer-Mediated Communication, 13* (1), article 11.

Cangialosi, A., Monaly, J.E. & Yang, S.C. (2007) *Leveraging RFID in hospitals: patient life cycle and mobility perspectives. IEEE Communications Magazine*, 45 (9), 18–23.

Chen, A., Harko, C., Lambert, D. & Whiting, P. (2007) An algorithm for fast, model-free tracking indoors. *ACM SIGMOBILE Mobile Computing and Communications Review*, 11 (3), 48–58.

Cheng, C.S., Chang, H.H., Chen, Y.T., Lin, T.H., Chen, P.C., Huang, C.M., Yuan, H.S. & Chu, W.C. (2009) Accurate location tracking based on active RFID for Health and Safety Monitoring. *Proc. 3rd International Conference on Bioinformatics and Biomedical Engineering, 11–13th June 2009, Beijing.*

Cheng, Y.C., Chawathe, Y., LaMarca, A. & Krumm, J. (2005) Accuracy Characterization for Metropolitan-scale Wi-Fi Localization. *Proc. 3rd International Conference on Mobile Systems, Applications and Services, 5th June 2005, Seattle*, pp. 233–245.

Culler, D., Estrin, D. & Srivastava, M. (2004) Overview of sensor networks. *IEEE Computer*, 37 (8), 41–49.

Delfos, J., Tan, T. & Veenendaal, B. (2008) Scale-dependency in IP-based positioning of network clusters. *Journal of Location Based Services*, 2 (1). 3–23.

Dorsey, J. (2006) Twitter. Twitter, Inc, San Francisco, CA.

Duckham, M., Nittel, S. & Worboys, M. (2005) Monitoring dynamic spatial fields using responsive geosensor networks. *Proc. the 13th annual ACM international workshop on Geographic information systems, 31 st October–November 05th 2005, Bremen*. pp 51–60.

eUniverse. (2003) MySpace. MySpace Inc., Beverley Hills, CA.

Gordini, C., Kealy, A.N., Grgich, P.M., Hale M.J. & Gordini, C. (2006) Testing and Evaluation of a GPS CORS Network for Real Time Centimetric Positioning – The Victoria GPSnet™. *Proc. the IGNSS2006 Symposium*, Gold Coast, Australia.

Gornall, S. (2005) HostIP. "hostip.info".

Hutchinson, M. & Veenendaal, B. (2005) Towards a Framework for Intelligent Geocoding. *Proc. Spatial Sciences Conference 2005, Spatial Intelligence, Innovation and Praxis: The national biennial Conference of the Spatial Sciences Institute, 12–16th September 2005, Melbourne*, Spatial Sciences Institute,.

Karl, H. (2004) Data Transmission in Mobile Communication Systems. In: Jochen Schiller, & Agnes Voisard, (ed.), *Location-based services*, San Francisco, Elsevier.

Kim, K., Summet, J., Starner, T., Ashbrook, D., Kapade, M. & Essa, I. (2008) Localization and 3D Reconstruction of Urban Scenes Using GPS, *2008 IEEE International Conference on Wearable Computers. Pittsburgh, United States*, IEEE. pp. 11–14.

LaMarca, A., Chawathe, Y., Consolvo, S., Hightower, J., Smith, I., Scott, J., Sohn, T., Howard, J., Hughes, J., Potter, F., Tabert, J., Powledge, P., Borriello, G. & Schilit, B. (2005) Place Lab: Device Positioning Using Radio Beacons in the Wild. *3rd International Conference on Pervasive Computing, Munich, Germany*, pp. 116–133.

Lee, D.L. & Chen, Q. (2007) A model-based WiFi localization method. *Proc. the 8th ACM SIGCOMM conference on Internet measurement, New York*, ACM. pp. 111–124.

Lee, Y. (1998) RFID Coil Design. Microchip Technology Inc., [Online] Available from: http://ww1.microchip.com/downloads/en/AppNotes/00678b.pdf. [Accessed 21 st June 2009].

Liu, Y., Guo, Q.H., Wieczorek, J. & Goodchild, M.F. (2008) Positioning localities based on spatial assertions. *International Journal of Geographical Information Science*, 23 (11), 1471–1501.

Koppen, B., Bunningen, A.V. & Muthukrishnan, K. (2006) Wireless Campus LBS: Building campus-wide location based services based on WiFi technology. In: Stefanakis, E. Peterson, M.P. Armenakis, C. & Delis, V. (eds.), *International Workshop on Geographic Hypermedia*, Berlin: Springer-Verlag.

Kupper, A. (2005) *Location-based Services: fundamentals and operation*. Chicester, John Wiley & Sons.

Manesis, T. & Avouris, N. (2005) Survey of position location techniques in mobile systems. *Proc. the 7th international conference on Human computer interaction with mobile devices & services, New York*, ACM. p. 111.

MaxMind. (2010) Geolocation and Online Fraud Prevention from MaxMind. MaxMind, [Online] Available from: http://www.maxmind.com

Padmanabhan, V.N. & Subramanian, L. (2001) *An Investigation of Geographic Mapping Techniques for Internet Hosts, Applications, technologies, architectures, and protocols for computer communications., San Diego, California, United States*, ACM Press. pp. 173–185.

Ratcliffe, J.H. (2001) On the accuracy of TIGER-type geocoded address data in relation to cadastral and census areal units. *International Journal of Geographical Information Science*, 15 (5), 473–485.

Spaccapietra, S., Al-Jadir, L. & Yu, S. (2005) Somebody, Sometime, Somewhere, Something, *International Workshop on Ubiquitous Data Management*, Tokyo, Japan, pp. 6–16.

Skibniewski, M.J. & Jang, W.S. (2009) Simulation of accuracy performance for wireless sensor-based construction asset tracking. *Computer-Aided Civil and Infrastructure Engineering*, 24, 335–345.

Taylor, G. & Blewitt, G. (2006) *Intelligent Positioning: GIS-GPS unification*. Chichester, John Wiley & Sons.

Völkel, T. & Weber, G. (2005) Location-based and Personalized Information Services for Spas. *Interact 2005*, Rome, Italy.

Zeimpekis, V., Giaglis, G.M. & Lekakos, G. (2002) A taxonomy of indoor and outdoor positioning techniques for mobile location services. *ACM SIGecom Exchanges*, 3 (4) 19–27.

Zuckerberg, M. (2004) Facebook. Facebook Inc., Palo Alto, CA.

*Advances in Web-based GIS, Mapping Services
and Applications – Li, Dragićević & Veenendaal (eds)*
© 2011 Taylor & Francis Group, London, ISBN 978-0-415-80483-7

The mobile web: Lessons from mobile augmented reality

Sylvie Daniel
Department of Geomatics Science, Laval University, Quebec City, Quebec, Canada

Robin M. Harrap
Department of Geological Sciences, Queen's University, Kingston, Ontario, Canada

ABSTRACT: The last few years have seen an explosion of mobile Web services and GIS tools with the advent of smart phones which offer fast data access, high resolution displays, and significant processing power. Within a few years, mobile devices will know where they are, what direction they are facing, and how they are moving. Those designing Web resources and sites will have to address issues related to the user's mobility, role in consuming or creating information, and focus of attention. Mobile Augmented Reality (MAR) system design studies have much to offer to developers of such Web-based GIS services and applications. Researchers in MAR have, for many years, examined how users can interact with local information and especially augment their understanding of their environment using graphics. Herein we review MAR and the mobile Web, discuss current examples of augmented interaction and discuss the future of augmented reality and the ambient Web.

Keywords: Urban, visualization, augmented reality, mobile, education, GIS, web based

1   INTRODUCTION

At the time of writing, the World Wide Web (Berners-Lee 2000) is about 15 years old, maps on the Web are about 12 years old, and true Web-GIS applications that expose significant query, display, and analytical functionality are less than a decade old. The Web-on-desktop context is so ubiquitous that those in society who do *not* have easy access to the Web are significantly challenged when doing such simple tasks as planning a trip, registering for city services, and paying bills. The shift from physical, mail-based, and phone-based services has been driven by convenience and rich discovery opportunities for consumers and by cost-saving and sales and marketing opportunities for organizations. Entirely new types of service-oriented organizations—such as Google—have appeared and traditional publishing of short-term materials such as newspapers have declined dramatically.

A similar if not more profound transformation has happened in the telecommunications arena: 15 years ago mobile phones were somewhat rare; now they are ubiquitous and many younger people don't have a traditional 'land-line' phone at all. They associate their identity and accessibility with their cell phones, which have become fashion accessories, status symbols, and core survival tools.

In the last few years we have seen the merging of these two 'revolutions'—mobile Web services are exploding with the advent of the newest generation of mobile phones which offer fast data access, generous displays, and significant processing power. With this merging we have seen the arrival of mobile Web and mobile Application map delivery and simple mobile GIS tools. Many people now turn to phone-based Web services to answer questions that others only recently began answering via desktop computers and the Web!

Mobile Web GIS is more than the sum of Web GIS and a mobile device. Our two premises, which we explore in this chapter, are that the ideas that have been explored in the field of

Mobile Augmented Reality (MAR) over the last decades have much to tell us about how the mobile Web may, and perhaps should, evolve, and secondly, that beyond the nascent mobile Web we should expect to see an 'ambient' mobile Web of semi-autonomous mobile services that surround users as they navigate the world and provide a persistent link to information resources and services, driven by geospatial methods and technology, and ultimately putting those users in charge of content creation.

1.1 The desktop context

The origins of the Web can be traced back to a seminal paper on information sharing by Vennavar Bush (Bush 1945). In 'As We May Think,' Bush convincingly argued that future technology would allow seamless access to large amounts of scholarly information, and support rich community annotations to provide an index and emergent commentary on and between documents. The World Wide Web, pioneered by Berners-Lee (Berners-Lee 2000) in the early 1990's, took these ideas and emphasized simple publishing of scientific documents using open standards and software. The explosion of Web content and the subsequent race to extend Web markup languages and server capabilities to support publishing of rich media, interactive Web-based applications, and search over vast collections of semi-structured content dominated the next ten years. The fundamental context for this period was of desktop computers connecting to the Web through dial-up or, increasingly, faster land-lines. Laptops were largely seen as desktop computers that happened to be portable; there was no particular focus on designing tools for laptop users although many content providers and service providers recognized 'traveling' users as a market niche. The desktop context is still, as of 2010, the dominant context for Web development.

Web map delivery and Web GIS emerged during this desktop phase; the XEROX map server and subsequent GIS tools such as MapGuide and ArcIMS, all served up maps or data to users who were assumed to be on reasonably fast connections, using fast computers with large displays. These applications were largely server-side; the analytical tools and in most cases even the interactive interface components, were handled by scripts on a Web server, although with the emergence of Java applets some GIS tools made the transition to being client-side.

The realization that the emerging Web information space was dominated by content poor in structure led to proposals for mark-up and information science techniques such as the use of ontologies to build a 'Semantic Web' (Berners-Lee et al. 2001). The core idea of the Semantic Web is to use specialized markup languages to add context to documents, although at the control of the information author and so subject to their intentions. This contrasts with efforts by organizations such as Google to brute-force mine the existing Web content to generate indices and so support discovery of resources by Web-search. The use of such tools as Google Search has become so common place that we now use 'to Google' as a verb in casual conversation and find it hard to envisage the Web as anything more than 'what we can access through Google.' Google, of course, responded to GIS by acquiring simple Web mapping technology—now Google Maps—and sophisticated 3D visualization technology—now Google Earth—and integrating these to some degree with their search tools. The depth of this integration is quite limited at the moment: Google has concentrated on adding more content types to the mix, for example with their StreetView camera systems, rather than exploring the semantics of space in any detail to support contextual search using local knowledge. We are only just beginning to see search options where the user's location is sensed, and where the local environment is mined for search context; the Google Goggles application is an example and is discussed later.

The last significant event in the evolution of the 'Desktop Context' has been the emergence of the social Web. Sites that support user-generated content, and especially user-generated linkages and markup, have emerged in the last decade and have led to a dramatic shift in content creation from organizations to individuals, and of focus from traditional media content to a mix of personal scrap booking, journaling, and free-form discussions. Such sites as MySpace, FaceBook, and Twitter dominate the mind-share of many, especially younger, Web users. Although much of the content generated by these users is ephemeral and of little interest or use to those outside the immediate social group of the creator, indirect uses are and will continue to emerge from mining of the general character of such personal content. Sites such as Facebook, Flickr and Wikipedia show just how powerful user communities can be in creating rich, widely useable content. The tendency of many users to publish anything and everything about their lives has led to a fundamental shift in the perception and legal aspects of privacy.

Part of the shift to the social Web has been driven by the emergence of Web 2.0 methods (O'Reilly 2005) where multiple web content providers are accessed by a 'Web Page' to generate a desired mixed display, traditionally referred to as a mash-up. The reason we use the name 'Web Page' carefully is that such pages provide client side interactivity with asynchronous updates based on an approach called Ajax (Garrett 2005) and so break the traditional 'click and load' metaphor of earlier websites; there is in fact no 'page' until the interaction happens, and the result does not exist as a distinct document, even an ephemeral one like a Google Search result page. Still rooted in the desktop context, Web 2.0 mashups extend user generated content to user generated interactivity. Early sites such as Chicago Crime Maps (EveryBlock 2010) provided highly interactive, local context maps that 'mashed' crime statistics with map data.

The current Web context is thus highly complex: we still see 'Web Pages' of fixed content, server-side applications, client-side applets, and Web 2.0 social media as relevant and important for the future of Web GIS. However, these are all deeply rooted in the desktop, which is no longer the only relevant context and may soon not even be the dominant one.

1.2 *The mobile context*

The 'mobile context' is any situation where Web resources are accessed at any level by a device which is fundamentally mobile. As noted above, the emergence of cell-phones has shifted telecommunications away from land-line dominance to mobile-user dominance in the last 15 years. Many of these devices are capable of interacting with the Web in one or more of four ways:

1. Many mobile devices can display Web pages. These may be the same pages as a desktop user would see, but many content providers have alternative pages designed for fast-download onto mobile devices and with content appropriately scaled for small displays (e.g. any Web browser on a cell).
2. Many mobile devices can locate themselves through integrated cell-tower triangulation or embedded GPS and so can interact with Web sites or Web mapping applications to focus mapping, guide discovery, and so on (e.g. mobile Google Maps).
3. Newer mobile devices are capable of hosting applications—they are effectively low-performance computers—which may use Internet or Web-based resources, location, etc. (e.g. iPhone apps).
4. Newer devices may be capable of pushing their location and other content to Web applications, allowing other users to see 'where you are' and 'what you are doing' to some degree (e.g. Google Latitude).

There are obvious privacy and security issues with these forms of information exchange, especially with those that might be automatic, such as location posting. For example, such functionality could be used in unwarranted surveillance or stalking, or to aid in theft or robbery.

The devices used are typically low-power-consumption, low-performance handheld phones with quite small display screens. Most have built-in low-resolution cameras. The newer models frequently have GPS accurate to a few meters, although in the complex urban or interior context these may not function reliably if at all. The newest models have accelerometers capable of rough orientation (tilt) estimation as well as bearing orientation (which way the user is facing). There is good reason to believe that within a few years mobile devices will know exactly where they are, exactly where they are facing, and how they are moving precisely and furthermore will likely interact with other carried items to create an ecology of information devices. This is driven to a large degree by the desire to have oriented and tagged photo capabilities from the camera built into essentially all new phones and the realization that such sensors would enable a broad range of new interaction styles. Note, for example, the highly vibrant photo geotagging community, as well as the increasing integration of geotagged photos with Google Maps and Google Earth. In addition, these geotagged photos can serve location based search engines or 3D modeling applications which require user community input to better populate their databases. The MARA project from Nokia (MARA 2010) and Los Ojos Del Mundo project (Los Ojos Del Mundo 2010) are two examples of such use of geotagged photos. The innovative, though not mobile, Wii video game system shows just how rich interaction can be with a device that can sense orientation and acceleration, and mobile phone developers are rushing to emulate the Wii remote by releasing similar motion based controls in their applications.

Those designing Web resources and sites are now faced with a dilemma: when developing content does one focus on the desktop context or include the mobile context in a significant way? When developing tools to operate specifically on mobile devices, what are the relevant issues? For example, all of the following might be significant:

1. Where is the user, and what are they near?
2. Is the user moving or not, and if so, how fast?
3. Is the user taking part in a scripted activity or accessing resources or ...?
4. Is the user looking for something local, and if so, can we sense this focus?
5. What can the user be expected to see from their current vantage point?
6. Is the user 'searching' for some specific thing to acquire, and if so, can we suggest advertised sources?
7. Is the user a 'consumer' or a 'producer' of information—are they trying to add some local information to a social or traditional Web collection?
8. Does the user want a simple answer as a response? A map? A photo? Spoken words? Vibration of the device? Music providing spatial cues?

Many other questions could be formulated. Many more questions are not obvious outgrowths of existing tools and techniques and will only emerge as new resources become available. Just as it would have been virtually impossible to predict the current social Web landscape 15 years ago, it is very difficult to imagine what new opportunities will arise from innovative combination of the capabilities of mobile devices, individually or collectively.

These are not new questions and concerns. For many years, the field of MAR has examined how users can interact with local information and especially augment their understanding of their local environment using graphics. We thus review mobile web and augmented reality applications and approaches in some detail below. Second, we examine whether the emergence of mobile and especially mobile-autonomous services means that beyond the emerging mobile web we may see an ambient web. We then examine how users might create content on their mobile devices.

2 THE MOBILE SPATIAL WEB—METHODS, EXPERIMENTS, AND LESSONS

As noted above, the Mobile Web is an emerging environment for interaction between the Web and mobile devices with an emphasis on cellular telephones such as 'smart phones.' The mobile spatial web refers to mobile web applications with a specific spatial components or focus. This focus might include using spatial data 'behind the scenes' or explicit presentation of spatial data in the form of maps or visualizations. Finally, either of these two approaches may be 'hosted' in a Web browser or in an application as noted above. For example, an application hosted on an iPhone might be executing on the iPhone, with no web browser evident, but may be accessing Web Services to access spatial data. Alternatively, an iPhone may host an application that exists within a Web page. We will detail both of these approaches below and provide lessons learned.

Given the shift to the social Web, we also have to ask whether mobile spatial data is largely *used* on a mobile device or also *created* on that device, and whether the spatial context is really relevant when creating content. A user might have a smart phone and routinely use spatial displays such as maps, but not create new content. Another user might routinely create content for a blog or other shared site that is not explicitly spatial. Posting photos online, for example, is implicitly spatial since photos are of spaces and places, but those photos might not have an identified location. Finally, a user might intentionally engage in creating spatial content, whether as geotagged photos, content for Web mapping or visualization sites, or some new form of media.

Much of the landscape of possible spatial interactions shown in Table 1 has yet to be explored in the marketplace in any significant detail. Recent Google applications illustrate some of the possibilities as described below.

On the Apple iPhone a dedicated application provides access to Google Maps. The application includes a button that zooms to the current geographic position of the user determined by cell triangulation or GPS. The user is thus accessing explicit spatial support services—a map, possibly with resources identified. The application shifts point of view as the user moves through space. Visiting the desktop-oriented Google Maps page with a phone browser provides a significantly different experience because the Web browser cannot access the location of the phone directly, and so the user must manually navigate to their current location. The tight linkage between phone hardware and software and the specific application—mobile Google Maps—results in a much richer user experience than the individual components would, in disconnected form.

The Google Latitude service allows a mobile user to publish their location on the Web; location of one or more acquaintances can then be visualized by a mobile or desktop user as point locations on the Google Maps base. This is an example of pushing spatial presence to the emerging spatial Web. Recently social experiments have been conducted where many users agree to publish their Latitude location and, as a group, spell words on the landscape, an example of spatial meta-presence.

One distinction made is between the implicit and explicit use of spatial data by services. We note that an opinion can be expressed overtly—by writing a comment, augmenting or modifying the augmentation of an environment—or implicitly, by simply being there. For example, the track of where thousands of users go for coffee is an implicit comment on the attractiveness of different venues in an area, though a variety of social factors may contribute to this implicit 'rating.' Noticing the convergence of many users of Google

Table 1. Accessing and creating implicit and explicit spatial content on portable devices.

	Accessing	Creating
Implicit spatial support services	Phone uses location to find nearby resources	Push implicit opinions about locale to services
Explicit spatial support services	Phone shows a map and local resources	Push spatial content: locations, 3d forms, opinions
Media components	See local media such as photographs, video, …	Push media created on phone to Web
Narrative content	Read composite media on Web page or via a specialized viewer	Push composite media, such as stories with photos
Spatial presence	Discover nearby friends	Push location and status
Spatial meta-presence	Discover emergent phenomena such as groupings of friends	Publish emergent phenomena as content or rules of discovery

Latitude at a particular park on a particular day of the week is an implicit judgment of suitability for recreation, or perhaps another social phenomenon. Other users can use such implicit information, and systems can mine it for content to create new, explicit, spatial information.

Going through a full typology of spatial content access and creation is not our intent. We merely mean to show that new forms of access and creation, quite distinct from those seen on the desktop, are emerging. We next examine the field of Augmented Reality in more detail, and then turn to examples of content creation and access we have built that are informed by the augmented reality point of view, and that point towards the emerging ambient Web.

3 MOBILE AUGMENTED REALITY—METHODS AND LESSONS

Augmented Reality (AR) overlays computer-mediated information on the real world in real time. This ability enriches environments for action and learning and offers the potential for new kinds of shared experiences. Unlike Virtual Reality (VR), where the user is completely immersed in a virtual environment, AR allows the user to interact with the virtual images using real objects in a seamless way (Zhou et al. 2008). According to Azuma's definition of AR (Azuma, 1997), augmented reality systems must fulfill the following three characteristics:

1. combine real and virtual objects in a real environment;
2. run interactively, and in real time;
3. register (aligns) real and virtual objects with each other.

The first AR interface was developed by Sutherland in the 1960's (Sutherland 1965). This first system involved head mounted display and movement sensor. The real development of AR started in the 90's with Bajura (1992) and State et al. (1996) work as new interaction and visualization capabilities in the field of medicine. AR applications usually relate to various research areas ranging from computer vision, computer graphics, and human-computer interaction that operate in conjunction with the aim of presenting an enhanced reality as well as allowing the user(s) to interact with it in a natural way (Liarokapis 2006).

One common paradigm for AR is the *magic lens* allowing the user to *see-through* to an image of the real world with added AR elements (Cawood 2008). Optical see-through augmentation is based on semi-transparent Head-Mounted Displays (HMD), superimposing the real environment using semi-transparent mirrors while video see-through displays show a captured video image superimposed with the virtual content. Recently, handheld devices such as tablet PCs, ultra mobile PCs, and mobile handheld devices such as PDAs and smart phones have become popular platforms for AR applications. These systems are less bulky than the head mounted displays usually worn for see-through augmentation. Handheld devices are also more widely spread outside the research community today than the HMDs fostering a better integration of AR in various applications fields (ex. tourism, automotive industry, games) and their adoption by the user community. Similarly to video see-through HMD, visual extension with handheld devices is typically done using a video camera. It provides the handheld display with a live video stream of the real world that can be augmented with synthetic graphics.

Figure 1 shows an illustration of what an environment augmented with a handheld platform would look like. In this example, the user is holding the handheld device in the direction of his choice. He can see on the handheld display a view of his real environment in that direction. The viewpoint is similar to what he would see if he looked in that direction with his own eyes. Context-specific information is then superimposed on this view. In this example, text providing information about the user current location (i.e. Central Park) is added on the display. Information about buildings in his field of view could be displayed as well.

As described above, AR overlays graphics and text on the user's view of his or her surroundings. To achieve this, AR systems track the position and orientation of the user's head

Figure 1. Example of MAR where geometric features and annotations are superimposed on live video feed (Phonemag 2010). (photo courtesy of Mac Funamizu http://petitinvention.wordpress.com/)

so that the overlaid material can be aligned with the user's view of the world. Through this process, known as registration, graphics software can place a three-dimensional image in the real world and keep the virtual image fixed in that position as the user moves around.

Vision is commonly used for tracking the user position and orientation. Unlike other active and passive technologies, vision methods can estimate a camera pose directly from the same imagery the user observes. The pose estimate often relates to the object(s) of interest, not a sensor or emitter attached to the environment. Registration may require appropriate additional technologies for tracking such information when a higher level of accuracy is necessary. Orientation information may be provided by inertial trackers while the user's current location can be provided by Global Positioning System (GPS) receivers.

AR systems generally require some indication of where exactly the virtual graphics should supplement the real world. This is most commonly accomplished with AR markers (Cawood 2008). Marker-less augmentation (Comport 2003, Genc 2002, Stricker 2001, Vacchetti 2003) would be the preferred method. With such an approach, you would not need to add elements in the scene to be able to augment it—you could simply rely on recognition of naturally occurring scene features. This approach is computationally expensive and requires some information about the environment; it is not yet advanced to the point where it is possible to provide a simple way to the user community to use AR without markers. As a result, artificial markers are used to determine position and orientation of the camera within the environment (Kato 1999, Naimark 2002). The simplest form of marker is a unique pattern, usually geometric symbols in a square, that is visible to the AR camera and is physically added to the real world. The AR system software recognizes the square markers, finds the pose and then sets the model view matrix so that subsequent rendering operations appear relative to the array and, therefore to the real 3D world (Cawood 2008).

To create the illusion that the virtual graphics is part of the real world, the shape of the surfaces of real objects in the physical environment needs to be known in addition to the position and orientation of the user. The virtual objects have to "behave" as real objects, i.e. when occluded by a real object, the corresponding parts have to be invisible to the observer. To achieve this effect, the rendering engine must know the exact position

191

of the mobile unit, and the position and the shape of the occluding real object. Therefore accurate 3d models of the real environment are generally required by AR systems. In addition, the 3d model can be used for the accurate positioning of the user in the real world; in principle, this is achieved by a matching procedure between features extracted from the video images (obtained from the mobile video camera generally embedded in the mobile devices) and features retrieved from the 3d model. Such 3d modeling of the real environment is especially needed by AR systems designed for outdoor use. Note that the 3d-like models of environments found in Google Earth and similar applications are too crude for these types of applications.

The main technologies used to build such 3d models are airborne, terrestrial or mobile (i.e. vehicle-based) photogrammetry, surveying approaches relying on total stations and GPS devices, and airborne, static terrestrial or mobile terrestrial lidar scanning. Even though 3d representations of urban environments are becoming more and more available, further developments are still needed to allow "anytime, anywhere" augmentation.

3.1   *Distinction between AR and mobile web GIS*

The main characteristics of mobile AR applications are mobility, awareness, context and personalization. The same characteristics can be attributed to mobile web apps. However, researchers in the two fields are not exploring these aspects from the same perspective and to the same extent.

In mobile web apps that include spatial content, representations of the world are generally provided by, and limited to, maps or photos. Therefore, the user deals with a partial, incomplete and abstract view of their environment. Furthermore, the user's attention is mainly focused on the device screen rather than on their surroundings, limiting social or physical interactions with this environment or other members of the user community. Collaboration in mobile web apps is strongly related to user location and social networking as discussed. The collaboration might be dialog-centered by messaging (e.g. chat), involve the exchange of media files (e.g. photos, videos, sound files), or indirect exchange via available upload/download functionalities on the web; localization to a specific environment is not guaranteed. For example, does local or even regional context affect a user's ability to exchange a file with another user? Further, in mobile web apps, users are generally restricted to using keyboards, touch screen, and stylii, constraining the possible styles of interaction. Context-awareness in mobile web apps, where present at all, consists in mapping who and what is in the user vicinity and what information would be of interest to the user according to his profile, his preferences, his location and his social network, and to date developers have not done a compelling job of showing why a user should use such awareness.

Mobile AR overcomes spatial, temporal, and social boundaries of conventional location-based applications by making the real environment intrinsic: the whole point is to interact *with and through the environment* rather than with an abstracted device-based representation of it: the device merely provides the 'window and wand' for interaction. Mobile AR apps thus represent distinctive location-aware applications—that is, they use AR technology to enhance or modify the user's real environment with virtual content. Users can interact with the world in a more meaningful and tangible way by adding annotation, graphic elements (e.g. buildings) or characters to their view of the real world. For instance, to initiate search queries about objects in visual proximity to the user, a user can point with the mobile device in a specific direction indicating their interest and browse through information available at that particular location. Once the system recognizes the user's target, either through an image-based recognition approach or knowing the user location and bearing and retrieving the corresponding item from a database, it can augment the view finder with graphics and hyper-links that provide further information, for example, the menu or customer ratings of a restaurant.

Such an approach provides a more natural and intuitive way to interact with the local environment. Interaction with features such as buildings and plants becomes intrinsically part of normal use of the application. Furthermore, instead of being represented by an icon or an avatar, as in a game, users are their own avatars moving in the application world by moving in their real environment, as are mobile elements and other people as well. Cooperation with other users might occur when sharing geographic space in a way similar to common real world activities. In mobile AR apps, interactions with entities or people happen in real-time to provide convincing feedback to the user and give the impression of natural interaction. A strong sense of "reality" and "feeling connected" comes from seeing and being seen by other users. Applications in mobile AR are therefore more vivid, immediate, direct and engaging, immersing the user in a physical and virtual world. However, getting the right information at the right time and the right place is key to all these applications which put a high demand on mobile device performance.

The difference, for mobile Web users, between indirect and implicit use of geography—Web GIS—and explicit use and especially augmentation—Mobile AR—will be highlighted with the specific case studies that follow.

3.2 *Mobile AR as a specific case of GIS*

With advances in tracking and increased computing power, researchers are developing mobile AR systems. These enable a host of new applications in navigation, situational awareness, and geolocated information retrieval. An outdoor AR system can be considered as a special case of a mobile Geographic Information System (GIS). If we compared these two systems from a component standpoint, mobile GIS integrates four essential components: a Global Positioning System (GPS), a handheld computer, GIS software and a mobile communication network for data access. The mobile communication network allows remote access to an entire GIS toolset and data library on a server. The mobile GIS user interface is relatively simple up-front, being very task-oriented and specific. The system relies on the server-based GIS to carry out complex queries and analysis. The user location, provided by the GPS, enables the software system to personalize the interface according to the user's preferences, and to display only data relevant to the user's vicinity.

Figure 2 proposes a description of the major components of an outdoor MAR system. Like mobile GIS, MAR system includes a positioning component, a computational platform and a mobile communication component for data access and storage. These components allow the mobile AR system to present geo-referenced information in real-time, based on the physical location of the user, user preferences, and other context-dependent information (Schmalstieg 2005). Similarly to mobile or web-based GIS, mobile AR applications generally present only a small subset of the information contained in geospatial databases, as determined by the location and context of the user. Efficient access to the database server is needed due to the strong real time constraint of mobile AR applications. Indeed, due to the user mobility, his location or orientation may be changing continuously: the displayed information should be updated at the same rate in order to keep the augmentation synchronized. Advanced technologies to efficiently store, query, and retrieve geo-referenced information from databases (data storage / access technologies in Figure 2) are a common focus of interest among mobile AR design and GIS communities.

Like GIS applications, mobile AR systems aim at providing powerful spatial analytical capabilities that will enable better assessment of natural situations through visualization support. In mobile AR applications, the computer augments the real scene with additional information that could provide visual support for special field based applications. The augmentation relies on the data provided by the camera, the orientation and positioning systems. These data are fed to the computational platform where the co-registration and rendering of graphics information according to the user viewpoint is computed (Figure 2). AR techniques improve visualization of 3D geo-data by communicating the distance to occluded objects (Furmanski 2002), improving the readability of text overlays (Leykin 2004), providing automated layout of presentation items (Höllerer 2001), filtering information (Julier 2000) or facilitating navigating through the 3D environment (Gruber 1995). Visualization techniques

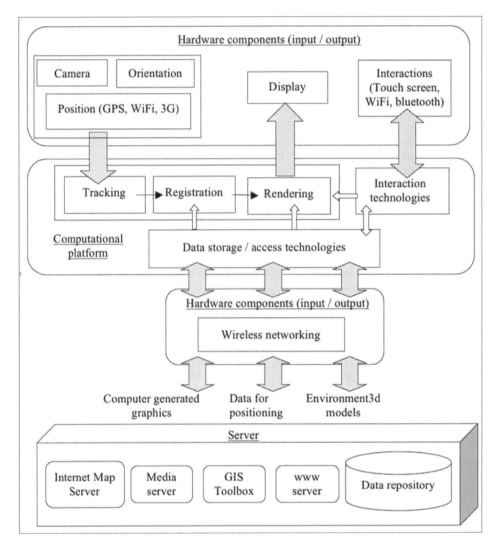

Figure 2. Major components of a mobile AR system.

for spatial data in GIS applications, especially mobile ones, are not yet as advanced. In recent years, some attempts have been made at exploring the use of perspective maps and 3d graphics to communicate geographic information on mobile devices for navigation purposes. These subtle modifications of the visualization help users to gain a better understanding of the structure of their environment. However, 2d maps are still largely used for orientation even if it requires a mental effort to switch between the egocentric perspective of the viewer and the geocentric perspective of the map.

Mobile AR applications change the nature of how we interact with and understand spatial data. The advanced AR techniques render the interface in a far more intuitive way than usual GIS solutions making it easier for users to match what they see in the display with their view in the real world. In addition tangible interactions with the 'real' world can be performed through multimodalities components (cf. Figure 2: interaction component). A connection is thus formed between the physical and the virtual worlds in which the users find themselves, and many layers of information are accessible at the same time.

4 CASE STUDIES—APPLYING AUGMENTED REALITY SENSABILITIES TO THE MOBILE SPATIAL WEB

We now present a number of short case studies highlighting different approaches to augmentation of the mobile user experience. These range from approaches on the line between Web GIS and AR to fully AR based tools.

4.1 *Situated storytelling: Massively parallel annotation of landscape*

As a first experiment in applying augmented reality sensibilities to the mobile Web, we have constructed a simple mobile Web application that supports rich annotation of landscapes in space and time (Harrap 2008).

Traditional augmented reality applications focus on augmenting a graphical view of space. Our goal in developing Situated was to explore, instead, augmentation of the semantic surrounds of a location, and to explore in particular whether semantic augmentation leads to different kinds of experiences than traditional graphics-based augmented reality. The long term goal is to merge the semantic augmentation described below with more traditional graphical AR. The core technologies in Situated are Web GIS based, but the philosophy of the user's relationship to their environment is closer to an AR perspective.

The Situated application is a mobile Web tool that allows individual users to locate themselves on a Google Maps backdrop, create locations for annotations of the landscape, roughly conceived as landmarks, and then situate stories at those landmarks. By story we mean any human description, discussion, or annotation that captures some sense of place and meaning. For example, a user might describe a park bench where they sat as a child; another might annotate a site that was once the home of a famous artist. Another might construct a tour of an urban space. Situated provides an iPhone-based interface for annotating landscapes with content, for accessing that content, and ultimately for augmenting our experience of local realities. Ultimately these augmentations are available to any user of spatial data, from desktop browsers through to full AR implementations.

We allow users to annotate landmarks comprising regions, neighborhoods, and local features; most annotations will be local features, and the neighborhood and city level stories are intended largely to provide framing information of general interest. Neighborhoods and Regions are currently implemented as point locations with large radii of interest, although in the future this could be replaced with polygon features created in a desktop or mobile GIS application. Users create landmarks as needed, and then create stories. Stories are personal experiences of place, and are explicitly located 'at' a landmark. Users in proximity to an existing Landmark can create stories there, or alternatively can create a new Landmark to differentiate the spatial context for stories. We allow, and provide intuitive interface tools for, modifying one's precise position on the map base, both to support refinement of position when using cell-based location services, and to allow annotation of features from some distance away. Representative views of the interface are shown in Figure 3.

Situated supports not only text notes but also the capture and editing of photographs taken with the iPhone's low-resolution camera. Photos are uploaded to the Situated server and become immediately available to other users.

Users are members of groups, which are mediated by leaders. New leaders are approved by the system controller. In Situated, user stories are submitted to group leaders who moderate the content before it appears to the community at large. This resolves a common problem with social Web applications where inappropriate use (advertising, objectionable content, etc.,) leads to controversy, though at the cost of imposing an administrative load on group leaders.

Limited personalization tools exist in Situated—a user can leave a message or 'story' for a specific group or user, though this raises the issue that that user would need to know that a story is 'waiting for them.' The exact interaction styles continue to be a subject for

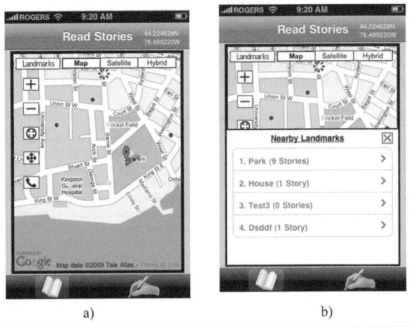

a)   b)

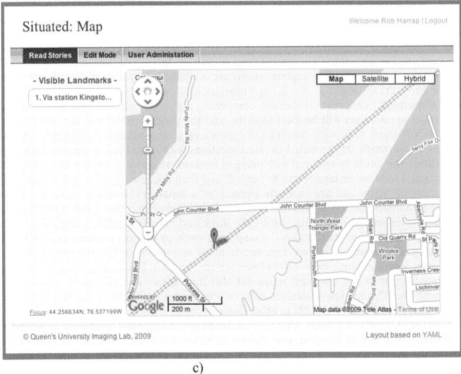

c)

Figure 3. The Situated mobile-Web augmentation tool running on an iPhone and a Web browser: (a) a view showing the location of the user (blue pin) and existing landmarks with and without stories (green and blue dots); (b) a view showing nearby landmarks; and (c) a web view showing the Situated Web application running in a desktop browser.

exploration, though at present they are limited by the capacity of the iPhone host to attend to multiple applications.

The Situated application is entirely Web-based. It comprises an APS.NET web engine built on the Microsoft .NET framework, and uses the SQLExpress database engine for data storage. The Situated 'engine' comprises a layer between the data storage and the user interface and provides objects for all the basic components of the application in the C# language. The user interface is coded in VB.NET since this language provides much better support for event-driven behaviours. Maps are generated using the Google Maps Applications Programming Interface (API). An iPhone 'look and feel' was achieved using a CSS framework. CSS—Cascading Style Sheets—provide a hierarchical approach to separating Web content from presentation. A tagged content stream interacts with a CSS style set to generate a specific display. The power of such an approach for Web GIS tools is that it allows a content set to be reconfigured for a new use without changing the content: the local styles determine what is presented and how.

The mobile Web nature of Situated is an advantage if not a requirement since our fundamental idea of having many users marking up the landscape simultaneously requires a centralized approach to data storage. The Web solution also makes use of the Google Maps base layers straightforward. A further advantage is that the use of a Web interface simplifies the process for a user: only one application, an augmented browser, is needed, and updates to Situated are transparent; the next time a user accesses the tool it may be significantly enhanced. We simply host our Web application on a commercially available enhanced Web browser, Big5, which allows traditional Web functionality but also access to the camera and location of the phone. Big5 also allows an application to access existing photos hosted on the iPhone and interaction with the accelerometer on the iPhone.

The main disadvantage of our mobile Web approach is look and feel: no matter how much work we do with CSS, the application does not look exactly like other iPhone applications, especially in how transitions between pages are accomplished.

Future development of Situated will focus on the creation of tours and other 'groups' of stories, and on development of simple game-related functionality to allow active exploration of situated landscapes. In the longer term, if applications such as Situated are to result in dense annotations of urban landscapes, tools to filter based on social and contextual preferences will be needed, as discussed in detail in Harrap (2008). These annotations are well suited for incorporation into other location-based tools, including those using full AR interaction styles.

Lessons learned in the Situated project are as follows. First, users must be able to augment their precise location since location technology is not always reliable or even available. Second, the application must be simple both to encourage wider use and also to minimize application lag following user interactions. Third, even such a simple application uses significant power on the mobile device and so long-term use in any one session is an issue. Fourth, the interface must be carefully designed to maximize clarity and to compensate for the small display size on mobile phones. Finally, a simple Web mapping interface can provide a mechanism for widespread generation of personal augmentations on real-world features for use in other AR applications.

4.2 Generation of simple 3d models in the field: Experiments and results

As a second experiment in applying augmented reality methods to the mobile Web, we developed an augmented reality application on the iPhone to support the fast and easy creation of 3d models. Augmented reality applications, whether taking place indoors or outdoors, require 3d models of the environment to be constructed as discussed above; these are used in the accurate positioning of computer-generated graphics in the real world, and the management of the occluding objects in the area between the user and the user focus of interest. If such objects exist, an accurate 3d model will help derive visible sub-structures of these objects.

Although 3d modeling of urban environments is becoming more and more common, 3d models are not yet available everywhere. Traditional surveying techniques such as airborne or satellite photogrammetry, lidar, and GPS-based capture of local environments are both time consuming and costly. We have therefore created an iPhone application, iModelAR, which allows a user to rapidly and easily create 3d models for use in local augmentation of the real environment. This application is used on a mobile device in that real environment. The application targets model creation for outdoor mobile AR games. The purpose is ultimately to allow anyone to build content for and play such games anywhere, especially in previously unprepared environments; this complements Situated, which aims to tag content about features, with geometric details about features. The goal is to foster massively parallel 3d model building in a similar fashion to how OpenStreetMap has encouraged and supported the construction of 2d georeferenced data. Users will contribute to the authoring of the world elements they want to interact with and around in AR games.

The motivation behind the development of an iPhone app rather than a web app concerns mostly the computation time and the access to the device accelerometers and camera; our goal is to capture geometry with a purpose-built tool and that geometry will then support mobile AR and mobile Web applications. The core idea of the application is to use the iPhone characteristics (GPS positioning, accelerometers, camera) as well as widely available Google Maps imagery as building blocks towards 3d geometry. The main steps of iModelAR consist of:

1. Pinning down 2d coordinates of the building's footprint in a Google Maps environment;
2. Taking pictures of the marker locations;
3. Computing the height of the building using the user's distance from each marker and the iPhone's vertical inclination;
4. Extruding the footprint to get the 3d model. The model is encoded in X3D format. X3D is the ISO standard XML-based file format for representing 3D computer graphics. The X3D geometry file can be transferred to a laptop or desktop computer for further visual inspection using the available X3D viewer.

Once the 3d model has been created, it can be used to augment the environment. Its geometry will help position computer-generated graphics according to a user's viewpoint, as is required by graphical AR applications. In addition, accurate 3d models provide support for dealing with occlusion. Representative views of the application are shown in Figure 4.

The application has been developed using the Objective-C language and the iPhone SDK 3.0 and iPhone OS 3.0 beta; it takes fully advantages of all the technologies offered by the current generation of iPhones. The interface tries to keep things as simple as possible to be synchronized with the iPhone interface style and to support novice users.

We have conducted some experiments to assess the accuracy of the 3d model created using iModelAR. First, we surveyed the corners of four buildings using a total station in order to have a ground truth data set. The total station was positioned on geodetic point with a known location. Then, we surveyed the same corners using iModelAR application. Several measurements were recorded for each of the corners in order to ensure good data redundancy and a sound analysis of the application accuracy. The experiments yielded to a spatial accuracy of 2.018 meters and vertical accuracy of 4.230 meters. The main source of inaccuracy stems from the iPhone GPS.

Future development of the iModelAR app will focus on further assessment of the model accuracy and relevance for AR mobile games. The purpose will be to determine the level of detail and accuracy that is sufficient to be able to augment the world and detailed enough to complement the annotation content recorded with the Situated app. To meet this goal, terrestrial lidar data with very high spatial resolution will be used. 3d models created from this data will be compared to the models provided by the iPhone app. They will help underline the benefit provided by extra detail and accuracy in terms of world augmentation. In addition, further functionality in iModelAR needs to be developed in order to tackle the 3d modeling of buildings with complex shapes.

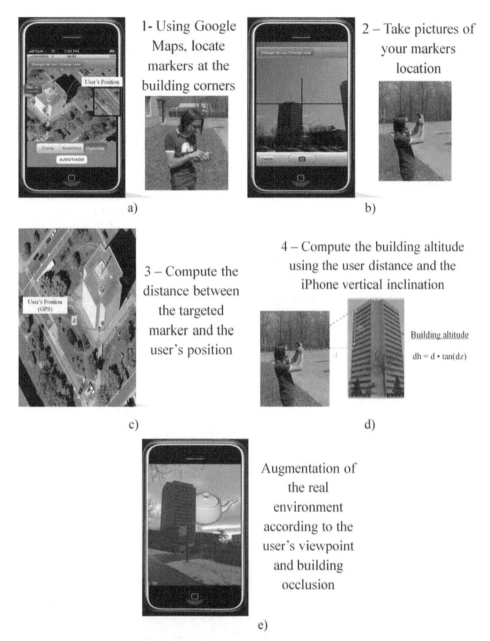

1- Using Google Maps, locate markers at the building corners

a)

2 – Take pictures of your markers location

b)

3 – Compute the distance between the targeted marker and the user's position

c)

4 – Compute the building altitude using the user distance and the iPhone vertical inclination

Building altitude

$dh = d \cdot \tan(dz)$

d)

Augmentation of the real environment according to the user's viewpoint and building occlusion

e)

Figure 4. iModelAR application steps: (a) marker computation; (b) pictures acquisition; (c) distance computation between the user and the building; (d) building height computation; and (e) augmentation of the environment with computer-generated graphics taking building occlusion into account.

Lessons learned in iModelAR creation are as follows. First, using current smart phones with similar functionality as those available on the iPhone, it is possible to build 3d models and rely on them to augment the environment and manage some occlusion contexts. The unique access to the user position, orientation, on site photo and web mapping allows extraction of geometries from the real environment and thus opens real possibilities in terms of world description by a wide range of users. Second, Google maps imagery is not ortho-rectified meaning there is inherent inaccuracy in the model creation process. Third, building roof or corner visibility is an issue in some contexts: it is required in order to be able to extract the footprint of the building. Finally, as in the Situated project, the application uses significant power on the mobile device and so long-term use in any one session is an issue.

4.3 *Experiments in mobile graphical AR*

The Situated case study highlights semantic augmentation of space. The iModelAR case study highlights generation of geometric models for use in subsequent graphical AR work, whether mobile-Web based or not. A number of mobile Web AR applications have been developed by other workers and we briefly discuss these here.

In December 2009, Google introduced an Android-phone based tool called Google Goggles. The application links the camera in an Android phone to a search engine: the camera GPS and compass data are combined with the picture and submitted to a Web search engine, and the system returns information about the local environment. For example, pointing a phone at a storefront overlays the store name on the live video view, and clicking on the name returns a web page with information about that business.

The best realized mobile AR tool for a smart phone currently is Layar, for the iPhone. Layar is a free augmented reality application that, for a select number of locations, allows overlay of information or entertainment graphics onto a scene. Figure 5 shows two representative examples: a building superimposed on the landscape of Rotterdam, and a whimsical representation of the Beatles crossing Abbey Road. These align with our showcased applications, iModelAR (architecture) and Situated (location information and tours).

Wikitude was one of the first world browsers. It is a predecessor of the Google Goggles tool and the first AR application used by the general public. Wikitude presents users with data about their surrounding, nearby landmarks and other points of interests by overlaying information on the real-time camera view. It uses Wikipedia and the online yellow-page Qype content as its main external content sources. GPS, compass and orientation sensors are required components for the augmentation to take place. Like Layar, Wikitude offers capabilities to augment the real world with 3d features in addition to 2d annotations. One Wikitude 3d application is a tribute to the World Trade Center in New York City. People in the area around Ground Zero can use Wikitude to see the World Trade Center as it looked pre-9/11. This visualization is aligned with our showcased application iModelAR although Wikitude does not involve 3d models and therefore does not manage occlusion (generated by trees for instance).

Figure 5. Screenshots of the LAYAR application, circa January 2010, showing Version 3 of LAYAR. At the left, an entire building is superimposed on the landscape in Rotterdam to highlight the form of a new building. At the right, the Beatles are shown crossing Abbey Road as part of a tour of 42 locations in London (Layar 2010).

Figure 6.   Screenshots of the Wikitude application running on Android. At the left, the picture presents the Wikitude world browser superimposing information about the landmarks and points of interest in the surroundings. At the right, Wikitude allows a user to see a 3d representation of the World Trade Centre when looking at a location in the Ground Zero area in New York City (Wikitude 2010).

Figure 7.   Screenshots of the Metro Paris application running on the iPhone. At the left, the picture presents the nearest metro stations and points of interest from the user position on Google Map within the application. At the right, the picture presents the same information but overlaid on the iPhone camera view.

Nearest Tube (Nearest Tube 2010) and Metro Paris (Metro Paris 2010) are two other world browser examples dedicated to subway stations but also providing information about other points of interest in the user's surroundings. Nearest Tube was one of the first augmented reality applications to 'go live' in the iPhone AppStore. The applications include cartographic visualization (i.e. Google maps) and augmented reality visualization. These align with our showcased applications, combining web-based tools and views (Situated) with 3d annotations superimposed on the user viewpoint (iModelAR).

5 DISCUSSION

The history, background, and case studies provided show a clear trend towards the widespread availability of local and global information on mobile devices, and the use of this information to support social, entertainment, and business tasks.

We emphasize that access to mobile spatial data from the mobile Web currently includes applications dominated by 2d data, such as Google Maps, but is rapidly expanding to include 3d information and ultimately the use of the mobile device as a lens to see the local environment augmented in new ways. These new approaches include roles for the mobile user as both the consumer of content and as the creator of content, and as newer and richer forms of interaction move to mobile devices, content creation tools are evolving to keep pace. We note as well that these applications are inherently social and local, and many involve new forms of interaction that make use of the hardware capabilities of mobile devices and of local geometric and semantic markup. As well, as has been seen with social Web applications, it is very likely that we will see new social structures that align with these new styles of world- and social interaction.

Our examples provide clear examples of how such content creation tools for mobile applications and the mobile Web in particular will proceed. First, Web- or mobile-device hosted applications will allow content creation roles previously seen on the desktop to migrate to the mobile realm. This is especially appropriate where the content being created is inherently situated, such as in photography, local history, tours, and local services. The 'Situated' application shows how local social Web content can be created on the landscape using a mobile Web GIS application, and of course the key question is what happens when many users start augmenting a landscape with a tool of this type.

Second, this migration of content creation to the mobile realm will result in new forms of content creation, where modalities unavailable on a desktop are used intuitively and effectively to create, modify, or simply interact with content. The 'iModelAR' application described above allows simple markup of Google Maps and photo data to create 3d geometries of urban features, ultimately driving the creation of rich urban 3d fabrics suitable for augmented reality use. This in turn raises the question of how we will drive the accuracy of such information capture up and how we will mediate between different 'versions' of the world as the same features are repeatedly captured by many different mobile users.

We envisage a world where all forms of content, with the requisite level of detail and accuracy, can be captured by simply walking, talking, and pointing in the local environment. Beyond story-based annotation is template-based creation of simple outdoor games. For example, spatial games are emerging where users interact with locations to compete for points (Foursquare 2010). Likewise, beyond pointing-based geometry capture is semi-automatic capture and re-capture of exterior and interior spaces by a host of different mobile devices with different capabilities.

With the rapid development of the 3d web-computing ecosystem, it is perfectly reasonable to imagine greater integration of work and play within present and representational environments through the use of MAR. The divisions between personal time and work time and between physical and virtual reality will be further erased for everyone who is connected, and the results will be mixed in their impact on basic social relations. Whether this blending of modes is a good thing, and what costs it imposes on personal and family happiness, requires significant social research.

In addition, the exploitation of mobile spatial data and MAR will deeply increase sensory experience, wayfinding, and orientation. Situated augmentation will be a powerful force to increase the sense of physical place. But before being able to move between the virtual and the physical worlds in seamless ways, the capacity of our current technology will have to be improved in order to be able to deal with the information overload.

Our work has also confirmed key challenges, primarily related to power consumption and the accuracy of positioning of existing devices. Designing effective interfaces on very small displays is also challenging. The technology of mobile devices is changing so rapidly that a product is out of date by release, though some of the location-sensing and power challenges persist between generations of devices.

In the longer term two challenges are identified. First, both of our applications capture and use content intentionally. Users must have taken out their device and opened the application. In the case of Situated, having ambient detection of proximity to stories of interest is not possible, since Situated only shows the user content when asked to do so. In the case of iModelAR, it supports only a 'survey' mode where the capture is focused; future tools might mine spatial content data from user positions, perform spatial analysis using available GIS data, and 'suggest' annotation targets to fulfill a mandate of a seamless 3d augmentation fabric. Working out how such continuous, ambient access to local semantic and spatial content, and continuous *contribution* to that content can work on limited devices requires research.

Second, the use of ambient transmission of location, as seen in Google Latitude for example, means that there will be serious questions about who can see what, and when. Ultimately stories, photos, games, and perhaps even geometry are personal, and both the social acceptability and legal framework for how such a mobile web functions will need to be determined. Consider this example: if Situated and iModelAR were used to build a model of the interior of a house, complete with personal details and photos, would this be public information? Should it be accessible, for example to emergency response crews? These issues get more difficult to assess when ambient creation of content, from the users daily path and activities, becomes possible.

In the future it is likely that devices such as mobile phones and cameras will interact with each other to create an ecology of devices around a user; this interaction will extend locally and temporally to nearby trusted resources such as desktop computers, network access nodes, and local content providers. The relationship between how such hardware can work, what is socially and legally responsible, and how the combined semantic and spatial model of this changing and complex environment can be constructed and maintained remains to be worked out.

6 CONCLUSIONS

The past 10 years have seen Web GIS evolving and being used to support a broad range of geographically related applications supporting user access to map visualization and simple mashups and user creation of spatial content. Such applications provide an easy way for anyone to participate, to share new experiences in terms of interaction and collaboration. They are giving rise to the notion of Volunteered Geographic Information (Goodchild 2007) which has seen millions of recorded observations made by neo-geographers. It underlines a pattern of sharing geospatial data on the Web that is interesting and promising for moving the spatial web to full geo-services on the web.

Location, mobility of the user and the growth of geographic based services are impacting many industries. There are now many ways to "consume" the geo-localization. Emerging opportunities on the mobile Web GIS include location-based gaming. In recent years, games for mobile devices such as smart phones and PDAs have gained popularity for providing personal entertainment on the go. With the number of mobile gamers around the world already reaching 220 million in 2009, the mobile gaming business is projected to expand to higher levels and constitute a bigger profit share of the gaming business. Currently emerging technologies enable the use of real world geographic environments for and within games. Within the domain of Location-Based Services (LBS) such games form a promising market for early adopters.

As the social Web evolves, there are also expanding opportunities for the mobile Web. Mobile phones and handheld devices have created a ubiquitous instrumented reality that goes much deeper than location awareness. Smart phones are powerful networked sensor devices that know a lot more about their users than just their location. With proximity, motion, accelerometers, voice, images, call logs, email content and so on localized in one device, much can be gleaned about who is doing what and where, with corresponding privacy challenges but also analytical and business opportunities. Analyzing these large amounts of mobile location data will help, for example, to drive relevant recommendation and personalization of the information. The mobile Web will therefore provide new insights into people behavior and networking. It will help better understand how to use the social web with mobile local context in order to provide people with relevant information according to their location and their network, like "where is everybody going right now".

The rapid evolution and growing interest of the mobile web will soon give rise to an ambient mobile web of semi-autonomous mobile services that surround users as they navigate the world. The ambient mobile web will interconnect people, devices, buildings, environment and content in an interoperable network. In a mobile environment, users will have access to additional ambient information that is not usually present in a fixed working location, and much of it will be content created on mobile devices by average users. The ambient web will provide an opportunity to improve the quality of information delivery service by utilizing ambient information. The nature of mobility provides proactive computing an opportunity to improve the quality of information delivery by identifying users' locations and their current tasks via pervasive or ubiquitous computing services.

There is an important market for such an ambient Web. The first target customers are "people with rich Internet lives," meaning people who can't bear to be untethered. The ambient mobile Web will give them active access to information even without the need for active search. Among the possible services offered to them, is whether there is anything near their current location of interest to them according to their usual behavior or their interest list. The determination of the content to be delivered to mobile users will be based on their current situation. Such capabilities could be exploited, for instance, for assessing the security level of a current situation, the ambient web resulting from the convergence of cutting edge communication and location-based technologies and decision support tools. The precise meaning of a situation will be vital for determining the appropriateness of information delivery. From a business perspective, the mobile ambient web can give rise to new market in terms of knowledge about consumer interest. Through pervasive ambient devices it will be possible to know who came near here and what was their behaviour; for example, did they pause when they passed a store display.

The history of Augmented Reality, and mobile Augmented Reality, is rich with examples of how to make such mobile applications possible, practical, and even enjoyable. We expect that mobile augmented systems that make use of ambient web content, rapidly expanding device capabilities, and rich interaction with the local environment will appear in the immediate future. Some of these will be Web-based, some will use Web services as part of their operation but be developed explicitly for a specific device, and some will blend all of the media and interaction forms available on current mobile devices in new ways.

ACKNOWLEDGMENTS

This work is funded by the GEOIDE Network Center of Excellence, Natural Sciences and Engineering Council of Canada Collaborative Research and Development grant, and the Department of National Defence, Canada. Thanks to editors Songian Li, Suzana Dragicevic and Bert Veenendaal for supplying us with the chance to write about MAR and mobile Web GIS with this volume. The authors acknowledge the support of all the partners and team members involved in the various projects related to this paper. We acknowledge especially David Ball for programming the Situated application, and Vincent Thomas, M.Sc. student, for developing the iModelAR prototype.

REFERENCES

Azuma, R.T. (1997) A survey of augmented reality. *Presence Teleoperators and Virtual Environments*, 6 (4), 355–385.

Bajura, M., Fuchs, F. & Ohbuchi, R. (1992) Merging Virtual Reality with the Real World: Seeing Ultrasound Imagery Within the Patient. In *Computer Graphics*, 26 (2), 203–210.

Berners-Lee, T. & Fischetti, M. (2000) Weaving the Web. *Collins Business*, p. 256.

Berners-Lee, T., Hendler, J. & Lassila, O. (2001) The Semantic Web. *Scientific American*, May, pp. 34–43.

Bush,V. (1945) *As We May Think*. [Online] Atlantic Monthly, July 1945, Available from: www.theatlantic.com

Cawood, S. & Fiala, M. (2008) *Augmented reality: a practical guide*. Pragmatic Edition, 978–1934356036. p. 328.

Comport, A.I., Marchand, É. & Chaumette, F. (2003) A real – time tracker for marker-less augmented reality. In *Proceedings of IEEE and ACM International Symposium on Mixed and Augmented Reality*, pp. 36–45

Everyblock. (2010) Everyblock.com. Newest realization of the ChicagoCrimeMaps site, [Online] Available from: http://chicago.everyblock.com/crime/, [Accessed 20th January 2010].

Foursqure. (2010) Foursquare: Check in, find your friends, unlock your city. [Online] Available from: http://foursquare.com, [Accessed 20th January 2010].

Furmanski, C., Azuma, R. & Daily, M. (2002) Augmented-reality visualizations guided by cognition: Perceptual heuristics for combining visible and obscured information. In *Proceedings of ISMAR 2002*, IEEE and ACM. pp. 215–224.

Garrett, J.J. (2005) Ajax: A New Approach to Web Applications. Online publication Available from: (http://www.adaptivepath.com/ideas/essays/archives/000385.php)

Genc, Y., Riedel, S., Souvannacong, F. & Navab, N. (2002) Marker-less tracking for AR: A learning-based approach. In *Proceedings of IEEE and ACM International Symposium on Mixed and Augmented Reality*, pp. 295–304.

Goodchild, M. (2007) Citizens as sensors: the world of volunteered geography. *GeoJournal*, 69 (4), pp. 211–221.

Gruber, M., Pasko, M. & Leberl, F. (1995) Geometric versus texture detail in 3D models of real world buildings. In *Automatic extraction of man-made objects from aerial and space images*, Birkhauser Verlag, Basel, pp. 189–198.

Höllerer, T., Feiner, S., Hallaway, D., Bell, B., Lanzagorta, M., Brown, D., Julier, S., Baillot, Y. & Rosenblum, L. (2001) User interface management techniques for collaborative mobile AR. *Computer & Graphics*, 25 (25), pp. 799–810.

Julier, S., Lanzagorta, M., Baillot, Y., Rosenblum, L., Feiner, S. & Höllerer, T. (2000) Information filtering for mobile augmented reality. In *Proceedings of ISAR 2000, Munich, Germany, 5–6th October 2000*, IEEE and ACM. pp. 3–11.

Kato, H. & Billinghurst, M. (1999) Marker tracking and HMD calibration for a video-based augmented reality conferencing system. In *Proceedings of IEEE International Workshop on Augmented Reality*, pp. 125–133.

Layar, (2010) *circa January 2010. Augmented Reality – Layar Reality Browser*. [Online] Available from: http://www.layar.com, [Accessed 30th January 2010].

Leykin, A. & Tuceryan, M. (2004) Automatic determination of text readability over textured backgrounds for AR systems. In *Proceedings of ISMAR 2004*, IEEE.

Liarokapis, F. (2006) An exploration from virtual to augmented reality gaming. *Simulation & Gaming*, 37 (4), December 2006, pp. 507–533.

Los Ojaos del Mundo. (2010) *The world's eyes*. [Online] Available from: http://sensable.mit.edu/worldeyes/ [Accessed 27th January 2010].

MARA, (2010) *Mobile augmented reality application project*. [Online] Available from: http://research.nokia.com/research/projects/mara/index.html, [Accessed 27th January 2010].

Metro Paris, (2010) *Subway iPhone and iPod Touch Application*, [Online] Available from: http://metroparisiphone.com [Accessed 30th January 2010].

Naimark, L. & Foxlin, E. (2002) Circular data matrix fiducial system and robust image processing for wearable vision-inertial self-tracking. In *Proceedings of IEEE and ACM International Symposium on Mixed and Augmented Reality*, pp. 27–36.

Nearest Tube. (2010) *Tells Londoners where their nearest tube station is via their iPhones*, [Online] Available from: http://www.accrossair.com/apps_nearesttube.htm [Accessed 30th January 2010].

O'Reilly, T. (2005) *What is Web 2.0: Design Patterns and Business Models for the Next Generation of Software*. Online publication, Available from: http://oreilly.com/pub/a/oreilly/tim/news/2005/09/30/what-is-web-20.html

Phonemag. (2010) *Concept encyclopedia offers augmented reality. Phonemag electronic media publication*, [Online] Available from: http://www.phonemag.com/concept-encyclopedia-frame-offers-augmented-reality-021225.php [Accessed 27th January 2010].

Schmalstieg, D. & Reitmayr, G. (2005) The world as a user interface: augmented reality for ubiquitous computing. In *Proceedings of the Central European Multimedia and Virtual Reality, 8–10th June 2008 Prague, Czech Republic*.

State, A., Livingston, M.A., Hirota, G., Garrett, W.F., Whitton, M.C., Fuchs, H. & PisanoEtta, D. (1996) Techniques for Augmented-Reality Systems: Realizing Ultrasound-Guided Needle Biopsies. In *Proceedings of SIGGRAPH '96, 4–9th August 1996, New Orleans, LA*. pp. 439–446.

Striker, D. & Kettenbach, T. (2001) Real-time and marker-less vision-based tracking for outdoor augmented reality applications. In *Proceedings of IEEE and ACM International Symposium on Augmented Reality*, pp. 189–190.

Sutherland, I.E. (1965) The ultimate display. In *Proceedings of IFIP Congress*, 2, pp. 506–508.

Vacchetti, L., Lepetit, V. & Fua, P. (2003) Fusing online and offline information for stable 3D tracking in real-time. In *Proceedings of Conference on Computer Vision and Pattern Recognition*, 2, pp. 241–248.

Wikitude, (2010) *Wikitude*. [Online] Available from: http://www.wikitude.org [Accessed 30th January 2010].

Zhou, F., Duh, H.B.-L. & Billinghurst, M. (2008) Trends in augmented reality tracking, interaction and display: a review of ten years of ISMAR. In *Proceedings of ISMAR 2008, 15–18th of September 2008, Cambridge, UK*.

Advances in Web-based GIS, Mapping Services
and Applications – Li, Dragićević & Veenendaal (eds)
© 2011 Taylor & Francis Group, London, ISBN 978-0-415-80483-7

A survey on augmented maps and environments: Approaches, interactions and applications

Gerhard Schall
Graz University of Technology, Austria

Johannes Schöning
DFKI GmbH, Campus D3_2, Saarbrücken, Germany

Volker Paelke
Institut de Geomàtica, Barcelona, Spain

Georg Gartner
Institute of Geoinformation and Cartography, Vienna University of Technology, Vienna, Austria

ABSTRACT: With the advent of ubiquitous computing infrastructures, Augmented Reality (AR) interfaces are evolving in various application domains. Classically, AR provides a set of methods to enhance the real environment with registered virtual information overlays, which has promising applications in the domains of Geographic Information Systems (GIS) and mobile mapping services. AR interfaces have a very close relation to GIScience and geo-visualization, most notably in that AR systems deal with large volumes of inherently spatial data. The development of scalable AR systems therefore draws on many infrastructures and algorithms from GIScience (e.g. efficient management and retrieval of spatial data, precise positioning). Moreover, the use of AR as a user interface paradigm has great potential for novel geospatial applications: most importantly, AR provides intuitive mechanisms for inter-action with spatial data and bridges the gap between the real environment and abstracted map representations. We overview the state of the art on augmented maps and categorize applications using augmented maps. In addition we intend to outline current developments and future trends of augmented reality interfaces over the next decade.

Keywords: Mapping, visualization, augmented reality, GIS, tracking, three-dimensional, mobile interaction, multisensor

1   INTRODUCTION TO SPATIAL AUGMENTED REALITY

For over 6000 years, humans have used maps to navigate through space and solve other spatial tasks. For the vast majority of this time, maps were drawn or printed on a piece of paper (or on material like stone or papyrus) of a certain size. Very recently, more and more maps are displayed on a wide variety of electronic devices, ranging from small screen mobile devices to highly interactive and enormous multi-touch walls. The field of Augmented Real-ity (AR) in combination with ubiquitous technology and their interface paradigms have large potential to close the gap between digital available spatial information and the real world with the help of pervasive environments and computing infrastructures. More impor-tantly, however, these maps and map applications have become extraordinarily interactive (unlike paper and stone maps). One example of such a produce is the Wikitude application which is available to the public, showing the strength of AR for the area of GIScience and geovisualization. Wikitude (Breuss-Schneeweis 2009) is a mobile travel guide for mobile devices running Android (Google Inc.) based mobile phones using location-based Wikipedia content. It was the first public available spatial AR application for mainstream mobile devices, such as the "G1" running the Android operating system. The mobile device can be used as a display like a *magic lens* or *tool-glas* (Bier et al. 1993) to explore nearby georeferenced Wikipedia features overlaid in the camera's field of view.

1.1 *Chapter outline*

This chapter is structured as follows. First, we introduce the fields of AR and ubiquitous computing and highlight the connection to the geospatial domain. Section two describes various AR displays and tracking approaches that allow the augmentation of real environments with registered virtual information overlays especially for GIS, mobile mapping services and a variety of other applications. In section three the role of AR as an interface for Web & GIS services is discussed. Next, section four summarizes key visualization and interaction techniques. The state of the art is discussed in section five in which the autors provide an overview of related work. Finally, section six provides some concluding remarks and outlines future augmented reality interfaces and technical and interaction challenges of the next generation AR interfaces in the geospatial domain.

1.2 *Augmented reality & ubicomp technology*

AR is defined as an extension of a user's perception with virtual information and requires the three main characteristics of combining real and virtual elements, of being interactive in real-time and of being registered in 3D (Azuma et al. 1997). This definition incorporates non-visual augmentation (e.g. audio AR) as well as mediated reality environments, where a part of reality is replaced rather than augmented with computer-generated information. Exploiting these features, AR offers various new approaches and interfaces, especially for geospatial information. The spatial information can be directly displayed "on the spot", and the interaction can take place in a simple and intuitive way (Azuma et al. 1997). In contrast to Virtual Reality, which completely immerses a user in a computer-generated environment, AR aims at adding information to the user's view and thereby allows to experience both real and virtual information at the same time. AR has close connections to the fields of Virtual Reality (VR) and Mixed Reality (MR), where the virtual augments the real, and Augmented Virtuality (AV) (see Figure 1). While the term "Augmented Virtuality" is rarely used nowadays, AR

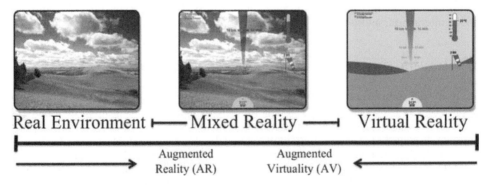

Figure 1. The Reality-Virtuality continuum of Milgram. It is a continuous scale ranging between the completely real (left), a mixed reality (middle), and the completely virtual (right) with a breakdown of the mixed reality segment. The area between the two extremes, where both the real and the virtual are mixed, is referred to as Mixed Reality. This in turn can be further subdivided into Augmented Reality, where the virtual augments the real, and Augmented Virtuality, where the real augments the virtual.

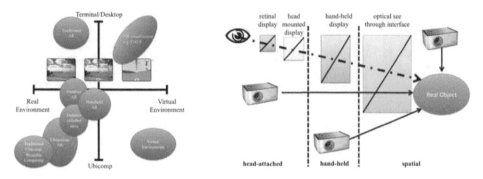

Figure 2. The Milgram-Weiser continuum of Newman et al. (left). Image-generation for augmented reality displays of Bimber et al. and different AR hardware setups (right).

and MR are now sometimes used as synonyms. However, to better understand the relationships between these fields, the Reality-Virtuality continuum of (Milgram and Kishino 1994) provides a good overview (see Figure 1) and describes the continuum in more detail and separating the MR section into the two subsections of AR and AV. Summarizing, the MR continuum describes a concept that there is a continuous scale between the completely virtual (in VR), and the completely unmodified reality. The Reality-Virtuality continuum therefore encompasses all possible variations and compositions of real and virtual objects. It has been somewhat incorrectly described as a concept in new media and computer science, when in fact it should belong closer to anthropology. Generally, maps—both paper maps and virtual maps—are a widespread medium deployed in many recent applications, especially location based systems (LBS) (Gartner et al. 2007). Analogous to Milgram's Reality-Virtuality continuum, maps can have a dimension of realness in the spectrum from reality to virtuality. This world in between, using maps to augment a visual representation of an area—a symbolic depiction highlighting relationships between elements of that space—is a challenging category of AR interfaces and applications. For realizing AR interfaces different methods are used. The most commonly used are the video-see-through-Display, the optical-see-through-Display, and the projection-AR-Display, which are explained in section two. To technically realize the different methods, various hardware components can be used as well. Ubiquitous computing (Ubicomp) is a post-desktop paradigm of human-computer interaction (Weiser 1999) in which information processing has been thoroughly integrated into everyday objects and activities. In the course of ordinary activities, users of ubiquitous computing engage many computational devices and systems simultaneously, and may not necessarily even be aware that they are doing so.

1.3 *Bringing "both worlds together" in the geospatial domain*

Weiser stated that ubiquitous computing is roughly the opposite of virtual reality (Weiser 1999). However, when one considers that Virtual Reality is merely at one extreme of the Reality-Virtuality continuum postulated by Milgram, then one can see that Ubicomp and VR are not strictly opposite one another but rather orthogonal as described and illustrated by (Newman et al. 2006). This new dimension was named by Newman the "Weiser's Continuum" and would have Ubicomp at one extreme and the concept of terminal-based computing at the other. The terminal is the antithesis of the Disappearing Computer; a palpable impediment to intuitive interaction between user and computing environment. Placing both continua, the Reality-Virtuality (see Figure 2) and the "Weiser's Continuum" at right-angles opens a 2D space shown in Figure 2, in which different application domains represent areas in this space. As mentioned before, we will concentrate on the third quadrant and describe

Figure 3. Augmented reality displays. Fixed projection-based AR display and the evolution of mobile AR hardware from backpack systems to Ultra-Mobile PCs, personal digital assistants and mobile phones (from left to right).

applications that enhance the real environment with registered virtual information overlays, especially for geographic information systems, mobile mapping services and a variety of other geospatial applications. There is increasing interest in linking AR with cartography or the geospatial domain. For example, Schmalstieg and Reitmayr describe how to employ AR as a medium for cartography (Schmalstieg & Reitmayr 2006). While ubiquitous computing aims on the computer becoming embedded and invisible in the environment, AR focuses on adding information to the reality. Thus new ways of interaction become feasible.

2 AUGMENTED REALITY DISPLAYS

Generally, AR displays can be split into Head-Mounted Displays (HMD), handheld displays and projection displays, the latter being stationary but potentially able to accommodate multiple users (Schmalstieg & Reitmayr 2006). Also, for image generation and merging with the real world, two approaches can be distinguished (Schmalstieg & Reitmayr 2006): optical see-through systems, which allow the user to see through the display onto the real world, and video see-through systems, which use video cameras to capture an image of the real world and provide the user with an augmented video image of her environment. As a result, five major classes of AR can be distinguished by their display type and their merging approach: optical see-through HMD AR, video see-through HMD AR, handheld display AR, projection-based AR with video augmentation, and projection-based AR with physical surface augmentation.

In the last decade augmented reality has grown out of children's shoes and is continuously showing its applicability and usefulness in today's society for various application domains, such as engineering, tourism or architecture. Along with the advance of mobile and wireless technologies, at the same time AR is emerging with web based technologies. The Web can provide extensive content with a location or geospatial component to serve for registered overlays in the user's view. In order to register, or align, virtual information with the physical objects that are to be annotated, AR requires accurate position and orientation tracking. Therefore, tracking devices are necessary to deliver all Six Degrees of Freedom (6DOF) accurately to determine the location of a user and the orientation she is looking. A wide range of tracking technologies exists. A widely adopted technique for AR is optical tracking, which uses video cameras and advanced computer vision software to detect targets, so called markers, in the camera image and calculates their position and orientation. Furthermore, markerless tracking approaches rely on detecting natural features in the environment and don't need any physical infrastructure. Optical tracking can be applied in both indoor and outdoor environments. For outdoor applications the Global Positioning System (GPS) is the predominant tracking system for delivering position estimates. Usually, GPS receivers are combined with inertial trackers and magnetic compasses also delivering the orientation of the user. While GPS is typically applied in ubiquitous applications for location information, careful GPS setup also allows using it for AR applications. Moreover, for indoor environments also infrared (IR) or Ultra Wide Band (UWB) and other sensors such as electromagnetic-trackers (e.g. Polhemus) can be used for tracking the user. But, these approaches require the physical preparation of the environment with sensors. However, the increasing availability of video cameras in today's computer devices has led to their use as means for tracking the position and orientation of a user as described by Klein in more detail (Klein & Murray 2007). Also various combinations of tracking technologies have been integrated in hybrid tracking approaches for AR (Schall et al. 2009).

This section briefly overviews different AR displays by describing possible hardware configurations. The categorisation of (Bimber & Raskar 2006) (see Figure 2 (right)) illustrates the different possibilities of where the image can be formed, where the displays are located with respect to the observer and the real object, and what type of image is produced (i.e. planar or curved). Our goal is to place them into different categories so that it becomes easier to understand the state of the art and to help to identify new directions of research.

2.1 *Projection-based AR displays*

Projection-based AR with video augmentation uses video projectors to display the image of an external video camera augmented with computer graphics on the screen whereas projection-based AR with physical surface augmentation projects light onto arbitrarily shaped real world objects. It uses the real world objects as the projection surface for the virtual environments. Ordinary surfaces have varying reflectance, color, and geometry. Limitations of mobile devices, such as low resolution and small field of view, focus constraints, and ergonomic issues can be overcome in many cases by the utilization of projection technology. Thus, applications that do not require mobility can benefit from efficient spatial augmentations. Projection-based AR with physical surface augmentation has applications in industrial assembly, product visualization, etc. Examples range from edutainment in museums (such as storytelling projections onto natural stone walls in historical buildings) to architectural visualizations (such as augmentations of complex illumination simulations or modified surface materials in real building structures). Both types of the projection-based AR are also well suited to multiple user situations.

The recent availability of cheap, small, and bright projectors has made it practical to use them for a wide range of applications such as creating large seamless displays and immersive environments. By introducing a camera into the system, and applying techniques from computer vision, the projection system can operate taking its environment into account. For example, it is possible to allow users to interact with the projected image creating projected interfaces. The idea of *shader lamps* (Raskar et al. 2001) is to use projection technology to change the original surface appearance of real world objects. A new approach to combine real TV studio content and computer-generated information was introduced in the Augmented Studio project. For this purpose projectors are used as studio point light sources. This allows the determination of camera pose or surface geometry and enables the real-time augmentation of the video stream with digital content (Bimber et al. 2006). An example of large spatially augmented environments is the *Being There* project, where a walk-through environment is constructed by styrofoam blocks and is augmented by projecting view-dependent images. Thus a realistic simulation of the interior of a building can be realized and since the user is able to walk around in the augmented building a strong sense of immersion can be provided. To allow the user to freely move around in the setup a wide area tracking system (*3rdTech's HiBall*) is used to track to head position (Low et al. 2001).

(Reitmayr et al. 2005) have implemented a flood control application for the city of Cambridge (UK) to demonstrate possible features of augmented maps, in which a map of interest is augmented with an overlaid area representing the flooded land at a certain water level. The overall system centers around a table top environment where users work with maps. A camera mounted above the table tracks the maps' locations on the surface and registers interaction devices placed on them. A projector augments the maps with projected information from overhead. Tracking and localization is done via visual matching of templates, which are stored for each map. Moreover, with the increasing compactness of modern projectors, new and more flexible possibilities of their usage arise. For example, miniaturized handheld projectors can be combined with mobile AR interfaces serving as output device.

2.2 *HMD-based AR displays*

Head-mounted displays are usually worn by the user on her head and provide two image-generating devices, one for each eye. Optical see-through HMD AR uses a transparent HMD to blend together virtual and real content. Prime examples of an optical see-through HMD AR system are various augmented medical systems (Azuma et al. 2001). Video see-through HMD AR uses an opaque HMD to display merged video of the virtual environment with and view from cameras on the HMD. By overlaying the video images with the rendered content before displaying both to the user, virtual objects can appear fully opaque and occlude the real objects behind them. The drawback of video-based systems is that the viewpoint of the video camera does not completely match the user's viewpoint (Schmalstieg & Reitmayr 2006). This approach is a bit more complex than optical see-through AR, requiring proper location of the cameras. For security reasons, these systems cannot be used in applications where the user has to walk around or perform complex or dangerous tasks, since judgment of distances is distorted.

Early work on mobile AR, such as the Touring Machine (Feiner et al. 1997) used backpacks with laptop computers and head-mounted displays. (Höllerer et al. 2001) built a series of *Mobile AR systems* (MARS) prototypes, starting with extensions to the *Touring Machine* from Feiner et al. Similar augmented reality prototypes have been built by Piekarsky et al. in form of the *Tinmith system* (Piekarski & Thomas 2001). The *Tinmith-Metro* application is the main application, demonstrating the capture and creation of 3D geometry outdoors in real-time, leveraging the user's physical presence in the world. Furthermore, systems such as Signpost, which is a prototypical AR tourist guide for the city of Vienna, have been built by (Reitmayr & Schmalstieg 2004) allowing for indoor/outdoor tracking, navigation and collaboration based on hybrid user interfaces (2D and 3D). For tracking the mobile user usually GPS, inertial sensors and marker based tracking is applied. However, these systems are rather cumbersome for mobile applications deployed over longer working periods.

2.3 *Handheld AR displays*

With the advent of handheld devices featuring cameras the video-see-through metaphor has been widely adopted for AR systems providing augmented or "X-Ray vision" views to the user. Consequently, handheld AR displays also use the video-see-through approach (Schmalstieg & Reitmayr 2006). However, they can be built from tablet PCs, Ultra-Mobile PCs, or even mobile phones and devices which are highly available, and have good technical and ergonomic acceptance. Therefore, recently handheld display AR becomes popular and can be potentially used in ubiquitous computing, such as Location Based Services.

2.3.1 *Ultra-mobile PC displays*

This alternative and more ergonomic approach based on a handheld computer was originally conceived by (Fitzmaurice & Buxton 1994), and later refined into a see-through AR device by Rekimoto (Rekimoto 1997). UMPCs are basically small mobile PCs running standard operation systems. A number of researchers have started employing them in AR simulations such as (Wagner & Schmalstieg 2007), (Newman et al. 2006) and specifically the *Sony Vaio*™U70 and UX180, as well as *Samsung*™Q1. (Elmqvist et al. 2006) have employed the wearable computer *Xybernaut*™Mobile Assistant, which, although shares some common characteristics with UMPCs, does not belong in the UMPC category. This has started a strong trend towards handheld AR (Wagner & Schmalstieg 2005). Handheld AR prototype devices of this category have been designed and built, for example, by (Veas & Kruijff 2008). The tracking approaches are similar to that of HMD-based AR setups. Moreover,

Reitmayr et al. have shown that even highly robust natural feature tracking from IMU/vision sensor fusion is possible on a UMPC, if a detailed model of the environment is available (Reitmayr & Drummond 2006).

2.3.2 *Cell phone displays*

Before the recent introduction of UMPCs or cell phones with CPUs of significant computing power, PDAs were the only true mobile alternative for AR researchers. PDAs now have enhanced color displays, wireless connectivity, web-browser and GPS system. However, a number of computational issues make the use difficult for AR due to a lack of dedicated 3D capability and floating point computational unit. Wagner et al. demonstrating the Invisible Train (Wagner et al. 2005) have employed them as handheld display devices for AR applications, whereas (Makri et al. 2005) allowed for a custom-made connection with a special micro-optical display as an HMD. Smart phones are fully featured high-end cell phones featuring PDA capabilities, so that applications for data-processing and connectivity can be installed on them. As the processing capability of smart phones is improving, this enables a new class of augmented reality applications which use the camera also for vision based tracking. Notable examples are from (Wagner et al. 2008), (Henrysson et al. 2005) and (Olwal 2006) utilizing them as final mobile AR displays. A promising approach was implemented within the Wikitude (Breuss-Schneeweis 2009) project, basically implementing a mobile AR travel guide with augmented reality functionality based on Wikipedia or Panoramio running on the Google G1 phone. The user sees an annotated landscape, mountain names or landmark descriptions in an augmented reality camera view. The user can then download additional information about a chosen location from the Web, say, the names of businesses in the local shopping center (Breuss-Schneeweis 2009). The tracking of the mobile device is done by the built-in GPS sensor and orientation sensor. Also the Nokia research team has demonstrated a prototype phone equipped with *MARA* (Mobile Augmented Reality Applications) software and the appropriate hardware: a GPS, an accelerometer, and a compass. The phone is able to identify restaurants, hotels, and landmarks and provides Web links and basic information about these objects on the phone's screen (Härmä et al. 2004). Latest research on smart phones focuses on vision based tracking of natural features allowing tracking the user in unprepared and unconstrained environments. Rohs et al. used smart phones for markerless tracking of magic lenses on paper maps in real-time (Rohs et al. 2007). Furthermore, Wagner et al. already made major advances in pose tracking from natural features on mobile phones (Wagner et al. 2008).

3 AR AS INTERFACES FOR WEB & GIS

The real-time delivery of maps over the Internet to mobile users is still in its infancy. Increasing interactivity requires that the web-based infrastructures enable the delivery of both 2D and 3D geospatial data to the mobile user. In this context, multiple representations of geospatial objects linked the ones with the others are desired to allow navigation at different levels of detail, representation or scales. Moreover, the representations of digitalized or independently captured data need to be consistent. Additionally, online processes, also called Web services, need to be available to enable the real-time delivery, analysis, modification, derivation and interaction with the different levels of scale and detail of the geospatial data. Current geospatial Web services are very often limited to those specified by the Open Geospatial Consortium (OGC) and standardized by ISO, namely the Web Map Service (de La Beaujardière 2002) (service for the online delivery of 2D maps), Web Feature Service (Vretanos 2002) and Web Coverage Service (Evans 2003) (services for the online delivery of respectively geospatial vector and raster data). However, according to Badard (Badard 2006), if these services constitute the essential building blocks for the design of distributed and interoperable infrastructures for the delivery and access to geospatial data, no processing, such as online analysis or creation of new information is possible. To overcome these shortcomings

Badard is investigating various geospatial service oriented architectures. On demand Web services for map delivery or services such as Google Earth provide maps of cities to mobile users. In addition, Internet GIS applications in planning and resource management have become more widespread in recent years. This allows for nomadic access of GIS services anyplace and anytime via the internet by using a simple web browser. Already a growing number of companies from various sectors, such as the utility or transportation sector, rely on web applications to provide their data to construction companies or customers. In this context, Internet GIS enables mobile field workers to consult the mobile GIS at the inspection site. For example, the Austrian utility company *Innsbrucker Kommunalbetriebe* provides a web interface where registered users can mark the target area on the map by drawing a polygon around the area of which they want to extract information about buried assets, such as sewer pipes, electricity or water lines.

AR as a novel user interface promises to go one step further and allows viewing geospatial content in relation to the real world on-site by overlaying the virtual information over the video footage. One essential question is how to generate such geospatial content or models. Here, the Web can serve as an important pool of geospatial data. Maps are highly stylized models of spatial reality. Since a map is a 2D scale model of the 3D reality, identification of reference features on the map in the real world and vice versa is a difficult task for a casual user. Three-dimensional representation and visualization of urban environments are employed in an increasing number of applications, such as urban planning, urban marketing and emergency tasks. For creation of interactive three-dimensional visualizations from 2D geospatial databases we have built a pipeline architecture. A procedure that is able to make use of such databases is called transcoding: a process of turning raw geospatial data, which are mostly 2D, into 3D models suitable for standard rendering engines (Schall & Schmalstieg 2008). Note in particular that users from companies like in the utility sector expect a reliable representation of the real world, so strict dependence on real-world measurements is necessary. To fulfill these needs, the models in our work are generated from data exported from geospatial databases in the standard Geography Markup Language (GML) (www.opengeospatial.org). There are derivatives of GML, such as CityGML (Kolbe et al. 2005), which is a specialization of the GML language for 3D visualization of textured architectural models. It is very efficient but requires a special browser. Instead, the work aims at using a standard scene-graph system. Typically, the transcoding from semantic attributes in the geospatial database into purely visual primitives necessarily implies information loss. With the flexible scene-graph structure the authors are able to preserve the semantic data from the Geo-database management systems (GeoDBMS) in the resulting 3D models (Mendez et al. 2008). This has the advantage that semantic information can be used to change the appearance of the 3D model in real-time. Storing the model data in a geospatial database provides us with all the advantages of a GeoDBMS, such as data access control, data loss prevention and recovery, data integrity of a geospatial database and the pipeline approach create a considerable added value from an economic point of view since a geospatial database can be used by many visualization applications (Schmalstieg et al. 2007). In addition, the tedious and error-prone management of static, one-time generated models stored in separate files can be avoided. Data redundancy and inconsistency among spatially overlapping models are eliminated since all models refer to a common data source. Models are always generated with reference to the most up-to-date data. Figure 4 shows an example of geospatial GIS data and the resulting 3D model delivered by the transcoding pipeline. Clearly, the transcoding process allows for a neat separation of model content and presentation. Temporary models are generated rapidly on demand from the available geospatial data source and can be easily reconstructed any time. For the long term, 3D models are stored as whole but only their underlying GIS data, the rules for model generation and the styles to be applied for visualization. Temporarily the pipeline is run as a semi-automated offline process. That is, the area of the map including the objects of interest are interactively selected by the user, exported and then uploaded to the client for transcoding. The resulting GML file consists of a feature collection containing multiple features giving an abstract representation of pipes, buildings and the like. Current work aims

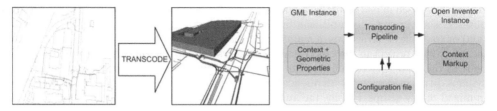

Figure 4.   Context preserving transcoding of data from a geospatial database (left). The pipeline trans-codes data into GML (right).

at an increased degree of automation of this process. By means of GPS the current position of the user in the field is determined. Using this information a Web service can be queried for online retrieval of the relevant data for the AR visualization. Future issues will include reconciling data that has been modified in the field with the database.

Figure 4 shows the transcoding process in more detail. Performing a transcoding pass means a change of the data format, in our case from a GML encoding consisting of context and geometric properties to a scene-graph (Open Inventor) visualization data including semantic context markup that can be applied by a variety of applications. A configuration file is used to hold a set of parameters for the transcoding process. Generally, web-based geospatial data sources present huge information stores, that can be leveraged by users. By sending a query to the Web service, a mobile user can access various geospatial information based on her location. This approach for content retrieval can advance both the Web-based GIS and the applications using it.

4   VISUALIZATION AND INTERACTION TECHNIQUES

Visualization and interaction techniques play a central role in mixed reality systems. The bulk of available augmentation information typically necessitates a selection of relevant content and requires effective visualization techniques to make the added information practically useful. Especially in systems that use head mounted displays, careful visualization design is essential as the virtual information integrated into the user's view may obscure important parts of the real-world environment or can distract the user significantly. Similarly, careful design of the interaction techniques in a MR system is required to ensure that the potential of MR systems to provide an intuitive and usable interface is realized (Gabbard et al. 2002). While new approaches to MR visualization and interaction are still emerging, there is also a growing body of knowledge regarding the applicability and usability of established techniques. The following sections aim to provide a brief overview of the available design space and introduce some common MR visualization and interaction techniques.

4.1   *Visualization techniques*

Visualization techniques for use in MR applications can be characterized by their spatial reference, their integration with the real-world environment and the amount of visual realism. The spatial reference frame describes how the 3D graphics objects that are rendered into the augmented graphics display are spatially bound to the real-world environment. Typical reference frames are the world, objects, the body of the user and the screen of the display device. Using the world as a reference frame, virtual objects are bound to a geo-spatial location and their visualization behaves like a physical object located at this position. Using this correspondence, many well known augmented reality applications can be implemented, e.g. the visualization of planned or historic buildings integrated into the current real-world environment or the visualization of hidden infrastructures. The use of objects as a reference frame defines a local coordinate system, where the visualizations move with the object. This reference frame is commonly used in marker-based augmented reality systems (where the absolute geo-spatial position is not known), to implement tangible user interfaces (discussed in the following section) or to display instructions in maintenance or assembly applications (where only relative locations with respect to the object under consideration are relevant). The use of the user's body as the spatial reference is used to make virtual tools easily accessible in virtual reality systems (in a fixed location with respect to the user, as in a virtual tool belt), but less common in mixed reality setups. Finally, the augmentation information can be bound to the screen of the display device, resulting in overlays that always appear in the same display location. A common use of this is the implementation of head-up-displays.

Regarding the integration with the real-world environment, MR visualization techniques can replace, enhance or mediate the real world environment. In the most simple case the graphics rendered from a MR visualization simply replaces the real environment in a part of the display. This approach prevents a tight integration of virtual content with the real-environment, but has the advantage that all existing visualization techniques can be embedded into an MR application in this way. More typical are visualization techniques where the added information is used to enhance the real-world view, which remains visible. By adjusting transparency the display can be seamlessly blended between virtual and real objects. Visualization techniques that mediate the real world environment filter information or objects from the environment; carried to the extreme complete real-world objects could be removed from the user's view in a setup known as "diminished reality". With respect to the visual realism of visualization techniques the design space spans a continuum from abstract to photo-realistic graphics. The central set of available design parameters are the depth-cues (occlusion, shading, shadows, parallax) (Hubona et al. 1999) but artificial meta-objects (illustration techniques) and visual abstraction techniques (e.g. NPR) are also possible (Furmanski et al. 2002). Drasic and Milgram (Drascic & Milgram 1996) examined the impact of stereoscopic vision in augmented reality displays and (Surdick et al. 1997) discuss the impact of various depth cues.

The central challenge in the development of visualization techniques for mixed reality applications is to design techniques that are perceptually easy to interpret for the user as well as efficient to model and render. Additional constraints can arise from the display device used, e.g. in optical see-through devices where the real background always remains visible (Rolland and Fuchs 2000). Depending on the application the level of realism has to be adjusted to either clearly convey the difference between virtual and real objects or to mix them as seamlessly as possible. The management of the amount of information to be displayed (filtering) (Julier et al. 2002) and the spatial layout of visualization objects are additional relevant issues (Bell et al. 2001).

4.2 *Interaction techniques*

Established Graphical User Interfaces (GUIs) are based on the WIMP concept (windows, icons, menu, pointer) in which a user interacts with a graphical interface representation using the mouse. This approach has become the standard means of interaction in desktop applications and has the key advantage that a limited set of standardized interaction hardware (mouse or similar 2D pointing devices) in combination with standardized graphical interface objects (widgets or controls) enables users to control arbitrary applications. The direct transfer of such techniques to mixed reality applications is possible (Szalavri & Gervautz 1997), but mixed reality applications are not limited to such techniques. A potential benefit of mixed reality interaction techniques is to enable more direct manipulation in which the user manipulates real world objects, exploiting his everyday physical manipulation skills. While conventional GUIs are limited to indirect manipulation of virtual objects using a 2D pointer, mixed reality applications offer a larger design space for the development of task specific interaction techniques in which user interactions with objects in the real world control the application. A simple example for a common mixed reality interaction technique in which a physical object is used both in direct interaction and to control the application are so called magic books, introduced by (Billinghurst et al. 2001). The user turns pages in a physical book (direct interaction). This manipulation is tracked by the system (e.g. using image recognition on special markers on the pages) and additional actions like the display of augmentation information are triggered by the application.

The use of physical objects to control an application is commonly referred to as Tangible User Interfaces (TUIs). (Ishii & Ullmer 1997) define TUIs as systems relating to the use of physical artifacts as representations and controls for digital information. A similar approach under the name of "Graspable User Interfaces" was introduced by Fitzmaurice et al. in the Bricks project (Fitzmaurice et al. 1995). TUIs remove indirections in the interaction and can exploit real world skills of users like bimanual manipulation. However, they require careful design and must be tailored to each application to exploit this potential. The need for application specific development and sensor hardware can be problematic in some application contexts. In the Bricks project small wooden pegs were assigned as physical controls to virtual objects. Using a number of Bricks users could manipulate the objects. The physical embodiments are not application specific and can be reassigned in another application context (Fitzmaurice et al. 1995).

A tangible user interface for the manipulation and query of spatial data was introduced in the "Tangible Geospace" application as part of the metaDESK project by (Ullmer & Ishii 1997). Several TUI prototypes have explored the use in urban and architectural planning. In the Illuminating Clay system (Piper et al. 2002) users were able to physically model a landscape on the table surface. The geometry was then acquired using a laser-scanner and used within the modeling application. Such an approach enables very intuitive modeling interaction, if the desired information can be specified by the clay model but is obviously highly specialized. A more general, but less direct approach was introduced in the "Magic Cup" system by Kato et al. in which physical marker objects that can be spatially tracked were assigned to virtual buildings and street furniture objects to enable simple spatial manipulation in a city planning application (Kato et al. 2003). Another important category of interaction techniques for mixed reality applications is based in the recognition of user gestures. In addition to approaches that are similar to general gesture recognition in that finger, hand or body gestures of the user are tracked and recognized (e.g. (Buchmann et al. 2004)) the use of camera gestures in an inside-out-vision setup is of special interest due to the proliferation of mixed reality on camera equipped mobile devices like smartphones and PDAs. In this setup the user moves the camera to signal the gesture that is interpreted from the video stream. (Reimann & Paelke 2006) have given an overview of how common interaction tasks like selection and quantify can be implemented in such an approach.

5 APPLICATIONS AND EXAMPLES

In the following we overview a broad range of AR applications in the geospatial domain. We present demonstration and research prototypes as well as real world applications. The table provides an overview of, form our view, important related work on AR interfaces in the geospatial domain. This section can not claim completeness in listing all noteworthy applications but aims to provide a representative overview of current work in this context. Our selection should be considered as good example of AR Interfaces in the geospatial domain.

In the following, we use the above five major AR types to categorize the AR Geo-applications. We categorize using the different interface types (optical see-through HMD, video see-through HMD, handheld display AR, projection-based AR with video augmentation, projection-based AR with physical surface augmentation); see also section 2) and the tracking system (GPS, marker based, marker less (optical tracking), IR, RFID, Inertial, UWB) that is used.

Table 1. Related augmented reality projects in geospatial applications (in alphabetic order).

Project name (Persons, Institutions)	Project description, reference
Augmented Maps Cambridge (Tom Drummond, Gerhard Reitmayr, University of Cambridge)	They developed a system to augment printed maps with digital graphical information and user interface components. These augmentations complement the properties of the printed information in that they are dynamic, permit layer selection and provide complex computer mediated interactions with geographically embedded information and user interface controls. Two methods are presented which exploit the benefits of using tangible artifacts for such interactions. The overall system centers around a table top environment where users work with maps. One or more maps are spread out on a table or any other planar surface. A camera mounted above the table tracks the maps' locations on the surface and registers interaction devices placed on them. A projector augments the maps with projected information from overhead (Reitmayr et al. 2005). `http://mi.eng.cam.ac.uk/~gr281/augmentedmaps.html`
AR PRISM (University of Washington and Hiroshima City University)	This system presents the user geographic information on top of real maps, viewed with a head-tracked HMD. It allows collaborative work of multiple users (via multiple HMDs) and gesture-based interaction. `http://www.hitl.washington.edu/publications//r-2002-63/r-2002-63.pdf`
ARMobile (VTT)	ARMobile technology is mobile software (Symbian, Java) that adds user-defined 3D objects to the camera view of the mobile phone. The application enables placing, for example, virtual furniture on mobile phone's camera image. `http://www.vtt.fi/liitetiedostot/innovaatioita/AR%20Suite_technology.pdf`
Enkin (Rafael Spring and Max Braun, Universität Koblenz–Landau)	Enkin displays location-based content in a unique way that bridges the gap between reality and classic maplike representations. It combines GPS, orientation sensors, 3D graphics, live video, several web services and a novel user interface into an intuitive and light navigation system for mobile devices. `http://www.enkin.net/`, `http://www.enkin.net/Enkin.pdf`
GeoScope (Volker Paelke, Claus Brenner, Leibniz University Hannover)	This application is a telescope like novel mixed reality I/O device tailored to the requirements of interaction with geo-spatial data in the immediate environment of the user. The I/O device is suitable for expert and casual users, integrates with existing applications using spatial data and can be used for a variety of applications that require geo-visualization including urban planning, public participation, large scale simulation, tourism, training and entertainment (Paelke & Brenner 2007).
Handheld Augmented Reality (project, 2003 - today) (Daniel Wagner, Dieter Schmalstieg, Graz University of Technology)	It aims at providing Augmented Reality anywhere and anytime. It mainly focuses on developing a cost-effective and lightweight hardware platform for Augmented Reality (AR). Based on this platform, they developed some applications. `http://studierstube.icg.tu-graz.ac.at/handheld_ar/`

(Continued)

218

Table 1. *(Continued)*

IPCity - Interaction and Presence in Urban Environments (EU funded Sixth Framework programme, 2004–2008) (Graz University of Technology)	The vision of the IPCity project is to provide citizens, visitors, as well as professionals involved in city development or the organisation of events with a set of technologies that enable them to collaboratively envision, debate emerging developments, experience past and future views or happenings of their local urban environment, discovering new aspects of their city. `http://www.ipcity.eu/`, `http://studierstube.icg.tu-graz.ac.at/ipcity/sketcher.php`
Localization and Interaction for Augmented Maps (2005) (Gerhard Reitmayr, Ethan Eade, Tom Drummond, University of Cambridge)	It augments printed maps with digital graphical information and user interface components. These augmentations complement the properties of the printed information in that they are dynamic, permit layer selection and provide complex computer mediated interactions with geographically embedded information and user interface controls. `http://mi.eng.cam.ac.uk/~gr281/docs/ReitmayrIsmar05.pdf`
MARA - Sensor Based Augmented Reality System Mobile Imaging (Application, Finished) (Nokia Research Center)	MARA utilizes camera equipped mobile devices for as platforms for sensor-based, video see-through mobile augmented reality. It overlays the continuous viewfinder image stream captured by the camera with graphics and text in real time, annotating the user's surroundings. `http://research.nokia.com/research/projects/mara/index.html`, `http://research.nokia.com/files/maraposter.png`
MARQ - Mobile Augmented Reality Quest (project, 2005–2007) (Daniel Wagner, Dieter Schmalstieg, Graz University of Technology)	It aims at developing an electronic tour guide for museums based on a self-contained, inexpensive PDA, that delivers a fully interactive 3D Augmented Reality (AR) to a group of visitors. `http://studierstube.icg.tu-graz.ac.at/handheld_ar/marq.php`
MOBVIS (EU funded Sixth Framework programme, 2004) (JOANNEUM RESEARCH, University of Ljubljana, Royal Institute of Technology (KTH), Sweden, Technical University of Darmstadt, Tele Atlas N.V.)	The MOBVIS project identifies the key issue for the realisation of smart mobile vision services to be the application of context to solve otherwise intractable vision tasks. In order to achieve this challenging goal, MOBVIS claims that three components, (1) multi-modal context awareness, (2) vision based object recognition, and (3) intelligent map technology, should be combined for the first time into a completely innovative system - the attentive interface `http://www.mobvis.org/index.htm`, `http://www.mobvis.org/demos.htm`
Overlaying Paper Maps with Digital Information Services for Tourists (Moira Norrie, Beat Signer, ETH Zurich)	It implements interactive paper maps based on emerging technologies for digitally augmented paper. A map of the Zurich city centre was printed using the Anoto pattern and a PDA used to visualise the supplementary digital information. It also developed an Interactive Map System for Edinburgh Festivals. `http://www.inf.ethz.ch/personal/signerb/publications/2005a-ns-enter.pdf`
Signpost (Vienna University of Technology)	It is a prototypical tourist guide application for city of Vienna covering both outdoor city areas as well as indoor areas of buildings. It provides a navigation model and an information browser mode. His low-cost indoor navigation system runs on off-the-shelf camera

(Continued)

Table 1. *(Continued)*

	phones. More than 2,000 users at four different large-scale events have already used it. The system uses built-in cameras to determine user location in real time by detecting unobtrusive fiduciary markers. The required infrastructure is limited to paper markers and static digital maps, and common devices are used, facilitating quick deployment in new environments (Mulloni et al. 2009). `http://www.ims.tuwien.ac.at/media/documents/publications/reitmayrauic03.pdf`
Situated Documentaries (Columbia University)	It is an experimental wearable augmented reality system that enables users to experience hypermedia presentations that are integrated with the actual outdoor locations to which they are are relevant. The system uses a tracked see-through head-worn display to overlay 3D graphics, imagery, and sound on top of the real world, and presents additional, coordinated material on a hand-held pen computer. `http://graphics.cs.columbia.edu/publications/iswc99.pdf`
The Touring Machine (1997) (Steven Feiner, Blair MacIntyre, Tobias Höllerer, Columbia University)	It presents information about Columbia university's campus, using a head-tracked, see-through, head-worn, 3D display, and an untracked, opaque, handheld, 2D display with stylus and trackpad. `http://graphics.cs.columbia.edu/projects/mars/touring.html`, `http://graphics.cs.columbia.edu/projects/mars/`
Timmi (2006) (University of Münster)	The main idea behind the Timmi application is that the camera image of the physical map is augmented with dynamic content, for example locations of ATM machines on the map. By moving a tracked camera device over the physical map users can explore requested digital content available for the whole space of the map by just using their mobile PDA or smartphone as a see-through device. For this purpose the mobile camera device has to be tracked over the physical map using AR Toolkit Markers (Schöning, et al.2006).
Tinmith (Wayne Piekarski, Bruce Thomas, University of South Australia)	It supports indoor and outdoor tracking of the user via GPS and fiducial marker. Interaction with the system is brought by the use of custom tracked gloves. The display of overlap is delivered by a video see-through HMD. The main application area of Tinmith is outdoor geometric reconstruction. `http://www.tinmith.net/`
Urban Sketcher (Graz University of Technology)	It describes how Mixed Reality (MR) technology is applied in the urban reconstruction process and can be used to share the sense of place and presence. It introduces Urban Sketcher, an MR prototype application designed to support the urban renewal process near or on the urban reconstruction site. (Authors: Sareika Markus, Schmalstieg Dieter, Appeared in CHI2008 Workshop. In Proceedings of 26th Annual CHI Conference Workshop, ACM SIGCHI, 2008)

(Continued)

Table 1. *(Continued)*

Vidente (Gerhard Schall, Dieter Schmalstieg, Graz University of Technology)	Vidente is a handheld outdoor system designed to support field staff of utility and infrastructure companies in their everyday work, such as maintenance, inspection and planning. Hence, hidden underground assets including their semantic information, projected objects and abstract information such as legal boundaries can easily be visualized and modified on-site (Schall et al. 2008) (Schall and Schmalstieg 2008). `www.vidente.at`
WalkMap (J. Lehikoinen and R. Suomela, Nokia Research Center)	WalkMap is targeted at a walking user in an urban environment, and offers the user both navigational aids as well as contextual information. WalkMap uses augmented reality techniques to display a map on the surrounding area on the user's head-worn display. `http://www.springerlink.com/content/x436r5602116jr88/`
WikEye & Wikear (Deutsche Telekom Laboratories and University of Münster)	In the WikEye project geo-referenced Wikipedia content is made accessible by moving a camera phone over the map. The live camera image of the map is enhanced by graphical overlays and Wikipedia content. The WikEar application use data mined from Wikipedia. It is automatically organized according to principles derived from narrative theory to woven into an educational audio tours starting and ending at stationary city maps. The system generates custom, location-based "guided tours" that are never out-of-date and ubiquitous - even at an international scale. WikEar uses the same magic lens-based interaction scheme for paper maps as Wikeye (Schöning et al. 2007).
Wikitude -AR Travel Guide (Application, continuously updated) (Mobilizy (Salzburg))	Wikitude is a mobile travel guide for the Android platform based on location-based Wikipedia and Qype content. It is a handy application for planning a trip or to find out information about landmarks in surroundings; 350,000 world-wide points of interest may be searched by GPS or by address and displayed in a list view, map view or cam view. `http://www.mobilizy.com/wikitude.php`

6 OUTLOOK

Augmented reality promises to combine the interactive nature of computer generated content with real world objects, thereby creating new forms of interactive maps. By using AR as an interface, maps can have a dimension of realness and interactivity in the spectrum from reality to virtuality. The past five years have seen an increasing use of maps in mobile and ubicomp applications, in which they are visually presented as realistically or as representationally as suits the users needs. More recent forms of maps are already building on online access to geographic information and leverage geospatial Web services. By using augmented reality such geographic information can be represented intuitively. The fast increasing demand from the general public for prompt and effective geospatial services is being satisfied by the revolution of web mapping from major IT vendors. Since maps are two-dimensional representations of the three-dimensional real world, maps obviously will continue to evolve towards more integrated, more realistic and higher-dimensional representations of the real world. Moreover, latest technological developments for handheld devices, such as smartphones, allow AR to become mobile and

Table 2. Geo AR examples categorized by interface type.

AR display type	Geo-applications: "Application name (Developer)"
Optical see-through HMD	• AR WalkMap (Nokia Research Center)
	• The Touring Machine (Columbia University)
	• AR PRISM (University of Washington and Hiroshima City University)
	• Situated Documentaries (Columbia University)
	• AR/GPS/INS for Subsurface Data Visualization (University of Nottingham)
	• Signpost (Vienna University of Technology)
Video see-through HMD	• HMD AR Tinmith (University of South Australia)
Handheld display AR	• Wikitude (Mobilizy)
	• MARA Sensor Based Augmented Reality System for Mobile Imaging (Nokia Research Center)
	• WikEye (Deutsche Telekom Laboratories and University of Münster)
	• Enkin (University at Koblenz-Landau)
	• MOBVIS (EU funded Sixth Framework programme)
	• MARQ - Mobile Augmented Reality Quest (TU Graz)
	• Handheld Augmented Reality (TU Graz)
	• ARMobile (Technical Research Centre of Finland)
	• Vidente (TU Graz)
	• Signpost (Mulloni, TU Graz)
Projection-based AR with video augmentation	• Urban Sketcher (TU Graz)
Projection-based AR with physical surface augmentation	• Augmented Map System (Cambridge University)
	• Digitally augmented paper maps (ETH Zurich), IPCity Interaction and Presence in Urban Environment (EU funded Sixth Framework programme)

ubiquitous for the general public. Both developments combine well and will also lead to social implications of where and how information is consumed by users. Mobile and location-based use of computers brings novel ways of presenting and interacting in-situ with information in general and geospatial information in special. Over the next decade this will lead to future developments of augmented reality interfaces and interaction techniques.

REFERENCES

Azuma, R. et al. (1997) A survey of augmented reality. *Presence-Teleoperators and Virtual Environments, 6* (4), 355–385.
Azuma, R., Baillot, Y., Behringer, R., Feiner, S., Julier, S. & Macintyre, B. (2001) Recent advances in augmented reality. *Computer Graphics and Applications, IEEE 21* (6), 34–47.

Badard, T. (2006) Geospatial service oriented architectures for mobile augmented reality. In *Proc. of the 1st International Workshop on Mobile Geospatial Augmented Reality*, pp. 73–77.

Bell, B., Feiner, S. & Höllerer, T. (2001) View management for virtual and augmented reality. In *UIST '01: Proceedings of the 14th annual ACM symposium on User interface software and technology*, ACM. pp. 101–110.

Bier, E.A., Stone, M.C., Pier, K., Buxton, W. & DeRose, T.D. (1993) Toolglass and magic lenses: the see-through interface. In *SIGGRAPH '93: Proceedings of the 20th annual conference on Computer graphics and interactive techniques, New York, NY, USA*, ACM. pp. 73–80.

Billinghurst, M., Kato, H. & Poupyrev, I. (2001) Magicbook: transitioning between reality and virtuality. In *CHI '01: CHI '01 extended abstracts on Human factors in computing systems*, ACM. pp. 25–26.

Bimber, O. & Raskar, R. (2006) Modern approaches to augmented reality. In *International Conference on Computer Graphics and Interactive Techniques*. New York, NY, USA, ACM.

Bimber, O., Grundhöfer, A., Zollmann, S. & Kolster, D. (2006) Digital illumination for augmented studios. *Journal of Virtual Reality and Broadcasting, 3* (8).

Breuss-Schneeweis, P. (2009) *Wikitude, an AR Travel Guide* [Online] Available from: http://www.mobilizy.com/wikitude.php [Accessed 16th April 2009].

Buchmann, V., Violich, S., Billinghurst, M. & Cockburn, A. (2004) Fingartips: gesture based direct manipulation in augmented reality. In *GRAPHITE '04: Proceedings of the 2nd international conference on Computer graphics and interactive techniques in Australasia and South East Asia*, ACM. pp. 212–221.

de La Beaujardiére, J. (2002) Web map service (WMS) Implementation Specification, Version 1.0.0. *Open Geospatial Consortium, Wayland, MA, USA.*

Drascic, D. & Milgram, P. (1996) Perceptual issues in augmented reality. In *SPIE Volume 2653: Stereoscopic Displays and Virtual Reality Systems III*, pp. 123–134.

Elmqvist, N., Axblom, D., Claesson, J., Hagberg, J., Segerdahl, D., So, Y., Svensson, A., Thoren, M. & Wiklander, M. (2006) 3DVN: a mixed reality platform for mobile navigation assistance. Technical report, Technical Report.

Evans, J. (2003) Web Coverage Service (WCS) Implementation Specification, Version 1.1.0. *Open Geospatial Consortium, Wayland, MA, USA.*

Feiner, S., MacIntyre, B., Höllerer, T. & Webster, A. (1997) A touring machine: Prototyping 3D mobile augmented reality systems for exploring the urban environment. *Personal and Ubiquitous Computing, 1* (4), 208–217.

Fitzmaurice, G. & Buxton, W. (1994) The Chameleon: Spatially aware palmtop computers. In *Conference on Human Factors in Computing Systems, New York, NY, USA*, ACM. pp. 451–452.

Fitzmaurice, G.W., Ishii, H. & Buxton, W.A.S. (1995) Bricks: laying the foundations for graspable user interfaces. In *CHI '95: Proceedings of the SIGCHI conference on Human factors in computing systems*, ACM Press/Addison-Wesley Publishing Co. pp. 442–449.

Furmanski, C., Azuma, R. & Daily, M. (2002) Augmented-reality visualizations guided by cognition: Perceptual heuristics for combining visual and obscured information. In *ISMAR'02: Proceedings of the 1st International Symposium on Mixed and Augmented Reality*, IEEE Computer Society. p. 320.

Gabbard, J., Swan, E., Hix, D., Lanzagorta, M., Livingston, M., Brown, D. & Julier, S. (2002) Usability engineering: Domain analysis activities for augmented reality systems.

Gartner, G., Cartwright, W. & Peterson, M. (2007) *Location Based Services and TeleCartog-raphy.* Springer.

Google Inc. *Android – An Open Handset Alliance Project*, [Online] Available from: http://code.google.com/android/

Härmä, A., Jakka, J., Tikander, M., Karjalainen, M., Lokki, T., Hiipakka, J. & Lorho, G. (2004) Augmented reality audio for mobile and wearable appliances. *J. Audio Eng. Soc, 52* (6), 618–639.

Henrysson, A., Billinghurst, M. & Ollila, M. (2005) Face to face collaborative AR on mobile phones. In *Fourth IEEE and ACM International Symposium on Mixed and Augmented Reality, 2005. Proceedings*, pp. 80–89.

Höllerer, T., Feiner, S., Hallaway, D., Bell, B., Lanzagorta, M., Brown, D., Julier, S., Baillot, Y. & Rosenblum, L. (2001) User interface management techniques for collaborative mobile augmented reality. *Computers & Graphics, 25* (5), 799–810.

Hubona, G.S., Wheeler, P.N., Shirah, G.W. & Brandt, M. (1999) The relative contributions of stereo, lighting, and background scenes in promoting 3d depth visualization. *ACM Trans. Comput.-Hum. Interact., 6* (3), 214–242.

Ishii, H. & Ullmer, B. (1997) Tangible bits: towards seamless interfaces between people, bits and atoms. In *CHI '97: Proceedings of the SIGCHI conference on Human factors in computing systems*, ACM. pp. 234–241.

Julier, S., Baillot, Y., Brown, D. & Lanzagorta, M. (2002) Information filtering for mobile augmented reality. *IEEE Comput. Graph. Appl., 22* (5), 12–15.

Kato, H., Tachibana, K., Tanabe, M., Nakajima, T. & Fukuda, Y. (2003, Oct.) Magiccup: a tangible interface for virtual objects manipulation in table-top augmented reality. In *Augmented Reality Toolkit Workshop, 2003, IEEE International*. pp. 75–76.

Klein, G. & Murray, D. (2007) Parallel tracking and mapping for small AR workspaces. In *6th IEEE and ACM International Symposium on Mixed and Augmented Reality, 2007. ISMAR 2007*, pp. 1–10.

Kolbe, T., Gröger, G. & Plümer L. (2005) CityGML–Interoperable Access to 3D City Models. In *Proceedings of the first International Symposium on Geo-Information for Disaster Management, Springer Verlag*. Springer.

Low, K., Welch, G., Lastra, A. & Fuchs, H. (2001) Life-sized projector-based dioramas. In *Proceedings of the ACM symposium on Virtual reality software and technology, New York, NY, USA*, ACM. pp. 93–101.

Makri, A., Arsenijevic, D., Weidenhausen, J., Eschler, P., Stricker, D., Machui, O., Fernan-des, C., Maria, S., Voss, G. & Ioannidis, N. (2005) ULTRA: An Augmented Reality system for handheld platforms, targeting industrial maintenance applications. In *11th International Conference on Virtual Systems and Multimedia, Ghent, Belgium*.

Mendez, E., Schall, G., Havemann, S., Junghanns, S. & Schmalstieg, D. (2008) Generating 3D Models of Subsurface Infrastructure through Transcoding of Geo-Databases. *IEEE CG&A, Special Issue on Procedural Methods for Urban Modeling 3*.

Milgram, P. & Kishino, F. (1994) A taxonomy of mixed reality visual displays. *IEICE TRANSAC-TIONS on Information and Systems, 77* (12), 1321–1329.

Mulloni, A., Wagner, D., Barakonyi, I. & Schmalstieg, D. (2009) Indoor positioning and navigation with camera phones. *IEEE Pervasive Computing, 8* (2), 22–31.

Newman, J., Schall, G., Barakonyi, I., Schürzinger, A. & Schmalstieg, D. (2006) Wide-Area Tracking Tools for Augmented Reality. *Advances in Pervasive Computing, 207*, 2006.

Newman, J., Schall, G. & Schmalstieg, D. (2006) Modelling and handling seams in wide-area sensor networks. In *Proc. of ISWC*.

Olwal, A. (2006) LightSense: enabling spatially aware handheld interaction devices. In *Proceedings of International Symposium on Mixed and Augmented Reality (ISMAR 2006)*, pp. 119–122.

Paelke, V. & Brenner, C. (2007) Development of a mixed reality device for interactive on-site geo-visualization. In *Proceedings of 18th Simulation and Visualization Conference*.

Piekarski, W. & Thomas, B. (2001) Tinmith-Metro: new outdoor techniques for creating city models with an augmented reality wearable computer. In *Wearable Computers, 2001. Proceedings. Fifth International Symposium on*, pp. 31–38.

Piper, B., Ratti, C. & Ishii, H. (2002) Illuminating clay: a 3-d tangible interface for landscape analysis. In *CHI '02: Proceedings of the SIGCHI conference on Human factors in computing systems*, ACM. pp. 355–362.

Raskar, R., Welch, G., Low, K. & Bandyopadhyay, D. (2001) Shader Lamps: Animating real objects with image-based illumination. In *Rendering Techniques 2001: Proceedings of the Eurographics Workshop in London, United Kingdom, 25–27th June 2001*, Springer Verlag Wien. pp. 89.

Reimann, C. & Paelke, V. (2006) Computer vision based interaction techniques for mobile augmented reality. In *Proc. 5th Paderborn Workshop Augmented and Virtual Reality in der Produktentstehung*, HNI. pp. 355–362.

Reitmayr, G. & Drummond, T. (2006) Going out: Robust model-based tracking for outdoor augmented reality. In *Proceedings of 5th IEEE and ACM International Symposium on Mixed and Augmented Reality (ISMAR 2006)*, pp. 109–118.

Reitmayr, G., Eade, E. & Drummond, T. (2005) Localisation and interaction for augmented maps. In *Proceedings of the 4th IEEE/ACM International Symposium on Mixed and Augmented Reality*, IEEE Computer Society Washington, DC, USA. pp. 120–129.

Reitmayr, G. & Schmalstieg, D. (2004) Collaborative augmented reality for outdoor navigation and information browsing. In *Proc. Symposium Location Based Services and TeleCar-tography*, pp. 31–41.

Rekimoto, J. (1997) NaviCam- A magnifying glass approach to augmented reality. *Presence: Teleoperators and Virtual Environments, 6* (4), 399–412.

Rohs, M., Schöning, J., Krüger, A. & Hecht, B. (2007) Towards real-time markerless tracking of magic lenses on paper maps. In *Adjunct Proceedings of the 5th Intl. Conference on Pervasive Computing (Pervasive), Late Breaking Results*, pp. 69–72.

Rolland, J.P. & Fuchs, H. (2000) Optical versus video see-through head-mounted displays. In *in Medical Visualization. Presence: Teleoperators and Virtual Environments*, pp. 287–309.

Schall, G., Mendez, E., Kruijff, E., Veas, E., Junghanns, S., Reitinger, B. & Schmalstieg, D. Handheld. Augmented Reality for underground infrastructure visualization. *Personal and Ubiquitous Computing*, 1–11.

Schall, G. & Schmalstieg, D. (2008) Interactive Urban Models Generated from Context-Preserving Transcoding of Real-Wold Data. *Proceedings of the 5th International Conference on GIScience (GISCIENCE 2008)*.

Schmalstieg, D. & Reitmayr, G. (2006) Augmented Reality as a Medium for Cartography Multimedia Cartography.

Schmalstieg, D., Schall, G., Wagner, D., Barakonyi, I., Reitmayr, G., Newman, J. & Leder-mann, F. (2007) Managing complex augmented reality models. *IEEE Computer Graphics and Applications*, 48–57.

Schöning, J., Hecht, B., Rohs, M. & Starosielski, N. (2007) WikEar- Automatically Generated Location-Based Audio Stories between Public City Maps. *adjunct Proc. of Ubicomp07*, 128–131.

Schöning, J., Krüger, A. & Müller H. (2006) Interaction of mobile camera devices with physical maps. In *Adjunct Proceeding of the Fourth International Conference on Pervasive Computing*, pp. 121–124.

Surdick, R.T., Davis, E.T., King, R.A. & Hodges, L.F. (1997) The perception of distance in simulated visual displays: A comparison of the effectiveness and accuracy of multiple depth cues across viewing distances. *Presence, 6* (5), 513–531.

Szalavri, Z. & Gervautz, M. (1997) The personal interaction panel - a two-handed interface for augmented reality. In *COMPUTER GRAPHICS FORUM*, pp. 335–346.

Ullmer, B. & Ishii, H. (1997) The metadesk: models and prototypes for tangible user interfaces. In *UIST '97: Proceedings of the 10th annual ACM symposium on User interface software and technology*, ACM. pp. 223–232.

Veas, E. & Kruijff, E. (2008) VespR: design and evaluation of a handheld AR device. In *7th IEEE/ACM International Symposium on Mixed and Augmented Reality, 2008. ISMAR 2008*, pp. 43–52.

Vretanos, P. (2002) Web Feature Service (WFS) Implementation Specification, Version 1.1.0. *Open Geospatial Consortium, Wayland, MA, USA*.

Wagner, D., Pintaric, T., Ledermann, F. & Schmalstieg, D. (2005) Towards massively multiuser augmented reality on handheld devices. In *Third International Conference on Pervasive Computing*, Springer. pp. 208–219.

Wagner, D., Reitmayr, G., Mulloni, A., Drummond, T. & Schmalstieg, D. (2008) Pose tracking from natural features on mobile phones. In *7th IEEE/ACM International Symposium on Mixed and Augmented Reality, 2008. ISMAR 2008*, pp. 125–134.

Wagner, D. & Schmalstieg, D. (2005) First steps towards handheld augmented reality. In *Seventh IEEE International Symposium on Wearable Computers, 2003. Proceedings*, pp. 127–135.

Wagner, D. & Schmalstieg, D. (2007) Artoolkitplus for pose tracking on mobile devices. In *Computer Vision Winter Workshop*, pp. 6–8.

Weiser, M. (1999) The Computer for the 21st Century. *ACM SIGMOBILE Mobile Computing and Communications Review, 3* (3), 3–11.

Collaboration and decision making

Advances in Web-based GIS, Mapping Services
and Applications – Li, Dragićević & Veenendaal (eds)
© 2011 Taylor & Francis Group, London, ISBN 978-0-415-80483-7

Map-chatting within the geospatial web

G. Brent Hall
School of Surveying, University of Otago, Dunedin, New Zealand

Michael G. Leahy
Department of Geography and Environmental Studies, Wilfrid Laurier University, Waterloo, Ontario, Canada

ABSTRACT: This chapter describes an approach that allows participants to enhance Web collaboration with maps and spatial discussions using Ajax, SVG/VML, JavaScript, and Open Source geospatial components. It provides a new means of encouraging interaction in the use of Geographic Information Systems (GIS) on the Web, by supplementing the now well known concept of 'geo-tagging' popularized by Google in its Google Earth and Google Maps applications with a new concept of 'map-chatting'. Map-chatting facilitates the creation of user-generated content and knowledge mapping, twinned with live discussion in the form of spatially linked and general text messaging to achieve dialog over issues of spatial relevance to communities of participants. The architecture of the software is discussed and its use in a case study in the South Island of New Zealand is presented. While the promise for the tool is considerable, the case study reveals that not all communities are necessarily ready or able to use the tools that are made feasible through the emergence of the geospatial web.

Keywords: Geospatial web, GIS, map-chatting, open source, web collaboration

1   INTRODUCTION

The map-chatting concept that is at the heart of the application discussed in this chapter has part of its origins in cell phone texting and computer-based instant messaging of the sort facilitated by software including Microsoft Live Messenger, Yahoo! Messenger, Google Chat and the open source tool Pidgin, among others. These applications comprise easy-to-use means of 'staying in touch' with friends, family, colleagues and even strangers over the Internet. Use of computer-based instant messaging is by no means new, as it has existed since the first appearance of Internet Relay Chat (IRC) in 1988. It was migrated to the personal computer over a decade ago by Mirabilis (http://www.mirabilis.com) in the form of ICQ, and later by Microsoft through the MSN Messenger Service. The appearance of more recent forms of social networking on the Internet, best revealed in the recent Facebook (http://www.facebook.com) and Twitter (http://twitter.com) phenomena, has added new forms of Web-based communication for staying in touch. Less social and more work-oriented forms of networking are also rapidly evolving and changing traditional means of person-to-person, person-to-and-from-group and group-to-group collaboration on tasks. For example, software from Adapx.com, which uses digital pen technology to facilitate markup of office documents as well as Geographic Information System (GIS)-based spatial data, facilitates real-time collaboration on matters of mutual interest by dispersed individuals and groups.

Collectively, these forms of communication comprise different yet related parts of the response to what Scharl (2007) describes as a call for a new generation of geospatial interfaces with simple yet powerful navigational aids that facilitate real-time access to and manipulation of geospatially and semantically referenced information. The appearance of these applications is closely associated with the introduction into mainstream Web parlance and operation of the concept of 'Web 2.0', which Musser (2007, 4) defines as 'a set of economic, social, and technology trends that collectively form the basis for the next generation of the Internet—a more mature, distinctive medium characterized by user participation, openness, and network effects'.

The so-called 'Architecture of Participation' that Web 2.0 encourages allows users to add personalized inputs to the content of generation 1.0 Web sites by contributing voluntary information (or user-generated content) to a site, as well as discussing the contributions of others. This interactivity transcends the traditional use of Web pages as passive sources of information, by allowing the Web to serve as a participatory platform and become, in fact, a 'participatory Web'. This concept and its underlying technologies have allowed applications that mimic conventional stand-alone computing desktops to be launched across the Internet, as well as facilitate the transfer of substantial volumes of information between Web servers and client computers, using only a fraction of the bandwidth that such transactions only recently required. In the geospatial domain both the transfer of voluminous data and the high degree of interactivity associated with the new technologies offer a broad horizon of possibilities within the participatory Web. In fact, the rate of growth of geospatial information on the Web reveals the potential for any point in space to be linked to commentary and media describing personalizations of its historical, environmental and cultural context (Goodchild 2007–2008).

This chapter builds upon the concept and programming environment of Web 2.0 to introduce the concept of map-chatting. Map-chatting allows flexible assemblages of co-located or dispersed individuals and/or groups on the Internet to select interactively existing map features, create new features on map layers, tag existing and new features with comments, and engage in synchronous or asynchronous, threaded or unthreaded, general or specific discussion about their own and other participants' linked comments, building discourse as they go. To operationalize the map-chatting concept, a Web map interface forms the medium of communication between specific chats and chatters. Technically, the MapChat software tool weaves well established Open Source Geospatial (OSG) components together through extensive customized programming to achieve this new approach to geo-collaboration. The second version of the software is the basis of the discussion in this chapter.

The chapter first discusses the concept of map-chatting as a form of volunteered or contributed geographic information and public participation GIS that is facilitated by the interactivity and social networking functions of Web 2.0. Section 3 presents a high-level technical description of MapChat version 2, where the components of the software and their interactions are described. Although software development is at the heart of the discussion, the broader issue of achieving effective software use by non-expert users is equally important. Section 4 presents a protocol used to deploy the software in an ongoing case study in New Zealand. The use of the tool by participants is discussed in Section 5. Although generally successful, the study reveals that while technology under the umbrella of Web 2.0 has made significant strides, especially in recent years, there remain difficult technical, logistical and human barriers to solve in achieving the objectives of broad-based collaboration among members of the public on issues of local spatial importance. This is discussed in Section 6 and the chapter is concluded in Section 7.

2   PARTICIPATORY GIS AND MAP-CHATTING

The concept of map-chatting is new to the emerging collection of Web 2.0 applications. Hence, it is also new to the participatory geospatial Web, where the process of adding user-provided geo-tagging of digital map features to on-line maps has a recent history of broad-based *public* use dating from Google's introduction of personalized Web maps in early 2007. Academic interest has quickly picked up on the nature and implications of the phenomenon of Web-based geo-tagging and other user contributions to on-line maps. Research has initially coalesced around several related themes, some of which have been under the investigative lens for more than a decade, while others are more recent and are closely associated with the new architectures that characterize Web 2.0 applications.

One of the new and provocative areas of public participation focuses on the concept of volunteered geographic information (for example, Goodchild 2008, Mummidi & Krumm 2008, Seeger 2008 and others that form a recent special issue of GeoJournal edited by Sarah Elwood). Clearly, software applications that foster the collection of volunteered *geographic* user-generated information on the Web have a specific geospatial focus. For example, sites such as Wikimapia (http://www.wikimapia.org - 'let's describe the whole world!'), or the open street map project (http://www.openstreetmap.org) fall into this category. Sites with volunteered *non-geographic* user-generated content, such as Wikipedia (http://www.wikipedia.org) fall into the same category of voluntary applications, albeit without the specific spatial content and relevance. As noted by Elwood (2008a, 133), 'these technologies and practices are dramatically altering the contexts of geospatial data creation and sharing, the individuals and institutions who act as data producers and users, and perhaps, most strikingly, geospatial data themselves'.

The widespread realization and uptake of Web sites that facilitate user contributions expose many profound and, as yet, unanswered questions about our understanding of the nature of volunteered or user generated content, both in the geographic and non-geographic domains. For example, there are significant ethical issues associated with the terms under which information is supplied. In this context, the tools of Web 2.0 allow as much if not more information about the actions of contributors as their contributions to be harvested, while they go about interacting with a Web-based digital map. Using event logging makes it possible to log virtually every action a contributor executes during his/her interactions with the site (such as mouse click events, pan events, zoom events on a Web map, the length of time taken to complete a task, etc.). Moreover, potentially more pernicious information can also be harvested, such as the general location of the contributor, their machine's Internet address and so on. This information can be gathered with or without the permission or realization of the contributor, depending on the ethical checks and balances that are in place. In the academic world, collection of such information would have to be made known to contributors in advance and informed consent provided for such 'hidden' information gathering. However, in the non-academic world attention is not normally paid to such detail. There are also important questions of the accuracy and reliability of contributed spatial data and information in the absence of any form of brokerage before contributions are committed to a site's database for the world to see and use (Flanagin & Metzger 2009).

The popularity of sites that facilitate the creation of Web 2.0 user 'produced' content on the Web has brought the above issues, and others, into sharper focus than they have been for some time. However, academic work exploring the linkages between maps and user supplied information has been on the research agenda for more than a decade (for example, Carver & Oppenshaw 1996, Carver et al. 1997, Carver et al. 2001, Kingston et al. 2000, among others). The initial interest in promoting inputs from members of the public through Web map interfaces to issues of local and even national importance (such as Carver & Oppenshaw's (1996) work on the location of nuclear waste disposal facilities in the United Kingdom) evolved into consideration of not just providing a means for individuals to 'have their say', but to facilitate *collaboration* between individuals and groups in the use of spatial data. This interest is especially evident in the area of spatial decision support (Jankowski et al. 2006, Maceachren et al. 2005, Nyerges et al. 2006). One primary focus of interest here has been on how best to generate interaction and contributions between individuals and groups discussing a spatial decision problem (Jankowski & Nyerges 2001, Jankowski 2009, Ramsay 2009). Another focus has been the development of techniques to support the identification and evaluation of decision alternatives through the medium of on-line digital maps (Malczewski 2006a, 2006b, Boroushaki & Malczewski 2009).

These themes of academic enquiry and accompanying examples of software application development have several common and binding threads. One that prevails upon the others is the concept of (public) Participatory Geographic Information Systems (PGIS). While there is a substantial literature on this topic, only a part of it is focused on the content of this chapter. This focus concerns the nature and role(s) of Web-based participation or collaborative use of geospatial data to discuss issues of local importance. Much of the discussion in the PGIS literature has sought to define and refine the concept (for example, Schlossberg & Shuford, 2005, Sieber 2006), and to clarify its role in the process of engaging members of the public in the use of GIS software and spatial data. Tulloch (2008) has intelligently questioned the extent to which user-provided content to Web map sites actually aligns with 'participation' in the sense conveyed in the PGIS literature, and the extent to which both fall within the umbrella of GIScience. In this context, he notes (p. 170) that 'one of the fundamental distinctions may turn out to be that VGI (Volunteered Geographic Information) is more about applications and information while PPGIS (Public Participation GIS) seems more concerned with process and outcomes' (parentheses added).

This chapter promotes the view there should be a convergence of applications and information *and* processes and outcomes, in that one aspect cannot be of much use without the other. While it is useful to conceptualize and refine our understanding respectively of the nature of user generated content on the one hand and PGIS on the other, these efforts may serve to blur the potentially more fertile intersections of the two (Elwood 2008b). In this context, the development of a Web 2.0-based application that facilitates user contributions to interactive digital maps of new spatial features as well as the selection of existing features of individual or group importance to an issue or theme, twinned with the ability to geo-tag such features with personalized comments or other information *and* engage in synchronous or asynchronous discourse with other participants on-line, offers a multi-faceted perspective on the intersection area.

Here, the emphasis is on contributions of individualized knowledge and discourse between contributors at the same time *and* within the same environment. The nature of participation is refocused somewhat in the sense that it is at least one step removed from users participating individually or *ex-post facto* collaborating by tagging information on a Web map for others to see, and/or leaving behind comments for others to respond to at a later date (c.f. Hopfer & MacEachren 2007 on the communication content of tags and annotations, Rinner et al. 2008, Sidlar & Rinner 2009, Simão et al. 2009). Instead, in this case the interaction is moved to a participant-website-participant dynamic in the sense that discussion can occur in real-time using the Web map as the medium of contact and communication. As suggested above, participants can be individuals and/or groups and both geo-tagging and discussion can occur synchronously in real-time (c.f. cell phone texting and instant messaging) or asynchronously through 'left messages'. Moreover, the locus of communication can be dispersed (for example participants working together from their own office or home computer) or co-located (working in the same room using individual computers connected to the Internet in a workshop setting). The objective of the dynamic in either case is deliberative discourse between participants through the Web map interface and the information gathered and its potential benefits at least equal those noted by Rinner & Bird (2008) in the context of on-line Web mapping tools.

Although still some way short of being able to contribute formally to supporting spatial decision making, the concept of map-chatting at the very least allows participants in a map chat to explore quite fully, based on their own experiences and knowledge, an issue of local or regional importance by informing each other, consulting with each other, involving each other in dialogue and collaborating in a way that other approaches do not allow. The innovative design of the MapChat tool is discussed in the following section, prior to describing its use in a preliminary case study.

3 MAPCHAT VERSION 2

The second version of the prototype MapChat tool improves upon the successful proof-of-concept that the first version provided. Both MapChat versions 1 and 2 use many of the same underlying OS components, although version 2 represented a break from the relatively constrained architecture underlying version 1 (Hall & Leahy 2008). Version 2 continues to use hypertext pre-programming language (PHP) on the Web server with the MapScript PHP library providing access to MapServer software functionality. It also continues to rely on the Open Source (OS) projects PostgreSQL for the backend database and PostGIS for spatial data storage and processing within the database. However, the entire architecture of Map-Chat itself was redeveloped from the ground up in version 2 to achieve greater flexibility and efficiency of both the server- and client-side components, as well as to improve the dynamics and usability of the client User Interface (UI).

Recent research using interactive Web mapping has placed considerable thought into interface design. For example, Simão et al. (2009) adopted a minimalist approach, but even with this principle at the forefront of their implementation they report challenges in achieving a simple, yet effective workflow. Although Simão et al. (2009) use proprietary software including ArcIMS, ArcGIS server, ArcSDE and Oracle, they note the last three components can be replaced with OS alternatives. However, in fact all components they use could fairly readily be replaced by OS projects. Beverly at al. (2008) used a combination of code written in Flash and PHP for their application which eliminated cursor-based map zooming, as this was thought to be too difficult for unskilled users of digital maps. In general, there is a discernable and general trend toward use of OS components in many other areas of Web mapping (for example, Anderson & Moreno-Sanchez 2003, Caldeweyher et al. 2006, Keßler et al. 2005, among others). In addition, Rinner and his associates (Rinner et al. 2008, Rinner & Bird 2009, Sidlar & Rinner 2007, 2009) draw upon the mapping Application Programming Interface (API) of Google and distributable code to produce a tool that is somewhat similar to, yet different from in several important ways, the MapChat application discussed in this chapter.

3.1 *MapChat software architecture*

Figure 1 illustrates the high level layout of the main components and resources used for MapChat version 2. The Zend Framework (http://framework.zend.com) provides a modular and extendible set of class libraries for PHP on the server. For the clients, the jQuery library (http://jquery.com) and its associated plug-in libraries provide a cross-browser JavaScript environment for dynamic Web pages. MapChat itself is essentially a set of class libraries that are built upon these PHP and JavaScript libraries. The mapping components of MapChat are based on the ka-Map PHP and JavaScript APIs for rendering tile images on the server, and displaying them in a dynamic Web map interface similar to the style popularized by Google Maps. This consistency in appearance is important as it maintains the look and feel of the popular Google application without 'being Google'.

One of the main features of the new design is the use of a front controller for directing all requests from Web browser client machines that connect to the MapChat Web server. The front controller receives incoming connections from client computers and forwards these to four types of specialized controllers depending on The Uniform Resource Locator (URL) requested. When a client browser first connects to the Web server, the request is passed to the 'index' controller by default. This controller initializes a PHP session on the server and returns an HTML document containing the main start-up Web page for the MapChat application. The client browser then renders the HTML document and loads the JavaScript and cascading stylesheet (CSS) resources that are required to initialize the client interface for the application. All JavaScript and CSS resources that are requested by the browser are directed through 'js' or 'css' controllers, which compress and cache the code (if it is not already compressed and cached), and then return it to the Web browser.

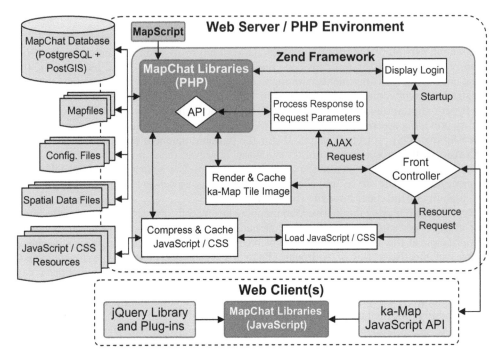

Figure 1. High level architecture of MapChat version 2.

After a client computer has loaded the initial MapChat interface, the majority of communication with the Web server is directed through an API controller. This provides access to functions for interacting with the core PHP libraries for MapChat through a customized set of API functions. API requests made by clients must name an API function, and supply any parameters required for that function. The controller loads and executes the corresponding API function, which generates an extensible markup language (XML) response that the controller returns to the client browser for client-side processing. Finally, the ka-Map controller handles requests for rendering image tiles for the map that is displayed in the client interface. This uses the PHP code from ka-Map to render, tile, and cache the map images (or retrieve previously cached images) that are loaded by ka-Map's client-side API.

3.2 *Speed and efficiency*

MapChat 2 implements several features that give it significant performance advantages in contrast to the first version. The most visible difference is achieved by the substitution of the Chameleon MapServer template system with the ka-Map API combined with a customized approach for defining the client UI using XML configuration files.

The Chameleon system used in MapChat 1 required that a single map image be rendered for the entire map image displayed in the browser every time a change was made to the map extents or displayed layers. This consumed significant bandwidth since each generated image was a unique file that needed to be downloaded from the server by the browser. In the case of digital orthoimagery this can amount to the consumption of significant bandwidth and result in slow response times. It also requires significant resources on the Web server, since all spatial data displayed in the map need to be processed in order to generate each new map image.

By using the ka-Map system in MapChat 2, map images are rendered in tiles at a set of predefined zoom levels. These tiles are cached by the server as they are rendered for each location, zoom level, and unique combination of map layers. They are also cached on the client-side, such that the browser only needs to download tiles that it has not already cached locally. In addition, groups of layers can be loaded into different tiles that are superimposed in the browser. This allows dynamically changing layers (such as user drawings) to be updated separately from other more static layers (e.g. aerial photography, roads, property boundaries, etc.). Individual tiles can also be updated when changes occur, instead of updating the entire displayed map. However, this approach does have a trade-off in that it requires more intensive processing by the browser, which must display and manipulate many individual tile images instead of one single image for the whole map.

Another major improvement in version 2 is the management of code resources that are loaded by the browser. In version 1, all JavaScript code resources used by the client-side interface are loaded every time a browser connects to a new MapChat session, regardless of whether a given JavaScript function is ever used in a given session. In MapChat 2, however, JavaScript code is loaded on-demand as it is needed. Configuration files assigned to each MapChat discussion are used to define the components in the client interface. However, rather than being interpreted by the server, the configuration is sent to the browser application. The client generates the UI elements (e.g. menu items, tool bar buttons) as defined by the configuration, without requiring additional code until the elements are used by a map-chatting participant.

Less visible improvements implemented in MapChat 2 include the use of gzip compression (http://www.gzip.org/) of all text-based data returned by the server (including HTML, JavaScript, CSS, and XML AJAX responses). JavaScript and CSS code is compressed on the server using the Minify library for PHP (http://code.google.com/p/minify), and this is cached similarly to the image tiles so that the code only need to be compressed once on the server and downloaded once by each client. While this has minimal impact on the actual functionality of MapChat, use of the application by users in rural or remote areas with low bandwidth connections is an important consideration, especially for the case study discussed in this chapter. Overall, the speed and efficiency improvements in MapChat 2 have proven effective for enabling a highly dynamic, Web-2.0 style application to run sufficiently well over dial-up connections.

3.3 *MapChat database and recording chat/event data*

The server-side of MapChat relies heavily on the PostgreSQL/PostGIS database environment for storing information. While the database structure was revised somewhat for version 2, its structure is similar to version 1. Figure 2 illustrates the general layout of the database, which essentially serves two purposes. First, it stores metadata about users, groups, and discussions. These metadata define users' membership and permissions in groups and discussions on the server. They also allow different discussions to be customized by linking each to its own MapServer mapfile (to define the data that are displayed for the discussion) and configuration file (to define the tools and behaviour of the client interface).

Second, the database records the chat history, events, and map data that are generated as users participate in MapChat discussions. This is accomplished using schemas that are created for each individual discussion. The discussion schema design is very similar to that used for version 1 (Hall & Leahy 2008). The main features in version 2 are two log tables used for storing records containing chat messages and event data, and two tables for storing map features that are drawn and/or selected by users during the discussion. Threaded discussions are achieved by storing unique identifiers for each message, and adding these identifiers as parents for any messages that are in reply to a previous message. Optional private messages are supported by adding records to two tables that link one or more users and/or groups to specific chat messages.

One of the more significant changes made in the database for MapChat 2 is the way that user drawn and/or selected features are recorded and linked to chat messages. In version 1, any features that were drawn or selected by a user were recorded in discrete sets of records in the selected features table. These selection sets were then linked to chat messages by storing

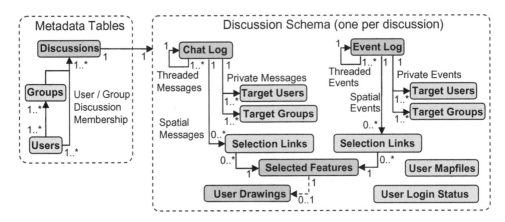

Figure 2.   General layout of the database for MapChat 2.

the selection set's identifier as a property of the chat message. This meant that each selection set needed to be retired any time it was linked to a message, since adding or removing features from it would also apply to any messages that were previously linked to the selection. In MapChat 2, however, this problem was avoided by using an intermediate table that stores records containing chat messages and selected feature identifiers. This makes more effective use of the capabilities of the PostgreSQL relational database, and establishes a flexible many-to-many relationship between chat messages and selected features.

The database is also used by MapChat 2 for storing login and mapfile 'states' for each user in a discussion. With the use of Chameleon in version 1, mapfile states were saved in a temporary folder associated with PHP session data. This made retrieval and analysis of the mapfiles difficult, given their temporary nature and a lack of any implementation in MapChat 1 for linking PHP sessions with permanent records stored in the database. Storing the user mapfiles in the MapChat 2 database also improves performance, as the mapfile text can be loaded directly into the PHP MapScript library without the need to read or write individual files to disk as an intermediate step. It also enables persistence of the map state for a given user (i.e. the on/off status of individual layers) across separate PHP sessions.

3.4   *Client events and synchronization*

Once MapChat is loaded in a browser, the Web server components outlined in Section 3.1 essentially serve as an interface between the client-side application and the MapChat database. Figure 3 illustrates this process, using the sending and receiving of chat messages between clients as an example. When a user sends a chat message from the client interface, an AJAX request containing the message text and any other relevant parameters (e.g. the ID of a parent message if the user is replying) is sent to the Chat_Send API command on the server. The Chat_Get API function then uses the corresponding internal MapChat libraries to interact with the database and record the new message. At the same time, all client computers that are currently connected to the same discussion will poll the server every five seconds to check for new messages. This involves sending an AJAX request to the Chat_Get API command, which queries for any new messages and returns them to the browser. Upon receipt of a new message, the client application will update any relevant content in the browser UI (e.g. the chat dialog UI).

The new approach in MapChat 2 to use an API controller for AJAX connections has also greatly improved the ability to transfer events between clients and the server. In the development of MapChat 1, a strategy for 'pushing' events to the client was implemented by appending additional data for one or more events in the XML responses to AJAX requests. In the browser, the XML responses are passed through a dispatch function that looked for

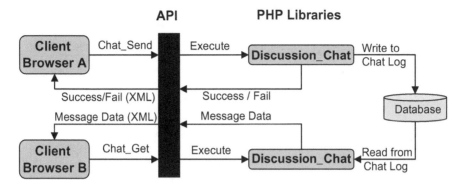

Figure 3.   Example of the role of **PHP Libraries** and **AJAX API** as the interface between client computers and the database in MapChat 2.

additional event data. Corresponding functions related to any received events are then triggered, regardless of whether they are directly related to the anticipated response from the original AJAX request.

In MapChat 2, the use of an API controller on the server complements this strategy by centralizing the sending/receiving of AJAX requests from client browsers. Figure 4 shows that as each AJAX request is made, the API controller can monitor the database for event information that is unrelated to the requested API function and appends additional data about these events to the response anticipated by the client-side function that initiated the request. In addition, it also allows incoming parameters submitted by the client to include additional event data that have been recorded in the browser between AJAX requests. The API controller checks each incoming request for event data and records them in the discussion database.

The benefits of this approach are greater efficiency of the application, as fewer connections need to be made to record important information, as well as more effective use of the original strategy for recording and dispatching event data. For example, events that describe a user's actions (e.g. panning, zooming, opening/closing dialogs) can be recorded as they take place in the client browser, then attached as event data to be logged on the server in the next AJAX request (usually as the browser polls for new messages every five seconds). Thus, detailed and important data that are useful for reconstruction and analysis of how users interact with the software during a discussion can be recorded with no impact on the application's functionality, apart from a negligible increase in bandwidth. As noted in Section 2, collection of this 'hidden' information should be revealed explicitly to each participant in a map chat prior to participation.

This is also beneficial for keeping multiple on-line users synchronized with each other. For example, when a participant user enters a discussion in MapChat, or draws new features on the map, any subsequent AJAX requests made by browsers used by other members of the discussion (e.g. when sending a message, selecting features, etc.) will automatically have additional data about the first participant's action appended to the anticipated AJAX response. On receipt of these additional data, appropriate events are triggered in the client browsers (e.g. a change in the first user's status icon, or refreshing tiles on the map to display new drawings, etc.).

3.5   *Client-side interface*

From an end-user perspective, MapChat 2 appears very different from version 1, as the UI presented in the browser has been completely redesigned for greater functionality and ease of use. In version 1, the interface used a conventional format with controls for interacting with the software were presented in a tool bar fixed across the top of the Web page. The map was presented in an area of static width and height dimensions. The chat history, user list,

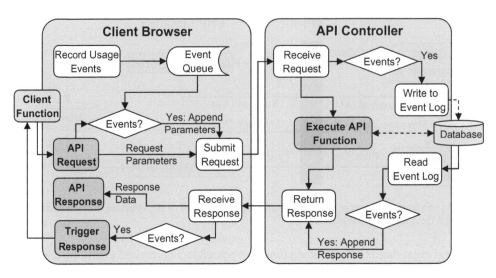

Figure 4. Process of logging and transferring of event information via AJAX requests processed by the API controller in MapChat 2.

and layer list were accessible as panels that could be displayed by selecting tabs across the top edge of the map area. Each of these panels was in a fixed location on either the left or right sides of the map. The version 1 design proved somewhat challenging to users, as the fixed dimensions and positioning of controls meant that interacting with the layer list and/or chat panels would temporarily leave much of the map area obscured. The fixed dimensions also could not take advantage of larger screen resolutions on many newer devices without adversely affecting older computers with smaller screen dimensions.

The adoption of the ka-Map JavaScript API, combined with jQuery and the jQuery UI plug-ins for user interface controls, has allowed MapChat 2 to overcome the deficiencies identified in version 1. Rather than positioning the map as a smaller element within a traditional frame of other elements in a Web page, the MapChat 2 canvas fills the entire browser window with a ka-Map object, as illustrated in Figure 5. The menus for accessing different controls are accessed by right-clicking the mouse anywhere on the map. Any controls that are made available from this menu, such as the chat history, layer list, or map navigation tools, are presented as objects within or on top of the map within jQuery UI objects. The jQuery UI objects allow these controls to be moved and/or resized within the browser window, so that users can position them to optimize the use of screen space for the task at hand. This layout for the interface has introduced a great deal of flexibility for configuring individual discussions in MapChat 2. Different tools can be included or excluded by configuring the options listed in the right-click menu. No modifications to the overall layout are needed, since each tool is contained in its own flexible dialog superimposed over the map.

Many other enhancements introduced in MapChat 2 stem from the greater performance and flexibility afforded by the use of jQuery combined with more effective use of CSS styles for fast and dynamic manipulation of HTML elements in the Web page. Figure 6 highlights the chat dialog for MapChat 2, which is much more advanced and intuitive than the first version. Version 1 simply provided a static text box that users could type messages into. To send a reply, a previous message needed to be selected, while the same text box was used to enter the reply text. In contrast, MapChat 2 dynamically places text boxes in appropriate locations, above the 'new topic' button in the dialog for new messages, or directly below a previous message when replying to it either in the chat dialog or in a callout bubble on the map canvas when replying. If users need to correct previous messages they have sent, an edit button next to the message can be clicked to replace the text dynamically with an editable box. These features alone have improved the intuitiveness and quality of contributions to discussions in

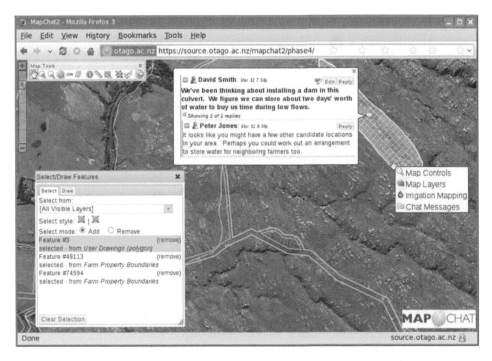

Figure 5. Overview of the Web interface for MapChat 2. Objects displayed include the map controls toolbar (top-left), the Select/Draw features dialog (bottom-left), a chat bubble for spatially-linked messages (top-middle), and the right-click application menu (right).

MapChat 2. For example, users are more likely to maintain threads in their discussions, since they must inherently choose to start a 'new topic' or 'reply' to previous messages and because they can ensure correctness in the content of their messages quickly and easily.

Spatial links attached to messages are also more flexible in MapChat 2. In version 1, any selected features or new features drawn were automatically linked to the next message sent by a user. This approach required careful deliberation on the part of the user to select/draw features prior to typing a corresponding message. It also prevented users from participating in a discussion if they wanted to make comments or reply to messages that were unrelated to the set of features they currently had selected or drawn. In MapChat 2, the linking of chat messages to features is optional, as toggle buttons are provided with the chat input text box. Three icons allow users to link their comments separately either to selected map features, drawn shapes, or the current map extent relative to the displayed zoom level. The links are also editable for previously sent messages, allowing a user to add, modify, or remove spatial links to any of their text messages at any point in time.

The utilization of Scalable Vector Graphics (SVG) in modern Web browsers (or Vector Markup Language (VML) for Microsoft Internet Explorer) has also allowed development of more interactivity with features on the map in MapChat 2. Version 1 relied on a library that emulated 2-dimensional graphics through styled HTML elements. This proved very inefficient when users chose to draw features with more than a small amount of detail, and limited the ability to manipulate the graphics. With SVG/VML, complex shapes can be drawn dynamically in the browser with a mouse or touch pad. This has greatly improved the usability of the drawing tools, while also allowing spatial features to be displayed dynamically on the map without the need to load or update image tiles. For example, clicking on icons associated with spatially linked messages in the chat dialog now retrieves feature coordinates from the server and flashes a vector graphic of the features on the map display. SVG/VML graphics can also respond interactively, for example when the mouse is hovered over a feature that a user has

239

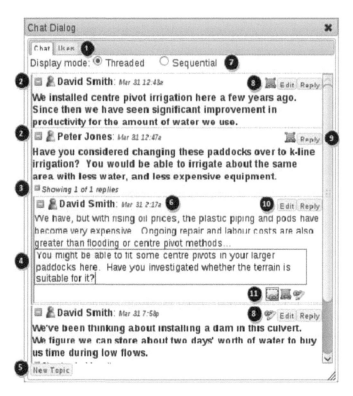

Figure 6.   Controls/features of the chat dialog in MapChat 2.
1. List of discussion members, private messaging;   2. User online/offline status icons;
3. Collapse/expand threads;   4. Message input textbox;
5. Start new thread;   6. Message timestamp;
7. Threaded/sequential display mode;   8. Zoom-to/display spatially linked features;
9. Reply to message;   10. Allows user to edit his/her own message;
11. Toggle message links to extent/selection/drawing.

currently selected, its corresponding record is highlighted in the list of features displayed in the selection dialog (see Fig. 5).

Overall, the new features of MapChat 2 not only resolve user-reported issues with the first version of the software, but provide much greater efficiency and functionality in virtually all aspects of the software. While the software is still some way short of satisfying the current 'wish list' of functions (for example, the ability to upload and geo-position user-provided digital photographs is not yet implemented; nor is the user-supplied addition of global positioning system co-ordinate points or tracks), the ability to facilitate on-line collaboration in discussions of spatially relevant issues and phenomena is satisfactorily implemented. The remainder of this chapter summarizes the use of version 2 of the software in an application focused on irrigation at the farm-unit and sub-watershed levels in an area of traditional dryland mixed sheep and beef farming in the South Island of New Zealand.

4   CASE STUDY AND PROTOCOL

As noted earlier, the basic intention of the MapChat tool is to facilitate real-time collaboration, using synchronous texting, geo-tagging and feature drawing as well as leaving asynchronous messages and/or geo-tags for *ex-post facto* responses from other participants in a MapChat discussion (c.f. the more limited tagging and asynchronous commenting approaches used by

Rinner et al. 2008, Seeger 2009, Sidlar & Rinner 2007, Simão et al. 2009). To stage an initial rollout and subsequent revision platform for the software, a case study was undertaken in a rural community in the Upper Taieri River catchment (UTC) in the south central area of the South Island of New Zealand. Water in the UTC is a scarce resource that is essential for the maintenance of agricultural practices dominated by dry-land mixed sheep and beef farming. This form of farming in the study area is currently faced with a number of challenges, including fluctuations in the global demand for wool and lamb, changing climatic conditions, new regulations on the lease of traditional high country pastures, impending changes in the historic rights of access to lowland water resources, and pressure from new water-intensive forms of farming, such as dairying. There are over 150 years of farming history in the study area and a pattern of inter-generational transfer of properties within families. Farmers in the UTC are currently seeking ways to cooperate at a community level to help ensure the security of their local water supply in order to preserve not only traditional farming practices but also continuity of inter-generational life-styles.

To address the issue of water access and use in the study area, local farmers have joined with participants from the New Zealand (NZ) Landcare Trust, the NZ Government's Department of Conservation, the Province of Otago's Fish and Game Department, the Otago Regional Council, and the Central Otago District Council to form the Upper Taieri Water Resource Management Group (RMG). The principal objective of the RMG is to develop a catchment-based water supply system to ensure effective future allocation of water through consolidation of mutual water access, simplification of compliance systems, and improved efficiency, flexibility, and consistency of management of water resources. It is intended that the system will be operated by a community-led management company.

While the stakeholders involved in the RMG support the overall objective of developing a community-led and managed system, a shared vision or plan of how such a system might be best implemented remains to be developed. In late 2007, members of the RMG contacted the researchers at the University of Otago to investigate whether this need could be served by a participatory GIS approach. The intention was to involve RMG members collaborating in the preliminary stages of forming a development plan for renewing the allocation of historic water access rights and implementing a water distribution system that meets the needs of the farming community. In principle, the MapChat tool is ideally suited for local farmers to share with each other their knowledge and perspectives on current and future irrigation practices and plans. The constant commitments of the farming calendar make it difficult for individuals to attend regular community meetings. Hence, it was thought than an online tool such as MapChat could replace community meetings with virtual collaboration using both synchronous and asynchronous messaging and map centred discussion of irrigation practices and options.

To facilitate this process a linked four-phase approach was developed. In Phase 1, ten local volunteer farmers, eight of whom were active members of the UTC RMG and two others who had attended RMG meetings, met one-on-one with a project researcher from the University of Otago. The purpose of this initial meeting was to identify on a digital, geo-registered map the land parcels that comprised each farmer's consolidated farm unit (many farmers in the study area own spatially discontiguous blocks of land). At the same time, information was collected from each participant regarding their computing skills and Internet access. Informed consent for participation was also obtained. The primary goals of this phase were to assess the comfort level of participants with computer use and to identify the types of spatial data layers each participant would like to use to review their own water needs and those of the broader community. The observations and information that were gleaned during this phase also guided the development of the functions that were ultimately programmed into version 2 of the MapChat tool, while advising the subsequent phases of the study.

Upon the completion of Phase 1, participants were initially partitioned into two groups of five members each for Phases 2 and 3. Group membership was based primarily on the access of participants to high speed Internet, followed by their level of comfort/experience using computers and digital mapping and also their general farm unit location within the overall

UTC study area. Following a control/experimental group design described by Jankowski (2009), Group A comprised a supervised/offline control group with dial up connection speeds and low computing comfort levels, and Group B comprised an unsupervised/online experimental group with either high-speed DSL or radio broadband Internet connections and higher levels of comfort with computer use.

Phase 2 spanned a period of approximately six weeks. During this time, all participants met, again, individually with a University researcher and were provided with a set of written instructions and a demonstration of the basic MapChat user interface. They were also instructed in the use of a MapChat irrigation mapping tool programmed specifically for this phase of the study. This tool allowed participants to create map layers representing different types of irrigation systems including water reticulation networks, areas for water storage, equipment, and other unclassified features on their farms. In general, important spatial features including water 'take' locations (points) from the Taieri River or it tributaries, irrigation channels (lines), and field-based irrigation dispersal locations (areas or polygons) were drawn to reflect their current and planned irrigation systems. All participants were given the task of creating map layers representing current systems, as well as layers representing plans for the relatively near future (up to 10 years) and longer term future (beyond 10 years). The control Group A participants worked alongside a researcher to complete their irrigation maps, while the experimental Group B participants were asked to complete their irrigation maps on the Internet in their own home after a preliminary demonstration and without a researcher present.

The primary intentions of Phase 2 were to give the participants experience in use of the map drawing/markup tools available in MapChat, to encourage them to think forward about their future on-farm irrigation practices, and to create a basis for dialog with other participants using MapChat's collaboration functions in Phases 3 and 4 of the study. The specific irrigation maps produced during this phase of the study are not discussed further in this chapter. At the end of Phase 2 each participant was asked to complete an interview to evaluate their use of the MapChat tool to date and to comment on general issues related to management of water resources within the catchment.

The initial intention of Phase 3 of the study was to involve group dialogue using MapChat in a researcher facilitated (control Group A) and non-facilitated (experimental Group B) co-located setting. Following from this, Phase 4 would combine the members of Groups A and B to review the outcomes from the Phase 3 co-located feature tagging and discussions, with the Group A and B maps and discussions merged into a combined map database. This process would take the separate control/experimental individual irrigation practices and plans of Phase 2 and the group collaborations of Phase 3, and combine them for a more embracing discussion of community-wide irrigation issues. Phases 3 and 4 would also end with further questionnaire-based surveys, intended to gauge what the participants had gained from the process in terms of consolidating their own and other farmers' irrigation plans into a community irrigation system. Also, the strengths and weaknesses of the MapChat tool and the general approach of Web-based collaboration would also be evaluated.

Hence, the incremental four-phase design, with control/experimental components in Phases 2 and 3, was expected to yield not only rich content in terms of discussions between participants, but also substantial information about the ease and success of use of the tool. However, due to a variety of factors, the design of the study following Phase 2 had to be modified. Foremost among these factors was the fact that Phase 3 fell at an extremely busy time during the New Zealand farming calendar and the time required to commit to the project, post-Phase 2, became a significant issue for virtually all of the participating farmers.

After the Phase 2 mapping was completed, all of the case study participants were contacted to coordinate their availability for the two group meetings that were planned for Phase 3. However, at this stage, two members of Group B from Phase 2 indicated that due to farming commitments they could not commit further time to the project. The reduced number of participants, combined with the logistical challenges of scheduling meetings that satisfied the geographically dispersed members of Groups A and B, led to the decision to break from the originally planned protocol. Instead, two groups of four were formed, and their members were primarily determined by geographic proximity to the rural towns of Ranfurly and Middlemarch. A third participant ultimately could not attend on the actual day of the Phase 3 workshop in Ranfurly due to unexpected commitments related to moving his stock.

Hence, the final outcome for Phase 3 was two groups, hereafter referred to as Group C (three members) and Group D (four members). A further adjustment was made to the Phase 3 by abandoning the use of the experimental/control approach used in Phase 2 and to treat both groups the same for the last two phases of the project. This decision was made in part because the original Phase 2 groups could not be sustained, but also because experience from a parallel case study using MapChat version 1 in British Columbia, Canada (Hall et al. 2010) showed that although simple in concept, the operational use of map-chatting, combined with map markup and feature chat linking was proving to be relatively complex for members of the lay public. Hence, initial and on-going moderation by researchers was required to maintain group discussions.

Another reason for the change to the Phase 3 plan was based on responses to the feedback from the participants during Phases 1 and 2. While the RMG and members of the Upper Taieri community in general understood the need to work towards a community-led water management approach, many were unsure about or had not agreed on the primary issues and key criteria that first needed to be discussed. Thus, to help ensure productive discussions during the Phase 3 workshops, both groups were given the same level of training in the use of MapChat to introduce them to the exploratory and chatting functions not used during Phase 2. The workload was reduced to focus only on discussing the collective irrigation maps produced during Phase 2, and broader group discussion on community-level strategies was left to Phase 4. Both groups were provided with a list of suggested topics to consider in Phase 3 as seeds for their discussion.

The format of Phase 4 of was adjusted to reflect the changes to Phase 3. The original intention was to have the participants discuss the practicalities of achieving a community-led water management scheme that would have been derived from the outcomes of Phase 3 combined with individual on-farm irrigation plans. However, since this level of discussion was not reached during Phase 3, a similar approach was taken in Phase 4 to seed the discussion with a list of related topics provided to the participants.

5 MAPCHAT USE

The protocol described above allowed Phase 1 of the case study to focus on data collection that was used for input into the MapChat application. The more direct use of MapChat's functions by the participants for creating maps and dialoguing on the irrigation issue took place during Phases 2 through 4. This section summarizes how the tool was used during these individual, group and distributed use phases based on data that were recorded in the application's database as geometries, chat messages and events.

5.1 *Individual use*

As noted earlier, in Phase 2 of the UTC study participants allocated to the control Group A worked alongside a researcher to produce their current, relatively near and more distant irrigation plans, while those in the experimental Group B were asked to use the irrigation mapping tool in MapChat online, after a preliminary demonstration without direct supervision from a researcher. One of the obvious differences between the Group A and Group B members is the amount of time taken with the software before the irrigation maps were completed. All participants in Group A completed the mapping within the two to three hours spent while a researcher was present in their home to assist with the use of the irrigation planner mapping tool. Also, participants in Group A were focused on completing the mapping tasks within the time allotted so they could return to other important work on their farm and also to ensure a second meeting would not be necessary.

It was expected *a priori* that Group B would take longer given that the participants could work on the tasks on-line at any time of their choosing. Table 1 summarizes the time spent using the MapChat tool by the five Group B participants during each week following the date they were given the Phase 2 instructions. In all five cases, more than two weeks passed before any of the participants completed their maps. After five weeks, three participants who had yet to finish were reminded that their maps were still incomplete, and were encouraged to finish them to ensure that the maps could be used for Phase 3. One Group B participant ultimately agreed to meet with a researcher in person in order to expedite the completion of the maps. Ultimately, four of the five participants completed their maps to their own satisfaction. However, one of the participants (User 5) did not finish the irrigation system mapping, and was unable to meet with a researcher to complete the work.

The participants in Group A required 2 hours on average to complete the mapping when working with a researcher, with the longest time being 3.37 hours. For Group B, the average time is much higher at 5.33 hours, but with a much wider range from 1.57 up to 9.75 hours. As noted above, the higher amount of time invested by this group is related to the fact that they did not need to complete the mapping within the time frame of a single meeting as well as the fact that they did not have assistance to complete the task. However, in each case, a variety of factors would likely have determined how much time each individual was willing or able to spend with the tool.

Table 2 provides a summary of the maps, features, and individual vertices that were created through use of the MapChat irrigation mapping tool in Phase 2. It is apparent that without a researcher present, Group B participants spent much more time with experimentation

Table 1.   Total hours spent by Group B participants using MapChat online for Phase 2.

User*	Week 1	Week 2	Week 3	Week 4	Week 5	Week 6	Week 7	Week 8	TOTAL
1	2.85	6.03	0.87	–	–	–	–	–	9.75
2	1.65	–	–	–	–	–	–	‡0.90	2.55
3	2.10	–	1.10	–	–	–	–	–	3.20
4	1.63	1.10	1.27	–	–	†3.55	†1.77	†0.23	9.55
5	0.40	–	–	–	–	–	†1.17	–	1.57

Calculations do not include times during which the participants appeared to be idle longer than 5 minutes at a time, and exclude login sessions during which no map edits were made.
* User numbers arbitrarily assigned;
† Time committed after being reminded;
‡ Participant met with a researcher to expedite completion of their mapping.

Table 2.   Summary of use of the MapChat irrigation mapping tool during Phase 2.

	Group A	Group B	Total
Maps			
Created	14	29	43
Discarded	0	14	14
Saved	14	15	29
Features			
Drawn	279	474	753
Discarded	12	108	120
Saved	267	366	633

(creating and discarding maps and features). With a researcher present for Group A, no maps and relatively few features were discarded.

Group A participants were usually more comfortable to have the researcher actually operate the mapping tool while they simply gave instructions verbally and directed the mouse for drawing features by pointing at map locations on the computer screen. In most cases, Group A participants would take control momentarily as they grew more comfortable with the tool, and/or when explaining finer details that were difficult to draw, though there was no explicit means to record this exchange. This generally explains the lower number of discarded features, as the researcher was more familiar with the workflow and therefore was able to use features such as the 'undo' function, while adding vertices to a drawing (reducing the need to delete and re-draw an entire feature to make corrections). Some of the users in Group B indicated they had encountered various connectivity problems as well as software issues while they worked with MapChat online, which may also have contributed to the additional experimentation and corrections required for them to complete their maps successfully. However, this input allowed bugs to be fixed and changes also to be made to improve the overall performance of the tool.

Another aspect of the mapping that can be compared is the level of detail and completeness of the features and attributes created for the irrigation maps. Table 3 lists some characteristics of the data recorded during the creation of the paddocks and irrigation network layers that the participants created to represent their current, near future, and long-term irrigation plans. These data include the number of vertices that were added, removed, and ultimately saved for representing the number of features created, and the percentage of possible attribute values that were completed. These two categories of features were the most frequently drawn (i.e. all of the farms had irrigated paddocks, and all but one included some form of water reticulation network features).

The use of MapChat described in Table 3 is generally consistent with the higher level summary in Table 2, as participants in Group B tended to remove or correct proportionally more vertices while producing their drawings than did those in Group A. In terms of the vertices that were used in the completed maps, there is proportionally more detail (i.e. more vertices per shape) for the paddocks that were drawn for Group A members versus Group B, while the reverse is true for irrigation networks. This is likely a result of Group A members focusing priority on producing the paddocks layers, as it was felt that this provided more useful information about what locations are being irrigated, and what types of irrigation were being used.

Table 3.   Complexity/completeness of Phase 2 irrigation map layers.

Time period	Current		Near Future		Long-Term	
Group	A	B	A	B	A	B
Paddocks						
Total drawings	37	66	44	34	31	55
Vertices added*	1187	1024	177	342	171	272
Vertices removed*	116	256	14	109	15	95
Vertices saved	1123	771	1182	546	603	755
Attributes saved	40%	34%	38%	23%	38%	20%
Networks						
Total drawings	20	42	21	12	19	16
Vertices added*	212	877	28	23	12	75
Vertices removed*	22	107	2	4	0	17
Vertices saved	196	607	202	154	138	198
Attributes saved	62%	55%	60%	56%	56%	58%

*Refers to recorded events associated with input/removal of individual vertices, and may directly correspond with vertices saved in final maps.

The paddocks layer also included thirteen optional attributes that participants were encouraged to complete in order to describe the irrigation on each paddock (e.g. irrigation type, amount of water used, frequency of irrigation, etc.). In this context, there was a noticeable tendency for Group A members to have higher proportions of attributes completed. With a researcher present, it was possible to ask/remind the participants about these attributes as each feature was created. The Group B participants seemed to focus more on completing the drawings first, and in some cases members did not supply many attributes without being reminded at a later time after the initial maps were inspected on-line by a researcher.

5.2 *Group use*

In preparation for the Phase 3 workshops, two combined digital maps were created, one for each group, that incorporated the group members' individual irrigation maps. While all of the participants agreed to share their maps in this setting, there was not the same uniformity in agreement in sharing all of the attributes. Some participants also either did not complete maps for future irrigation plans, or opted not to share them. Three did not create long-term maps, and of those one also did not create a near future map, while another did not want to share the near future map with other participants. In those cases, irrigation maps for missing future time periods were duplicated from the latest time period for which data were created based on the premise that no changes were planned.

The MapChat tool was configured to include selection lists for farms and time periods that could be used to filter the discussion in the chat dialog, and update the map to display the corresponding features. Three additional maps were included in this stage that highlighted areas of change where land was newly irrigated, or no longer irrigated in future plans relative to the current maps. Throughout the workshops, the participants were encouraged to review these data and maps as they discussed each others' farms. As they used the controls to switch between farms and irrigation/change maps, their comments were flagged with categorization for subsequent analysis. The change maps revealed information more directly to participants and allowed them to see more clearly what would be new and what would be decommissioned in time on each farm. However, these data are not discussed further in this chapter.

Table 4 provides a summary of the comments made by the participants using the MapChat chat function during the Phase 3 workshops. Messages were classified by examining their properties within the MapChat database (e.g. links to spatial features), as well as interpreting

Table 4. Summary of Phase 3 workshop discussions.

Group	Total			Percent		
	C	D	Both	C	D	Both
Total chat messages	57	150	207	100	100	100
Topic threads	16	43	59	28.1	28.7	28.5
Irrigation-related messages	44	146	190	77.2	97.3	91.8
Off-topic messages	12	1	13	21.1	0.7	6.3
Questions posed	28	61	89	49.1	40.7	43.0
Answers given	17	53	70	29.8	35.3	33.8
Unanswered questions	4	9	13	7.0	6.0	6.3
Messages indicating user confusion	3	0	3	5.3	0	1.5
Erroneous messages	0	2	2	0	1.33	0.97
Messages with spatial links	8	13	21	14.0	8.7	10.1
Messages linked to drawn features	4	11	15	7.0	7.3	7.3
Messages linked to selected features	5	2	7	8.8	1.3	3.4
Messages linked to map extents	0	0	0	0.0	0.0	0.0

the message content. The most notable difference between the two workshops is the total number of messages. While Group D had one extra participant and lasted for approximately the same length of time, it had nearly three times as many comments made using the MapChat tool. This is likely due to the fact the members of Group D all generally had relatively higher levels of experience and comfort with the use of computers and the Internet and three of them had used the tool independently in Phase 2. In contrast, two of the three members in Group C (who had not used MapChat independently before) had relatively less computer-related experience, particularly with typing, which limited the amount of comments that could be made during the time period of the workshop.

The differences between Groups C and D also stem from other factors that relate to the group dynamics in a face-to-face environment, as well as the participants' perception of the purpose for the workshop. At the end of the Group C workshop, some members indicated that they had believed the purpose of the workshop was to 'practice' for Phase 4, rather than to have any serious discussion. This is evident in the relatively high proportion of off-topic messages in their discussion versus that of Group D. Another major difference is the level of verbal (non-computer) discussion observed during the workshops. The members of Group C appeared to be much more focused on operating the MapChat tool and typing their messages, and as a result spent relatively little time talking verbally to one another. In contrast, Group D used verbal discussion frequently throughout the workshop. While this did not result in less discussion recorded via MapChat, it did allow the participants to coordinate their efforts better. In this respect, these members avoided confusion, for example by asking each other which map layers they were currently viewing.

Some minor differences are visible in terms of how frequently chat messages were linked to map features in MapChat (see Fig. 7). While Group C overall had fewer unlinked and spatially-linked messages, the proportion of spatially-linked messages was higher overall than Group D. This is perhaps due to the lower number of participants, and lower volume of chat messages and higher level of verbal discussion between participants. In this context, the two researchers present during the workshops were better able to encourage and provide instructions to the three participants in Group C for using the spatial tools available to them.

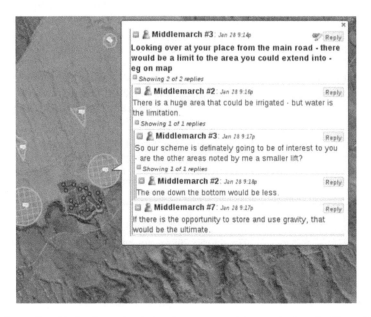

Figure 7. Example of irrigation-related chat between participants during the Phase 3 workshops (white callout bubbles within polygons represent comment/discussion tags).

5.3 Dispersed use

The same eight participants that had originally agreed to be involved with Phase 3 work (including the one that could not attend) agreed to participate in Phase 4. A two week period was allotted for the group discussion in this phase, with the option for participants to continue beyond this time frame if they desired. Participants could log into the project on-line and engage in discussion synchronously, depending on the presence of other participants, or leave comments for asynchronous responses by other participants at a later time. Workday commitments dictated that collaborations would most likely occur between 7pm and 11pm on weekdays, or potentially any time during the day on weekends. Table 5 shows that the time invested by the participants during this phase varied significantly. One participant did not login at all, while another only did so briefly without making any significant contribution to the discussion. In contrast, three users spent substantially more time online and logged in more frequently. While these users also made most of the contributions to the discussion, their time on-line did not translate directly into active discussion. Much of their time spent on-line was relatively inactive apart from viewing/interacting with the map display, as they would passively monitor for activity from other participants.

Examining the characteristics of the Phase 4 discussion relative to Phase 3, discussions showed more variability in terms of their focus. Table 6 summarizes the group discussion for Phase 4. A noticeable difference stems from having the discussion in an on-line-only setting. In this context, support from the researchers could not be obtained verbally, but instead was provided virtually on-line via the MapChat discussion itself. Thus, the numbers in Table 4 are organized to reflect the data recorded for everyone involved (E), for just the participants (P), and for only participant's messages that did not involve discussion with a researcher (OP).

Comparing just the total numbers of chat messages in the Phase 4 discussion shows that nearly half of all messages recorded were between a researcher and a participant, and that they originated with a researcher. This form of communication was primarily conducted using the private chat function of MapChat. The purpose of most of these private chats was for support when requested by a participant, or when a participant became stuck with the use of a specific aspect of the tool. In some cases, participants were pre-emptively contacted to notify them of problems that were observed with use of the software, to make recommendations for using the spatial tools relative to the discussion, or to ask if a participant was in need of assistance (e.g. when a participant signed on for the first time, but appeared to be inactive).

In total 40 messages were conveyed between participants during the two week period. This is much lower than the on-line messaging that was achieved during either of the Phase 3 workshops, despite having a considerably longer time frame for participating in Phase 4.

Table 5. Total minutes spent by participants during the Phase 4 online discussion.

User*	Day													
	1	2	3	4	5	6	7	8	9	10	11	12	13	14
1	127.3	1.4	0.4	137.5	3.6	7.3	26.9	4.8	23.4	70.9	1.3	0.3	–	–
3	–	39.4	–	–	–	–	–	33.7	78.9	–	10.9	–	–	–
4	–	–	–	–	29.0	180.8	208.6	–	–	–	–	27.4	–	21.0
5	–	5.9	–	–	38.7	1.6	–	–	–	–	–	–	7.0	–
2	7.3	2.6	29.3	1.7	–	0.2	–	–	7.1	–	–	–	–	–
6	–	–	8.8	–	–	–	–	32.0	5.9	–	–	–	–	–
7	–	22.1	–	–	–	–	–	–	–	–	–	–	–	–
8	–	–	–	–	–	–	–	–	–	–	–	–	–	–

*User numbers arbitrarily assigned.

Table 6. Summary of Phase 4 online discussion.

Groups*	Total			Percent		
	E	P	OP	E	P	OP
Total chat messages	144	74	40	100	100	100
Private messages	104	36	2	72.2	48.7	5
Topic threads	38	21	15	26.4	28.4	37.5
Irrigation-related messages	32	32	31	22.2	43.2	77.5
Community scheme related messages	6	6	6	4.2	8.1	15.0
Water supply related messages	12	12	12	8.3	16.2	30.0
Water take related messages	3	3	3	2.1	4.1	7.5
Weather related messages	4	4	4	2.78	5.41	10
Off-topic messages	3	2	2	2.1	2.7	5.0
Questions posed	32	25	16	22.2	33.8	40.0
Answers given	23	21	18	16.0	28.4	45.0
Unanswered questions	1	1	0	0.7	1.4	0.0
Messages indicating user confusion	20	20	2	13.9	27.0	5.0
Erroneous messages	14	8	6	9.7	10.8	20.0
Messages with spatial links	10	8	7	6.94	10.81	17.5
Messages linked to drawn features	5	4	3	3.5	5.4	7.5
Messages linked to selected features	2	1	1	1.4	1.4	2.5
Messages linked to map extents	4	4	4	2.8	5.4	10.0

* E = messages from everyone including the researchers; P = chat messages from participants; PO = chat messages from participants, excluding any replies related to private discussion with the researchers.

A variety of factors are likely contributors to the lower rate of input during this stage. First, the Phase 4 discussion did not have a specific set of tasks for the participants to follow (i.e. commenting on each individual farm's current irrigation use and plans). Instead, the participants were encouraged to discuss issues they felt were important to the development of the community-led water resource management scheme. However, as mentioned earlier, it was not evident that there was consensus among the participants about the nature of this scheme despite there being widespread agreement that it was a good idea. While a list of recommended points of discussion was provided with the instructions given to each participant, by and large these were not addressed.

Overall, 77% of the 40 messages were related to irrigation in general, but only 15% appeared in any way related to the broader issue of creating a community managed irrigation scheme. It may be that the individual participants did not feel they knew enough to address this issue, or that they did not feel confident about addressing it on-line relative to individual farm irrigation use. Finally, the function of linking spatial features to chat messages that was expected to be an important aspect of the interactions in this project, was not utilized by the participants to any great extent. Of the 40 chat messages in the group discussion, only 7 had links to spatial features. This issue is considered further in the following section.

Feedback provided by the participants at the end of Phases 3 and 4 indicated that all participants acknowledged the potential for MapChat to facilitate community interaction in exchanging information and dialoguing on the use of irrigation and reviewing future plans. However, at the same time several indicated that they were not sure how this approach would serve them personally. This may be due to the lack of consensus of what issues need to be addressed to begin with in the upper Taieri River catchment area, and it remains unclear as to the ability of software such as MapChat to capture this level of understanding for this community. Hence, in the absence of any significant discussion on the community irrigation issue throughout the four phases of the case study, it is difficult to assess the extent to which the integration of user generated content, in conjunction with the variously structured discussions that were facilitated through MapChat, will impact on the immediate irrigation problems faced by this community. Despite this, a number of important lessons were learned that have implications for future uses of the tool and the concept of map-chatting. These lessons are noted in the next section.

6 DISCUSSION

In the case study discussed above MapChat was used in a closed (password-controlled access) collaboration between a small number of members of a distinct farming community. The four-phase protocol for information collection and collaboration was implemented that lead participants incrementally first through learning use of the tool in a control/experimental design proved to be useful subject to the time constraints that affected the participating farmers. This process was intended to follow in a general sense the first four of the five levels of participatory GIS use identified by Jankowski (2009), derived originally from the International Association of Public Participation (2006), namely inform, consult, involve and collaborate. While there was generally agreement on the part of participants that the MapChat tool offered considerable advantages over the more conventional or traditional forms of discussion and information sharing, such as public or community meetings, the end product of the engagement process and use of the tool was somewhat less satisfying than had been hoped at the outset.

Before the end of the second phase of the case study's protocol was completed it was clear that the planned approach required adjustments. As the study process unfolded it became evident that there was less clarity and considerably less agreement than was initially understood to exist between participants on the types of irrigation solutions that were relevant to them individually and collectively for the UTC area as a whole. This lack of clarity and agreement was compounded by what can best be described as a general difficulty for the participants to be able to devote enough time to learn how to use the MapChat tool fully and confidently. Even although the initial design allowed for facilitating the control group through their use of the tool in the early stages of the study, this approach had to be abandoned and the control/experimental nature of the approach revised.

The outcomes of the study revealed that participants struggled generally when left to complete tasks by themselves from Phase 2 onward. While they appreciated the usefulness of the tool and the process, they generally had insufficient time available to use tool fully to complete tasks related to geo-tagging map features, establish spatial feature-chat linkages, and generate discourse between each other. This failure in sufficiency is evident in various aspects of the tool's use. Hence, this user group experienced difficulties in utilizing the basic functions of the software. Similar initial participant difficulties with using the software were experienced in an unrelated parallel study using MapChat version 1 in British Columbia, Canada. However, in this case participants have through time become able to use the tool successfully to geo-tag, comment on and discuss a substantial number of landscape features of value in their community (Hall et al. 2010). It is likely that this success is due largely to the significant and on-going involvement of local volunteers who serve as facilitators in the use of the tool, plus a community of users who do not have the same time commitments and constraints that affect the farming community in the UTC.

More generally, it is important to note that other case studies reported in the literature encounter similar constraints. For example, the 12 active participants reported in the study by Rinner & Bird (2009), despite being 'experienced computer users', registered only 22 replies to 26 contributed comments and found 'the handling of the online neighbourhood map difficult and did not use the option to link their comments to the map' with the software in question. Several other related studies use either somewhat ambiguous examples of *public* involvement in on-line map markup and attached commenting (for example, Hopfer & Maceacheran 2007), undesignated participants in experimental use of tools (Sidlar & Rinner 2007,

Simão et al. 2009), explicitly contrived usage (Cai & Yu 2009), or members of a University community (Rinner et al. 2008, Sidlar & Rinner 2009). Moreover, in all cases the number of participants when explicitly reported is always small, not reaching more than approximately 20. While understandable in the early stages of prototyping software, these numbers and some of the groups involved (e.g. student participants) are unlikely to encounter the same encumbrances and constraints that affect members of the general public.

Logistically, it is much easier to prototype and test in controlled environments with readily available participants than with the general public. However, in no instance is a case study reported in the literature where more than a relatively small number of participants are involved. Hence, the complexities of multiple simultaneous discussion threads and a large number of users creating and annotating map features and, in the case of MapChat, chatting at once remains substantially unexamined and unknown, but based on the current literature it is likely to be highly complex and highly prone to confusion on the part of participants. This chapter has shown that the technical barriers affecting MapChat version 1 relating to speed of data transmission across the Internet and, in principle, flexibility and simplicity of map-based software use on the Web can be addressed with creativity and good software design and implementation. However, the human end use barrier remains a significant obstacle for all such tools and is likely only to be resolved through greater familiarity with and ease of use among the public of digital mapping and computer-based interactivity.

Some of the benefits gained through the case study discussed in this chapter are somewhat less easily quantifiable than those enumerated in the previous section. While MapChat's functionality appeared to be underutilized during the case study discussions, six out of seven participants that provided feedback for the Phase 3 workshops indicated that using the digital map layers aided them during their discussions. Others noted that their participation benefitted them by improving the interaction between community members, despite the fact that important issues had yet to be addressed. There has been ongoing use by participants of the application as a generic mapping resource (e.g. for measuring/mapping irrigation-related and other information about individuals' farms), as well as interest expressed in using MapChat for other purposes not directly related to the case study (e.g. as an irrigation monitoring or roster management tool, or for mapping irrigation for larger groups of community members). Thus, despite not achieving the specific goals originally set for the case study, there appear to several indirect positive outcomes from the use of MapChat in a PGIS context.

7 CONCLUSION

This chapter described the design and use of a new approach and new tool for facilitating individual and group collaboration in the discussion of issues of spatial importance to local communities. The software at the heart of the chapter, MapChat version 2, utilizes open source components, new Web-based computing techniques, and a new conceptualization of the Web as a platform for participation. In the 'participatory web' the traditional focus on static pages that provide only producer-generated content has shifted to focus on interactive Web sites that contain both producer- and user-generated content. In fact, the trend is for developers now to produce Web sites that provide users/visitors with facilities to input personalized content that may be entirely whimsical, that may be of inherent value and interest to other users, and that can comprise important compilations of locally relevant knowledge that would otherwise remain invisible and unknown. This shift is especially evident within the geospatial domain, as revealed in the concept of citizens as sensors (Goodchild 2007) who provide volunteered geographic information (Goodchild 2008) through social tagging (Behar 2009) to create participatory knowledge repositories (Scharl 2007). While the nature and role of these topics is still being debated, their application is actively being taken up with a rapidity characteristic of other new ideas in academic enquiry.

Work will continue to refine aspects of MapChat 2, not so much in terms of function or user interface design but in terms of the protocols that need to be developed around this and other such tools to explore how best to introduce them to local communities. This will require further case studies and further experimentation in the deployment of the software. However, one thing is clear and this is that such deployment must take place through a long term process of general community engagement that may have to last for many months and involve many facets of continued interaction between local participants and researchers. In this context, there remains a gulf between the relative difficulty with which a tool such as MapChat can be used quickly and transparently by members of the public compared to the extremely popular uptake of Web 2.0 applications such as Google Maps and Google Earth. One of the needs for future work is to understand better why this is so, in order to achieve the full potential of the tool in assisting communities to dialogue over issues of importance to them.

ACKNOWLEDGEMENTS

We would like to acknowledge financial support for Version 1 of MapChat from the Canadian GEOIDE Network of Centres of Excellence. The University of Otago partially funded the development of MapChat version 2. We also wish to thank the farmers from the UTC, especially Geoff Crutchley, for their inputs and Gretchen Robertson of the New Zealand Landcare Trust for assisting with establishing the case study. Albie Thomson of Careys Bay provided very useful coding improvements for version 2.

REFERENCES

Al-Kodmany, K. (2001) Bridging the gap between technical and local knowledge: tools for promoting community-based planning and design, *Journal of Architectural and Planning Research*, 18 (2), 110–130.

Anderson, G. & Moreno-Sanchez, R. (2003) Building Web-based spatial information solutions around open specifications and open source software, *Transactions in GIS*, 7 (4), 447–466.

Behar, K.E. (2005) Capturing Glocality – online mapping circa, *Parsons Journal for Information Mapping*, 1 (1), 1–22.

Beverly, J.L., Uto, K., Wilkes, J. & Bothwell, P. (2008) Assessing spatial attributes of forest landscape values: an Internet-based participatory mapping approach, *Canadian Journal of Forestry Research*, 28, 289–303.

Boroushaki, S. & Malczewski, J. (2009) ParcitipatoryGIS.com: A WebGIS-based Collaborative Multicriteria Decision Analysis, Unpublished paper available from the second author at jmalczew@uwo.ca.

Cai, G. & Yu, B. (2009) Spatial annotation technology for public deliberation, *Transactions in GIS*, 13 (s1), 123–146.

Caldeweyher, D., Zhang, J. & Pham, B. (2006) OpenCIS - open source GIS-based web community information system, *International Journal of Geographical Information Science*, 20 (8), 885–898.

Carver, S. & Oppenshaw, S. (1996) Using GIS to explore the technical and social aspects of site selection for radioactive waste disposal facilities, Working paper 96/18. Available from: <http://www.geog.leeds.ac.uk\wpapers\96 - 18.pdf> [Last accessed 28/05/09].

Carver, S., Blake, M., Turton, I. & Duke-Williams, O. (1997) Open spatial decision making: evaluating the potential of the World Wide Web, In: Kemp, K. (ed.), *Innovations in GIS 4*, Taylor and Francis, London. pp. 267–278.

Carver, S., Evans, A., Kingston, R. & Turton, I. (2001) Public participation, GIS and cyberdemocracy: evaluating on-line spatial decision support systems, *Environment and Planning B: Planning and Design*, 28 (6), 907–921.

Elwood, S. (2008a) Volunteered geographic information: key questions, concepts and methods to guide emerging research and practice, *Geojournal*, 72, 133–135.

Elwood, S. (2008b) Volunteered geographic information: future research directions motivated by critical, participatory, and feminist GIS, *Geojournal*, 72, 173–183.

Flanagin, A.J. & Metzger, M.J. (2008) The credibility of volunteered geographic information, *Geojournal*, 72, 137–148.

Goodchild, M. (2007) Citizens as sensors: the world of volunteered geography, *Geojournal*, 69, 211–221.

Goodchild, M. (2008) Commentary: wither VGI?, *Geojournal*, 72, 239–244.

Hall, G. Brent. & Michael, G. Leahy. (2008) Design and implementation of a map-centred synchronous collaboration tool using open source components: the MapChat project, In: Brent Hall, G. & Leahy, M.G. (eds) *Open source approaches in spatial data handling*, Berlin Springer.

Hall, G. Brent, R. Chipeniuk, R.D. Feick, M.G. Leahy. & Deparday, V. (2010) Community-based production of geographic information using open source software and Web 2.0 *International Journal of Geographic Information Science*, 24 (5), 761–781.

Hopfer, S. & MacEachren, A. (2007) Leveraging the potential of geospatial annotations for collaboration: a communication theory perspective, *International Journal of Geographical Information Science*, 21 (8), 921–934.

International Association of Public Participation (2006) Spectrum of public participation, [Online] Available from: <http://www.iap2.org/associations/4748/files/spectrum.pdf> [Last accessed 20/05/09].

Jankowski, P. (2009) Towards participatory geographic information systems for community-based environmental decision making, *Journal of Environmental Management*, 90, 1966–1971.

Jankowski, P. & Nyerges, T. (2001) *Geographic information systems for group decision making: towards a participatory, geographic information science*, New York, Taylor and Francis.

Jankowski, P., Nyerges, T., Robischon, S., Ramsay, K. & Tuthill, D. (2006) Design considerations and evaluation of a collaborative, spatio-temporal decision support system, *Transactions in GIS*, 10 (3), 335–354.

Keßler, C., Wilde, M. & Raubal, M. (2005) An argumentation map prototype to support decision making in spatial planning', In: *Proceedings 8th Agile Conference*, Estoril, Portugal.

Kingston, R., Carver, S., Evans, A. & Turton, I. (2000) Web-based public participation geographical information systems: an aid to local environmental decision making, *Computers, Environmental and Urban Systems*, 24, 109–125.

MacEachren, A., Cai, G., Sharma, R., Rauschert, I., Brewer, I., Bolleli, B., Shaperenko, S., Fuhrmann & Wang, H. (2005) Enabling collaborative geoinformation access and decision making through a natural multi-modal interface, *International Journal of Geoinformation Science*, 19, 293–317.

Malczewski, J. (2006a) Multicriteria decision analysis for collaborative GIS, In: Balram, S., Dragićević, S. (eds.) *Collaborative Geographic Information Systems*, Hershey, Idea Group Publishing. pp. 167–185.

Malczewski, J. (2006b) GIS-based multicriteria decision analysis: A survey of the literature, *International Journal of Geographical Information Science*, 20 (7), 703–726.

Mummidi, L.N. & Krumm, J. (2008) Discovering points of interest from users map annotations, *Geojournal*, 72, 215–227.

Musser, J. (2007) *Web 2.0: Principles and best practices*, O'Reilly Media.

Nyerges, T., Jankowski, P., Tuthill, D. & Ramsay, K. (2006) Collaborative water resource decision support: results of a field experiment, *Annals of the Association of American Geographers*, 96 (4) 699–725.

O'Reilly, T. (2005) What is Web 2.0? Design patterns and business models for the next generation of software, [Online] Available from: <http://www.oreillynet.com/pub/a/oreilly/tim/news/2005/09/30/what-is-web-20.html>, [Last accessed 30.05.09].

Ramsay, K. (2009) GIS, modeling and politics: on the tensions of collaborative decision support, *Journal of Environmental Management*, 90, 1972–1980.

Rinner, C. & Bird, M. (2009) Evaluating community engagement through argumentation maps – a public participatory case study, *Environment and Planning B*, in press (advance publication online)

Rinner, C., Keßler, C. & Andrulis, S. (2008) The use of Web 2.0 concepts to support deliberation in spatial decision making, *Computers, Environment and Urban Systems*, 32, 386–395.

Scharl, A. (2007) Towards the geospatial web: media platforms for managing geotagged knowledge repositories', In: Scharl, A. & Tochtermann, K. (eds), *The Geospatial Web – how geo-browsers and the Web 2.0 are changing the network society*, London, Springer. pp. 3–14.

Schlossberg, M. & Shuford, E. (2005) Delineating "public" and "participation" in PPGIS, *URISA Journal*, 16 (2), 15–26.

Seeger, C.J. (2008) The role of facilitated volunteered geographic information in the landscape planning and site design process, *Geojournal*, 72, 199–213.

Seeger, C.J. (2008) The role of facilitated volunteered geographic information in the landscape planning and site design process, *GeoJournal*, 72, 199–213.

Sieber, R. (2006) Public participation geographic information systems: a literature review and framework, *Annals of the Association of American Geographers*, 96 (3), 491–507.

Sidlar, C.L. & Rinne, C. (2007) Analyzing the usability of an argumentation map as a participatory decision support tool, *URISA Journal*, 19 (1), 47–55.

Sidlar, C.L. & Rinner, C. (2009) Utility assessment of a map-based online geo-collaboration tool, *Journal of Environmental Management*, 90, 2020–2026.

Simão, A., Paul. J., Densham. & Haklay, M. (2009) Web-based GIS for collaborative planning and public participation: an application to the strategic planning of wind farm sites, *Journal of Environmental Management*, 90, 2027–2040.

Tulloch, D.L. (2008) Is VGI participation? From vernal pools to video games, *GeoJournal*, 72, 161–171.

Advances in Web-based GIS, Mapping Services
and Applications – Li, Dragićević & Veenendaal (eds)
© 2011 Taylor & Francis Group, London, ISBN 978-0-415-80483-7

Jump-starting the next level of online geospatial collaboration: Lessons from AfricaMap

Benjamin Lewis & Weihe Guan
Center for Geographic Analysis, Harvard University, Cambridge, Massachusetts, USA

ABSTRACT: This article discusses an opportunity to engage in new forms of collaborative research made possible by widespread access to the web and mapping technologies. While throughout history there have been references of various types to locations on the earth, very little has been organized in a way to support even simple spatial search or visualization on maps, let alone collaboration. Factors which hinder such geospatial collaboration will be suggested along with thoughts for moving forward. AfricaMap, a public online mapping project under development at Harvard, will be explored for ideas and lessons. Finally, a series of new tools is envisioned to push online collaboration to a more productive state of practice for researchers in all disciplines.

Keywords: AfricaMap, geospatial collaboration, mapping technologies, GIS, open source

1 DEFINITION OF THE PROBLEM

"Whatever occurs, occurs in space and time" (Wegener 2000). It follows that historic documents should include many references to geography (and time) that could be displayed dynamically on a map. Indeed, a study by the University of California of 5 million library catalog records found that half contain place name references (Petras 2004). Other studies which examine the contents of individual documents have confirmed the existence of an even larger percentage of place references.

Consider the potential significance of place references in light of the first law of geography which posits everything is related to everything else, but that near things are more related than distant things (Tobler 1970). If place references are converted to numbers corresponding to latitude and longitude a computer can understand, all documents become spatial datasets which can be discovered, visualized, and analyzed using geospatial technologies. To tackle the difficult problems of recognizing and then making productive use of geographic references within documents, a combination of online human collaboration and automated tool development is required.

Online collaboration occurs whenever someone adds new information to the web and shares it with others. Until recently technical skills were required to add content to the web. Now only literacy is required. The common and simple activity of writing on the web is changing the way we communicate and, in so doing, changing the way we do research. To give a sense of the breadth of this phenomenon, consider that it includes such diverse activities as creating a book review on Amazon, adding new code to an open source software project, tagging a photo in Flickr, editing an article on a wiki, loading a video to YouTube, adding a comment to a Facebook wall, as well as blogging, and tweeting. Even using a search engine writes information (not generally made public) which is used to shape future search results.

The use of a geographic reference is optional in most of these activities. In the past, few such references were added to information on the web because it was difficult, time consuming, and costly. With the advent of free tools the cost is now close to zero, and the volume of geospatial writing is increasing. A document describing Google Earth's collaborative geospatial format KML stated, "the current geo-mass market operating environment consists of millions of users ... using tens of millions of existing and indexed KML files and resources ... which is growing rapidly" (Wilson 2009).

Collaborative Geographic Information Systems have the potential to change the way we organize and share scholarly materials. By using latitude and longitude in addition to traditional search methods such as keyword, title and author, new kinds of questions can be asked such as: "Show me all items of type 'basket' from the Peabody collection which were found within 100 miles of Lake Tanganika" or "Display all ethnographic regions which intersect the travel routes of Livingston." The Ushahidi project built an application using the Google Maps Application Programming Interface (API) that allows users to post information on a map associated with the violence that followed the 2008 elections in Kenya. The application allows users from anywhere in the world to build a detailed picture of the timing and distribution of events (Zuckerman 2008).

"For true media innovation to have human impact it must effect the imagination" (Scharl 2007). Displaying information visually and referencing it to a familiar landscape may help people to better interpret information. For example, a photograph of an insect that causes crop failure is more useful when accompanied by a latitude/longitude coordinate. The process of adding geographic coordinates to information on the web represents a new form of geospatial communication, which is a particular instance case of the broad phenomenon of online collaboration.

Until recently the tools which enable map-based collaboration were not practical because most users did not have access to bandwidth necessary to display maps or large images quickly. By contrast, ASCII text-based collaboration had been available since the early days of the Internet. In 1988 the Internet opened to the public, and it became possible for anyone to obtain an internet connection and set up a server. In those days ASCII naturally carried the bulk of communication content because of its expressiveness and small memory requirements. By the mid-90's, as the cost of memory and bandwidth decreased, the traffic volume of other forms of media requiring more memory, such as images, sound, and video, increased (Hobbes 2010).

In 1996 MapQuest was released and quickly became a popular web tool for finding driving directions. MapQuest was not collaborative as it did not allow people to add their own information to the map, and one could not use it to define new geographic coordinates. Requiring less bandwidth and memory, text-based collaboration was able to start earlier. In 1999 the first blogs were created and mass-scale, text-based web collaboration began (Blood 2000). Before blogs, one had to be able to write HTML to publish simple text messages on the web. It would not be until 6 years later in 2005 with the introduction of the Google Maps API, that map-based collaboration would be positioned to reach a mass audience. As with blogging, the uptake was rapid and exponential, jumping from a few geospatial web users to millions in a couple years. To the surprise of many in the geospatial field, the commercially funded Google Maps platform fulfilled many elements of the geospatial collaboration vision GIS professionals had been prototyping for years (Miller 2006).

Important improvements in web infrastructure occurred between the appearance of MapQuest and the arrival of Google Maps, which made the success of the latter possible. First, the speed of the average internet connection went up 50 fold during this period (Nielsen 2008), and processing power increased even more, doubling every 2 years as had been predicted by Moore's law (Wikipedia, Moore's Law 2010). At the same time browsers became more sophisticated application deployment platforms, able to process XML that, when combined with increases in bandwidth, made possible the first global base map delivery service. Google took advantage of infrastructural improvements to remove a fundamental barrier to geospatial activities on the web, specifically, access to fast, global, high quality mapping.

Google Maps represented the first widespread use of AJAX, a new combination of the existing technologies XML and JavaScript. The mashup was born, and along with it, new forms of data sharing. By this time, the mass use of Global Positioning Systems (GPS) in conjunction with collaborative geospatial platforms like OpenStreetMaps (OSM) began to demonstrate the power of crowdsourcing for organizing geospatial knowledge. In a similar way, but starting earlier because of its text orientation, Wikipedia had begun to enable large scale organization of knowledge in the form of text and images. A significant difference between OSM and Wikipedia is that Wikipedia includes a large historical dimension. OSM and much of the geospatial web does not—yet.

The introduction of Google Maps and Google Earth in 2005 established many aspects of the technical foundation that exist today. Elements of Google's architecture such as the use of the Web Mercator projection and pre-cached map tiles were adopted by competitors. From a user perspective, the new elements were: 1) A detailed, current base map of roads and satellite imagery for the globe. 2) Free, fast delivery of base maps via an open Application Programming Interface (API) which can be incorporated into custom applications. 3) Tools for annotating the base maps. KML files created in Google Earth are one exampl of such tools. Mashups made with the Microsoft Bing Maps API are another.

Much has been built on top of this framework since 2005; however, most developments have been incremental. An important characteristic of Google's approach is the implicit distinction made between the two primary geospatial data structures, raster and vector. Raster formats (often JPEG, PNG, or GIF) requiring large amounts of storage space are used to display static satellite and roads base maps, while vector formats (primarily flavors of XML) requiring little storage space, support the addition of new geographic information to the web and serve as annotation of the base maps. The base maps, which if downloaded would comprise many terabytes, are delivered via a web service. The fundamentally necessary, always on, current, raster base map became the geospatial dial tone of the web. Against this reference system anything else can be referenced, comments can be made, new data created by creating XML annotations in the form of hyperlinks, geotags, GeoRSS, GeoJSON, KML, etc. For the first time, people with no professional background in geography or GIS began actively engaging with these technologies. The term VGI (Volunteered Geographic Information) was coined to describe the new kinds of data being created by this emerging class of geographers (Goodchild 2007).

Google's contributions, and the various efforts and innovations it spurred on the part of Microsoft, Yahoo, MapQuest, and others, moved geospatial collaboration forward. From the vantage point of a few years later, we can get a better sense of the size of the geospatial collaboration iceberg. We understand that most geospatial collaboration involves the creation of annotations of base maps that depicts current conditions. Little georeferenced historic information is available on the web, with the consequence that most annotations lack historical depth. Today makes sense as the place to start, but what about History? How should we handle the past geospatial annotations people have created in the form of cartographic (mostly paper) maps? Interestingly, people have been geospatially annotating the world for thousands of years (Harley 1987), often with more accuracy and consistency than web-born annotations being created today (Fig. 1).

Modern colonial powers have been drawing detailed maps of the territories they control or wish to control for hundreds of years. Since the development of the theodolite in 1790, a significant portion of historic maps are accurate enough to reference well to modern base map. (Wikipedia Ramsden 2010). Historical maps of lesser accuracy can also be profitably georeferenced. Examples of these are the David Rumsey maps prior to 1800 available in Google Earth.

From the perspective of our current geospatial collaboration paradigm in which current satellite and road map services are (arguably) the culturally neutral, interpretation-thin base against which other information is collated and discussed, existing paper maps are a treasure trove of interpretation-laden, historic commentary about places on the planet, yet to be brought into the light of the web and made available for the general user or professional researcher to explore and use. Historical maps constitute a rich source of historic places and

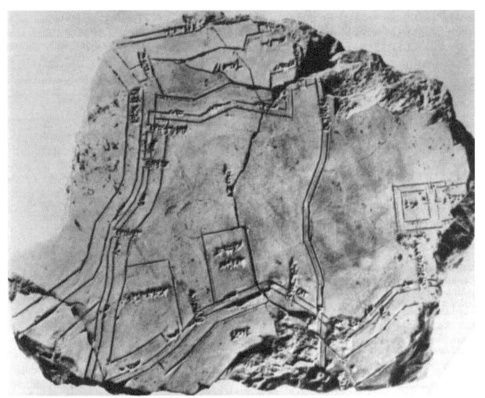

FIG. 6.7. PLAN OF NIPPUR, CA. 1500 B.C. Possibly the earliest town plan drawn to scale, this shows the temple of Enlil in its en closure on the right edge, city walls, canals, strochouses, and a park.

Size of the original: 18 X 21 cm, By permission of the Hilprecht Collection, Friedrich-Schiller-Universitat, Jema.

Figure 1.   A detailed plan drawn to scale depicting a portion of the ancient city of Nippur with temple, city walls, canals, storehouses, and a park that was created 3500 years ago (Harley 1987).

names against which texts, photographs, documents, datasets, and video can be manually and automatically referenced.

There are important unaddressed issues which hinder the widespread adoption of online mapping tools to support historical research. One problem is that most existing historical materials have not yet been digitized or brought online, let along georeferenced. Another is that there is no central place to find such information. Eventually search engines may be that place, but that is not the present case. If everyone used KML to tag web pages containing spatial data we might have the beginnings of a solution (Schutsberg 2007).

Google has current base maps and few historical maps. Google's search engine does not offer a good way to find old maps, especially ones which are georeferenced. (An exception to this is the excellent Google Earth interface for finding historic satellite imagery.) Other historical materials online are scattered across map libraries and private collections. Very few are georeferenced. Though the capabilities of online mapping systems have grown tremendously, it is still difficult for researchers to discover even the little historic spatial data there is on the web.

Barriers to spatial collaboration exist in the United States, but the situation is more severe elsewhere in the world. The problem is multifaceted, and solutions in this area must address a wide variety of areas including data storage, archiving, tool development, data integration, georeferencing, data distribution, interoperability, access to software tools, and cost (Mannak 2008). For many parts of the developing world, even basic scientific datasets are not available to researchers (Ogodo 2009). Based on our recent experience developing the

AfricaMap system, we will attempt to disect the problem and put forward some ideas for moving the nascent domain of collaborative mapping forward.

The AfricaMap project developed at the Center for Geographic Analysis represents an initial attempt to address a part of the collaboration problem by building a framework to make some high quality data available on the web in a way that researchers can use across disciplines. AfricaMap allows users to search against large amounts of data from across disciplines and display result on a map. Based on our experience with AfricaMap, we will attempt to define the factors which help or hinder collaboration; we will discuss in detail, placing the factors in the context of current and developing technologies and initiatives.

We perceive three basic limiting factors which hinder online, spatially oriented collaboration:

1. The availability of historical or current mapping other than current satellite and roads for most parts of the world.
2. The availability of a public host to which organizations can load materials they wish to share and make materials available to others permanently for discovery and visualization.
3. The availability of tools to allow interested users to collaborate, mark up maps, hold discussions, and create workspaces.

These areas are presented roughly in order of importance. Without (1) basic mapping, there would be little incentive for users to add materials to the system and store it in (2) permanent storage. Without dependable storage it would not be possible to build (3) tools for collaboration to support researchers. Taken together, these limitations make it difficult to even imagine the possibilities in the area of collaborative map building. It is similar to trying to imagine a collaboratively built encyclopedia before Wikipedia.

The AfricaMap project starts to address problem (1), but it still lacks many features. Based on our experiences with AfricaMap and feedback from users, it is clear that there is a strong need for solutions to problems (2) and (3). These three core areas contain sub-topics. We will unpack and examine them in detail. Finally, will look at the current AfricaMap project for lessons learned and attempt to describe a way forward.

2 AFRICAMAP BACKGROUND

AfricaMap supports research and teaching on Africa that servesstudents and faculty across African studies at Harvard University and increasingly internationally. Much public data, current and historical, exists for Africa, but they are difficult to discover, let alone obtain; many researchers with projects on Africa spend critical time and funds gathering cartographic data from scratch. People in Africa have an even harder time accessing mapping of their own regions. When researchers gather data it is sometimes lost again because there is no place to store and reference it. The AfricaMap project represents a framework for organizing African data, allowing it to be retrievable, map-able, comparable, and downloadable. The data is served live from various systems inside and outside of Harvard over the Internet, and brought together within a web browser to form an integrated map for the user. Data that is stored on the AfricaMap servers are made available to other applications outside AfricaMap as map services.

At its core, AfricaMap consists of a set of public digital base maps of the continent, viewable dynamically at a range of scales and composed of the best cartographic mapping publicly available. Behind the scenes a gazetteer provides rapid navigation to specific locations. As more detailed mapping becomes available it is added to the system. Because of its decentralized architecture, there is in theory no hardware or software limitation on the amount of data that can be incorporated.

The idea for AfricaMap was developed in the Center for Geographic Analysis (CGA) at Harvard (see Acknowledgement). The project is overseen jointly by professors Suzanne Preston Blier and Peter Bol, and designed and managed by Ben Lewis. The CGA has the dual aim of supporting Harvard research that involves mapping and location-based investigation, as well as making data created in the course of research available to others. In November of 2008 the Phase I release (Beta version) of AfricaMap was launched.

Basic AfricaMap features attempt to serve broad needs of researchers in multiple disciplines interested in Africa. These features include:

- A common web accessible set of current and historic maps for Africa. One of the greatest problems facing researchers is the lack of a commonly available base map for the continent upon which one can build one's own research materials and share them with others. Wherever possible AfricaMap is making base map data available without copyright restrictions.
- A comprehensive (and continually growing) gazetteer for African place names. Currently there are over a million names, in an array of spelling alternatives and languages in addition to English (French, Arabic, Chinese, Japanese and Russian etc.). This is not important to international users Furthermore, by overlaying maps at different resolutions and showing different features, users can potentially "translate" place terms into these different languages.
- A repository for spatial datasets, both those compiled from other sources and new datasets created by researchers.
- A clearinghouse for research projects on Africa, allowing users to discover what regions are studied in current and past projects. Users can view the data or metadata relating to those projects and contact the researchers involved.

3 FACTORS WHICH HINDER COLLABORATION

During the inception, design, development and support of AfricaMap, the authors have been exposed to user demands for spatially oriented collaboration tools, as well as the technical and institutional obstacles which exist in meeting these demands. The continued development of AfricaMap is aimed at overcoming these obstacles. In this section we will examine key elements based on our experience which hinder collaboration and propose ways of addressing them.

3.1 *Availability of mapping materials*

3.1.1 *Availability of historic base mapping*
Historical base mapping provides orientation for many geographic investigations. At a minimum base maps generally contain roads, town names, major geographic features, and political boundaries. In many cases historical mapping made in the past 50 years or so contains much more detailed information about a given place than one can identify on a Google satellite or roads base map, especially for rural parts of the world (Figs 2a-b). Historical base mapping, when coupled with Google's current satellite imagery, provides a powerful, hitherto underutilized way to visualize change.

Determining globally how much historical mapping is missing from our current online geospatial environment is difficult. Since the early 19th century, major world powers including Britain, France, Spain, Portugal, Russia and the U.S. have systematically mapped most of the habitable earth multiple times. Most of these maps exist in paper form in library drawers. Many are in private collections. A few libraries are in the process of scanning their maps, and a few of these are engaged in georeferencing them. East View Cartographics, a major private map seller which has a sizable collection of colonial maps, claims to have over 500,000 maps. Books are also a large source of valuable maps. John Orwant, the manager of the Google Books metadata team, estimates the number of book titles in existence as 168,178,719 from more than 100 metadata sources (Orwant 2009). If we assume just 1% of these books each contain 5 maps, that would be 8 million additional maps in books. By contrast, the number of historical georeferenced paper maps online which can be downloaded or incorporated into an application via web services today must certainly be less than 1000, and there is no easy way to find them. The latter estimate is based on extensive searches by the authors.

Making such maps available in a web-based mapping system is more of an organizational and financial challenge than a technical one, though technical breakthroughs will be as important. For example, a free online georeferencing tool for anyone to use could have a major positive impact by reducing the cost of this work and allowing the work to be shared. Such a tool could help crowdsource the work of georeferencing historic materials. An online tool has an advantage over a desktop tool because as the work is done, the newly georeferenced maps and other materials are in turn available as a reference against which additional materials can be referenced. Such an approach could allow users to comment on and repeatedly georeference the same object until the community feels it is finished. The versioning history of a given map object could be made available to the community in wiki fashion.

3.1.2 *Availability of georeferenced materials*

Historical base mapping is only one type of georeferenced material. If one considers that nearly all materials in the corpus of human knowledge, including maps, books, manuscripts, letters, photos of objects, photos of people, sound recordings, and video recordings, either describe places on the surface of the earth or are tied to a location in some manner, it becomes apparent that only a small portion of historical knowledge is currently available online, let alone georeferenced.

The universe of potential materials to be included in a collaborative mapping system may be divided into two broad areas: materials such as satellite imagery and GPS data, which are

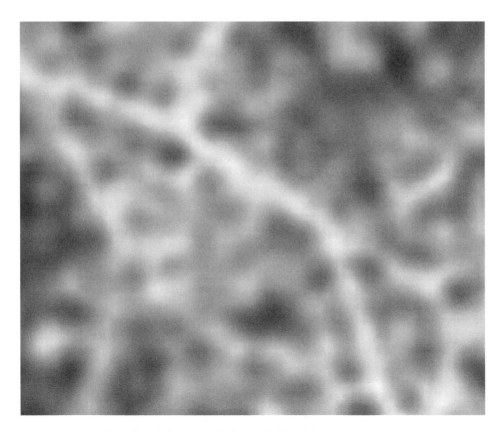

Figure 2a.   Image from Google for most of Africa (Ife, Nigeria).

261

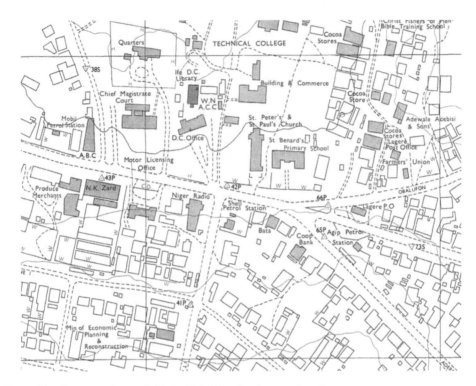

Figure 2b.   Base mapping available in AfricaMap for the same location.

born digital and spatial, and information which must be converted to digital form and then georeferenced. Much of the material that is born digital is currently well-handled: Google Maps and other commercial providers are providing imagery via its Google Maps API, and Open Street Maps is providing a crowdsourcing platform for uploading GPS tracks to build high value datasets such as roads for all parts of the globe. AfricaMap can easily incorporate these types of materials.

The material which offers perhaps the greatest opportunity for expanding the nature of geospatial collaboration because of it both breadth of subject area and its temporal depth are paper maps in map collections and books, as well as references to places in books, datasets, images, and documents. As we have stated, much of the recorded history over the past 5 millennia describes events that happened somewhere on the surface of the earth; it is information that can be organized spatially in addition to being organized by subject and author.

Georeferencing is the establishment of relationships between information and locations using names or times or events, in order to improve searches against and visualizations of collections (Hill 2006). One can think about georeferencing as an additional method of organization that can be applied to just about any type of information. Geographic element definitions are included in many metadata formats such as Dublin Core (DCMI), providing opportunities for map collections. In many cases these fields have not been populated; that may change with the improved availability of tools for generating accurate coordinates.

Perhaps the important question for our purposes is not whether a given piece of information has a spatial component, but how much it costs to extract the spatial information and represent it in a meaningful way. To answer this we need to know what we wish to extract and how we want to represent it on a map. The choice made will vary according to information types: an ancient document might be best represented by a point while a travel account in a book might require multiple lines. Irrespective of spatial information is extracted from an old map, a text document, or an old photograph, there are many considerations which must be taken into account, including what precisely it is that is being located in space and how the spatial information is likely to be used. Is the place the document describes to be mapped, or is it the place where the document was found to be mapped, or both?

3.1.3 *Integration of data with public commercial layers such as Google's*
Google and other commercial providers are making very detailed datasets available to the world and the terms are liberal. Data which a few years ago would have cost millions of dollars to access are available for free and can be incorporated into non-commercial, web-based systems easily. These datasets have immediately become, for many parts of the world, the best geographic information available to the public. But the commercial layers lack the historical depth and detailed place name information that are available on historic topographic maps. Because the commercial and historic maps each have their strengths and weaknesses, combining the two is logical. By doing so, suddenly it is possible to see change over time, and annotations on the historic mapping fill in for the missing labels on the commercial satellite mapping.

Commercial datasets have additional value because they are constantly being improved, are always on, and are well hosted on fast servers. Access never seems to be a problem, though that could change. For this reason it is important not to be dependent on the mapping of one commercial provider alone. There is a danger along with opportunity. These data may vanish at some point or become very expensive. The glib solution is to assume that they will disappear in a few years and therefore use them to the maximum degree to georeference other materials while they are still accessible. The other solution is to be aware of and support free data sources such as Open Street Maps and public domain satellite data which will always be available.

3.1.4 *The use of gazetteers*
A gazetteer is a list of place names with associated geographic coordinates which helps a user navigate a large spatial domain. Gazetteers can be used in a variety of ways to make content more accessible and comprehensible across space and across representations in multiple languages and historic variations. A gazetteer is very useful when combined with a historic base map from scanned paper maps. The scanned paper map is generally not searchable by place name, while the gazetteer is searchable but provides no cartographic information. As with the combination of satellite and scanned maps, the two combined are greater than the sum of their parts.

3.1.5 *Intellectual property considerations*
Collaboration requires the sharing of information. Often researchers do not want to give away their data until they have published their results. But sometimes it is more valuable to get feedback and find potential collaborators than to keep a dataset private. With the right tools, sometimes a researcher is willing to share data publically during research in controlled ways. Today such tools exist but they are limited in their power. One of the best is Google MyMaps but it is constrained in the number of features which can be displayed simultaneously. As there is no easy way in the geospatial world to make materials available to others and control accessibility on a day-to-day basis (as people are accustomed to doing with Facebook and other collaborative tools) many researchers default to not sharing.

Another reason that materials are not more frequently shared is that there may be no apparent, simple way of defining and enforcing the type of sharing that the user would like to implement. Perhaps a user would like to make materials available for only non-commercial use and would prefer to require attribution if data is used in a publication. The Creative Commons project (http://creativecommons.org/) provides legally defensible sharing-oriented licenses which cover the basic areas of attribution, modification, and commercial use and can be adapted to fit most requirements.

3.2 *Availability of a place to put mapping materials*

3.2.1 *The cost of serving data*

It is expensive to set up a GIS server, which at the very least costs thousands of dollars in hardware, software, bandwidth, and technical expertise to put together. Providing a low cost or free hosting solution for scholarly spatial materials would be an important step towards making more of such data available. Universities have an interest in supporting initiatives which increase the access of their own scholars to materials and encourage collaboration between their scholars and those of other institutions. Many universities already spend considerable resources to serve non-spatial digital collections to the world.

3.2.2 *The use of web services*

In addition to traditional web-based viewing and direct download of data, the use of web services, particularly open ones, makes many new kinds of data sharing possible. Web services allow interactive maps which reside on one system to be embedded in another organization's web or desktop application. Google Maps is such a service and can display on the Google Maps site or can be incorporated into other applications. Given the large file size of GIS downloads and the cost of hardware and software necessary to host a system, the use of web services promises to be an important part of geospatial collaboration.

3.2.3 *Permanence and citations of materials*

Researchers are often willing to share their materials provided the cost is not high, their materials are easily discovered, and credit or citation goes to the owner or author. In academia, citations are the coin of the realm, and improved citeability could be used as an incentive to scholars to make their materials available. There are no map-oriented systems which offer facilities to both store geospatial information and make them permanently citable. There are systems which meet this need for tabular social science datahowever, such as the Dataverse Network (King 2007).

3.2.4 *Data discovery, metadata and spatial indexing*

How can one find the best available online data for a particular purpose? This is currently very cumbersome when it comes to spatial materials and standard search engines. One reason is that search engines are oriented toward finding pages, not data, let alone the range of types of spatial data. Even within sites which specialize in data this task can be arduous. There is often no easy way to get a sense of the overall contents of a particular data warehouse. A user gets a list back, but if there are a large number of results in the list and the system does not have a good way to prioritize results, the chance of finding the materials searched for is small.

A possible solution is to provide a holistic view of the contents of a system using a map to show the footprints of all materials, a tag cloud view to show the concentration of keywords, or a time graph to allow one to find all materials which deal with a particular period in history. All of these approaches to some extent bypass the problem of discipline-based vocabularies or ontologies; instead they use discipline-independent terms such as space (lat/long), time (date range), and tag (user defined tag). Combining multiple such holistic views (map, tags, time) in a faceted search will provide a powerful way to sift through large collections without much a priori knowledge of the contents.

Many types of information describe a place on the earth and have an implied geographic footprint which defines the area of interest. Creating geographic footprints for non-spatial data is an important aspect of georeferencing. In AfricaMap, there is a Project Layer which is composed of the geographic footprints of research projects completed, on-going, or planned for in Africa. Such footprint layers are a spatial form of metadata, describing and linking to the project materials from the map and making non-spatial project information searchable spatially.

3.3 Usability of the system

3.3.1 Interface design
Many GIS systems are cumbersome and difficult to use. There is often a large tool bar that the user is expected to learn. Because the intended users of such data sharing systems are often not professionals in information technology fields, the system's interface design should be simple and conform to de facto standards wherever possible. Advanced functionality can be present, but should be ideally behind the scenes. Simple design promises a system easier to implement across disciplines and across cultures and languages.

3.3.2 The importance of performance and scalability
Many geospatial systems that are oriented toward scholars or developed by non-commercial efforts are slow. Slowness isn't just a problem of wasting a user's time; it is often a show stopper. People generally do not come back to unresponsive systems and are unlikely to recommend them to others. The problem of speed is related to its scalability. A system might be fast with a moderate amount of traffic but very slow when traffic increases beyond a certain point. In order to handle gracefully increasing amounts of traffic a system must be scalable.

A web services architecture supports scalability because it provides a way to decentralize computationally intensive functions. The same technologies can also support collaboration by allowing various organizations to provide data services which can be organized for the end user as a single mapping system.

In addition to technical approaches which support scalability, software cost is an important factor for allowing a system to scale in order to meet demand. When additional servers are needed to help with performance, the cost of software to support more servers can be prohibitive. Open Source GIS software is highly scalable in terms of cost; however there are many factors that must be considered when calculating cost, such as the skill set of the organization.

3.3.3 Interoperability with other systems
Web services, open standards, open source software, open data formats, open metadata formats, web access, operating system platform independence—all these are important elements of interoperability and are important to consider when designing systems that attempt to reduce barriers to data access and improve the chances of successful collaboration between scholars.

3.4 Availability of a place to collaborate with mapping materials

3.4.1 Research audience should be defined broadly
We are interested in improving the ability of academic researchers to collaborate, but we are also interested in building systems which bring a wide variety of users together, whether from government, non-profits, private sector, or the general public. The problems and opportunities are too great and crucial for interested people not to be able to work together.

The research audience differs from but overlaps with a general audience, and the success of Wikipedia is a vivid example of this. A system that only aims to serve the academic community but does not build common cause with other groups interested in mapping knowledge is missing an opportunity to create a more significant system with greater longevity.

3.4.2 Multiple language support
For both the research audience and the general public, multiple language support is essential for collaboration. Most of the open source components used in the AfricaMap framework are able to support multiple languages without further development. For example, one can search by place names in any language and will be able to find the place if it is recoded in that language in the gazetteer. However, some challenges exist in making the system compatible with a variety of client operating systems internationally; these are issues that which go beyond language support.

3.4.3 *Collaboration tools*

True collaborative research requires more than data searching, viewing, uploading and downloading. It should also include tools for defining and inviting collaborators, controlling access to some materials, commenting on materials, tracking changes to editable layers, and more. AfricaMap currently lacks many of the tools we deem to be important. Those we envision are described in the next section.

4 AFRICAMAP LESSONS LEARNED AND SOLUTIONS RECOMMENDED

4.1 *Data currently in the system*

The project started with publicly available continental scale data, and used that as a framework against which to start to bring in more detailed materials. In addition to Google's satellite and physical maps, AfricaMap includes continent-wide layers to support place-based search and materials georeferencing for even the most remote regions of the continent. Other important layers in AfricaMap include the Operational Navigation Charts (ONC), Tactical Pilotage Charts (TPC), Joint Operations Graphics (JOG), and Army Map Service (AMS) sets of mapping from the U.S. National Imagery and Mapping Agency (NIMA). In addition, a full set of seamless Soviet mapping at 1:500,000 scale for the continent was acquired. Other important thematic layers in AfricaMap useful for georeferencing include current and historical administrative boundaries, historic maps, ethnographic boundaries, and language boundaries. There are additional critical datasets that are gradually being added to AfricaMap such as Russian 200,000 scale mapping, census data, and health data. Table 1a and 1b list the major data layers in the current AfricaMap release.

The gazetteer (place name) layer in AfricaMap was derived from GeoNames (http://geonames.org) which itself is composed of several gazetteers. The main source for GeoNames is the GEOnet Names Server (GNS) maintained by the U.S. National Geospatial Intelligence Agency (NGA). The vocabulary of place types in GeoNames follows the NGA's place types, and GeoNames has added a few new types. Of the various public global gazetteers, none are complete for all feature types, and there is no widely agreed upon vocabulary for place types. GeoNames is perhaps the closest to providing a de facto standard for gazetteers, and because it is being improved wiki fashion by a global user community, seems most likely to become the most complete public gazetteer. One weakness of GeoNames is the lack of historical depth. Even though it records place name aliases, they are not marked by the time when the alias was in use.

Table 1a. Major raster data layers in AfricaMap phase 1 beta release.

Series name	# Map sheets	Coverage	Scale	Year	Size on disk GB
U.S. ONC	47	95% continent	1:1,000,000	1979–1998	13
U.S. TPC	142	75% continent	1:500,000	1974–2002	40
U.S. JOG	1039	50% continent	1:250,000	1965–2005	58
U.S. Burundi	40	100% continent	1:50,000	1994	3
Soviet	550	100% continent	1:500,000	1962–2003	6
Soviet	3250	80% continent	1:200,000	1962–2004	40
Nigeria	1039	50% country	1:100,000, 1:50,000	1960–1978	58
French 1898	62	100% continent	1:2,000,000	1898	12
Sierra leone	190	100% country	1:50,000	1964–1973	3
Freetown	108	100% city	1:2,500	1941–1965	5
Historical maps	10	100% continent	varies	1612–1930	2
Health maps	4	100% continent	1:10,000,000	2000	1
TOTAL	6477				240

Table 1b. Major vector data layers in AfricaMap phase 1 beta release.

Map name	# Features	Coverage
HRAF ethnographic atlas	847	100% continent
People's ethnographic atlas	1927	100% continent
Scholarly maps on history	150	100% continent
Harvard map collection index	260	100% continent
Projects index	45	NA
Place name gazetteer	~1,000,000	100% continent
Admin boundaries	17,155	100% continent
Population surface	1 km grid	100% continent
Population centroids	109,177	100% continent
Lakes and rivers	150,572	100% continent
Soils great groups	4,909	100% continent
Surficial geology	11,977	100% continent
Land cover	5 km grid	100% continent
Power plants	1637	100% continent
TOTAL	1,298,506	

4.2 Current AfricaMap system architecture

4.2.1 Hardware

AfricaMap currently runs on a single Linux server running the Red Hat Enterprise 5 operating system and is hosted at the Harvard-MIT Data Center (HMDC) with a one gigabyte per second connection to the Internet. AfricaMap currently accesses about half a terabyte of raster and vector data on disk.

4.2.2 Software

AfricaMap uses the Open Source PostgreSQL relational database for storage of vector (point, line polygon) spatial features. Geodatabase functionality is enabled within PostgreSQL by the PostGIS spatial libraries. PostGIS supports a wide variety of spatial SQL operators and is a platform for querying, analyzing, extracting, and converting spatial data. For rendering and symbolizing map data out of PostGIS, MapServer is used. MapServer connects directly to PostGIS and provides an Open Geospatial Consortium (OGC) compliant Web Map Service (WMS) interface for AfricaMap (Fig. 3).

AfricaMap uses a JavaScript-based map client to display map tiles from a local tile cache as well as tiles from remote servers such as Google's. All map data is rendered as tile caches including the gazetteer (place name database), political boundaries, and historic maps. Tile generation occurs based on WMS (Open Geospatial Consortium Web Map Service) requests.

Caches are generated from the original GIS data by map renderers MapServer and GeoServer. All queries against vector (point/line/polygon) GIS datasets are handled by PostGIS, a spatial relational database. Tiles are stored in an open tile format called TMS (Tile Map Service).

Data can be stored in any source projection and re-projected on-the-fly to Spherical Mercator for overlay with commercial map services such as Google Maps. Google Earth versions of the raster layers are also made available as KMZ files using superoverlays.

4.2.3 Web services

Using WMS for map rendering within AfricaMap makes it possible to provide web service access to AfricaMap data from applications anywhere on the web. The maps in AfricaMap can be used by systems other than AfricaMap without their needing to download and configure the (often very large) datasets. Since AfricaMap consumes open services it is also able to consume map services hosted by other organizations.

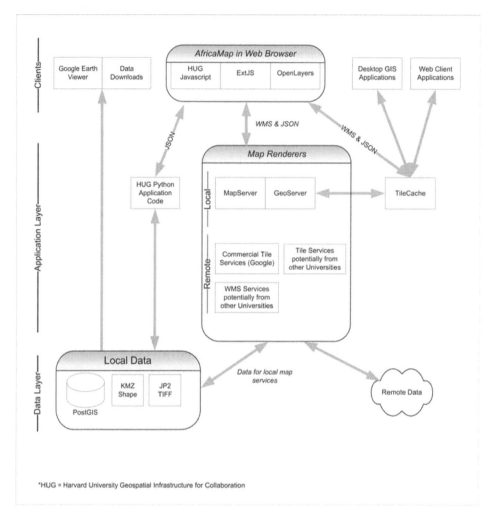

Figure 3.   Current AfricaMap phase I architecture.

To make AfricaMap's map services scalable, TileCache is used which generates an Open Source Geospatial Foundation (OSGeo) compliant Tile Map Service (TMS) from a WMS. In order to integrate TMS layers and Google Maps layers within a common application, AfricaMap uses the OpenLayers JavaScript library. OpenLayers also contains connectors to proprietary mapping services such as Google Maps and Microsoft Virtual Earth. Unlike Google Maps mashups, OpenLayers-based platforms are not dependent on Google Maps, but can use many different commercial or non-commercial services.

To support rapid query and map display against datasets with millions of features such as the AfricaMap place name database, full text indexing using the Tsearch2 PostgreSQL extension is used. This approach allows a user to search multiple layers at once and display the results of all matches from all layers on the map. Unlike many web GIS applications, there is no browser-side limit to the number of query results that can be displayed. This is accomplished by using an image-based service instead of an XML or vector-base approach for rendering features.

AfricaMap uses open standards, open web service protocols, and open data formats. Open standards are the basis of the web, and AfricaMap is a web-centric infrastructure which benefits greatly from the technical and social benefits accruing from open standards used in the Open Source software from which it is built.

4.2.4 *Data formats*

GIS image files used by AfricaMap conform to the open GeoTIFF standard for spatial imagery. Vector data conforms to the open Shapefile specification and Well Known Text (WKT), and large images are made available in the open JPEG 2000 format and KMZ format. All of these file types represent leading open geospatial data formats. The core GIS libraries used are the Open Source OGR Simple Features Library and the Geospatial Abstraction Library (GDAL), which are capable of reading and writing a wide variety of formats.

In AfricaMap all map data are projected to a common spatial reference system, Spherical Mercator, in order to overlay with commercial layers such as Google Satellite. Because Spherical Mercator is not commonly used in desktop GIS systems, data made available for download are provided in the more commonly used Geographic space WGS84 space. Web services in AfricaMap support both Spherical Mercator and Geographic. AfricaMap uses the OGC Well Known Text (WKT) standard for describing and communicating projection information in web services and GIS data.

4.3 *Functionality of the system and what's missing*

AfricaMap supports an initial implementation of a solution to some of the problems identified in this section. The areas for which AfricaMap does not provide solutions are critical and need to be addressed in order to jump start the next level of online geospatial collaboration envisioned here.

The AfricaMap project has made a small start toward addressing the problem of availability of basic mapping and other kinds of spatial data; AfricaMap presents some large datasets to users via a simple interface that supports intuitive comparisons between mapping layers. However as groups and organizations have been exposed to and become users of AfricaMap, many have expressed the desire for tools that would allow them to load their own materials to the system as well as tools to support various forms of online collaboration.

In addition, many groups have asked for versions of the system which focus on other geographic regions, and these groups of course need data loaded for those regions. In fact there have been so many requests for additional data that we have not been able to keep up with the requests and are working on a tool to allow users to upload their own materials. There have also been many requests for the ability to control access to datasets. Because everything in AfricaMap is public at this point, there are many datasets which we have not been able to add to the system because they cannot be made completely public.

Furthermore, there have been requests from various groups for tools to use to georeference collections of materials such as map sheets and digital objects representing objects in museums. The availability of georeferencing tools will be key to the creation of an environment in which a range of scholarly materials can be searched and resulting distributions visualized on a map.

5   PROPOSED ENHANCEMENTS TO AFRICAMAP

Based on our experience, the following are the key areas that must be addressed in order to move the AfricaMap framework forward in the direction of a general purpose collaborative platform:

1. Assemble fundamental sharable geographic materials such as major map series for globe developed by the United States and Soviet Union at a minimum. Include integration with commercial providers, gazetteers, and best global layers such as the LandScan population distribution surface and GADM administrative boundaries.

2. Provide a place (ideally a central index to materials on various servers) to store materials, control access, and make map materials usable. Discovery of any portion of the data should be possible via search engines as well as from a central mapping application.
3. Tools for georeferencing materials of various types need to be provided for anyone to use.
4. Tools for collaborating and engaging in scholarly exchanges with regard to map materials should be provided.

The current Beta release of AfricaMap delivered only a small portion of these functions; however, its architectural framework is designed to accommodate further enhancements in this direction.

5.1 *Proposed collaboration tools*

Users need to be able to create their own logins to the system and upload, georeference, and control access to their materials. One simple type of georeferencing tool would involve the referencing of URLs. Being able to define a geographic location and basic metadata for a persistent URL which points to valuable information is a potentially useful enhancement. Such URLs could point to digital objects in collections or to valuable online resources which describe a given place. The geographic "footprint" of the URL might consist of a single or multiple points or lines or polygons.

A bulk version of this tool could allow the user to upload a table of URL information and specify fields in their table likely to contain references to places. These could be automatically parsed and matched against a gazetteer to determine preliminary lat/long coordinates. After performing a preliminary automated coordinate lookup, the owner would be able to check and correct the coordinates assigned within an interactive mapping environment. Coordinate values will be stored in the database in the OGC standard Well Known Text format which could then be output to any spatial format. It is important that users be able to export out any materials they create for use in other systems.

A critical feature of this proposed system is the ability for users to control in a fine-grained way how their materials area shared. Users must be able set permissions such that their materials are visible to only themselves, visible to those they choose to share access with, or made public. Users should be able to control whether others can edit their materials or only view them. Such an approach supports the typical research materials life-cycle in which sources at early points in a project must be confidential; after publishing, the materials can be made public. There may even be an active desire on the part of the researcher to make materials available to support a published article.

For a map-centric collaboration platform, a tool to georeference scanned map images is critical. Historic maps are a key resource in this system which will be used to georeference other historical materials. The system should allow the user to upload a scanned map image, and then georeference the scanned image by identifying common points between the uploaded image and already projected mapping in the system. Once a sufficient number of common points are defined, the system would reference the scanned map to geographic space. The system could be designed to allow users to perform this task as frequently as desired to obtain a good fit between the map and the "real" world. Users would be able to comment on one another's referencing work and even revise referencing performed by others. The creator would be able to choose whether to accept a corrected version of their material.

A tool to facilitate scholarly exchanges and conversations relating to spatial objects could allow such a system to become a platform for communication at various levels of geographic resolution. Users will be able to comment on mapped materials by placing notes in the form of point, line, or polygon features on the map and annotating them. Such a system should automatically track the user, date, location, and map reference for each comment, and the user should be able to email their map markup to colleagues. A key element here could be a URL, automatically generated, and included in the email along with the comment. Upon clicking on the URL, the email recipient would be taken into AfricaMap and the comment displayed in the context of any maps that were displayed when the email sender created the comment.

Using this approach any comment created by a user could be syndicated using the syndication standard for blogs, RSS. Users would be able to choose to make any comment appear in their RSS feed which anyone would be able to subscribe to. This type of feature could make map comments made in AfricaMap usable by any blog in the world. As with the emailed feed, a subscriber reading a feed will have a URL that will take them directly into the AfricaMap system to view the comment in the map context in which it was made. We envision that syndication could be an important means of dissemination and could bring new users into the site where they would be positioned to contribute to and continue a discussion by adding comments.

Links pointing to content on other servers are subject to breakage whenever the linked-to site changes the URL or disappears. An important feature of a system would be an administrative tool that could be set to check periodically all links in the system and generate a report for the administrator. After a defined number of failed attempts to access a given link, the system would notify the creator of the link through email.

In addition to latitude and longitude, dates are a powerful metadata type that cut across disciplines; by using maps to represent time, time-enabled materials can be presented visually. To attain these ends, users need to be able to precisely define time ranges (start date and end date) for any feature in the system. To support interoperability with other systems ISO 8601, the international time standard, could be used to store date ranges in the database and in data output from the system. ISO 8601 is compatible with Dublin Core used by OAI and with KML used by Google Earth.

A simple way to enable AfricaMap to support the visualization of time enabled materials would be to output a version of the features in KML format for exploration using the Google Earth time bar. In addition, the query interface could support searches using start date and end date, and a time graph view could display the temporal distribution of all materials in the system for which time ranges are defined.

In addition to the standard query result list, we propose three interactive graphic views be available to help the user characterize a collection or the results of a query: map view; time graph view, and word cloud view. The map view will display a heat map showing the distribution of materials across the geographic extent of the selected set of materials. Areas with greater concentrations will show up in a different heat map color. The time graph will display a graph of the materials distributed across the time frame they represent. The word cloud will use the approach of a standard tag cloud except that it will be possible to choose certain fields such as Language and Author to create field-based, focused clouds in addition to the more traditional unstructured tag clouds.

The relationship between the four pieces (results list, map, time graph, word cloud) should be dynamic. Zooming in on the map to show a subset of the results on the map would filter the rest of the views, and the results will change to reflect the new spatial filter. Changing the time range will filter the other three views. Selecting a word from the word cloud will cause the other views to adapt to the refined query. In this way a user will be able to explore large collections of georeferenced materials with a better a priori sense of the distribution of materials in the collection.

5.2 *Proposed new architecture of AfricaMap*

Figure 4 presents the existing architecture in black and the proposed in blue. Highlighted are areas of significant software development.

The proposed architecture builds from the current AfricaMap and adds some significant new features to support ongoing collaboration and time-enabled GIS dataset development between geographically distributed participants.

An important enhancement is the creation of a mechanism for leveraging the wealth of historic geographic information in the world's paper map collections by making it easier for

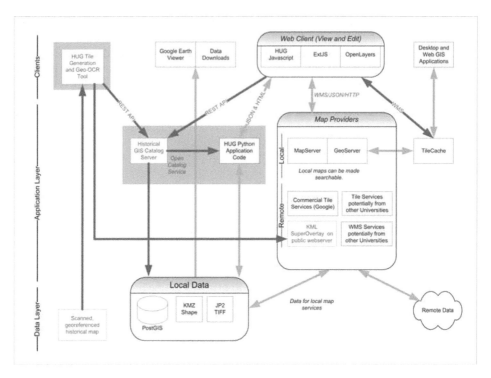

Figure 4.   Proposed AfricaMap phase II architecture.

organizations to publish their maps in a useful way. This approach uses tile caches as in the current AfricaMap but with some twists. First, the list of available tile caches is put on the web in the form of a database (**GIS Catalog Server**) which can be updated by those who wish to contribute new caches to it, and which can be queried by users searching for maps. These caches do not need to reside on a central server, but can reside anywhere, including on the server of the organization that owns the original paper map. No special software is required in order for the organization to serve them efficiently. The caches just need to be stored in a web accessible space.

The next phase of the AfricaMap project proposes the development of an open source tool for these organizations to use to create tile caches from their georeferenced maps and register metadata for the caches along with the cache location **URL** on the central Catalog Server. In order to be able to process large numbers of tile caches the central server should farm out the tile generating jobs using a grid technology like **BOINC** (BOINC 2010). Once a set of tiles has been create for a given georeferenced map and metadata defined for it, the tile cache should be uploaded to the data owner's web space and registered on the central catalog server index. The Catalog Server will make its contents searchable via an open **API** so that others will be able to build a client that can query the catalog and retrieve tiles from wherever the tiles reside.

In addition to a framework for distributed sharing of historic map data, tools will be provided (AfricaMap Python Application Code in diagram) for collaboratively editing GIS layers based on the tile cache backdrops. Tools will allow users to add features to existing layers with defined schemas. A key layer will be the time-enabled gazetteer (place name database). This layer will evolve and improve as more and more historic maps are uncovered and more place name information is extracted from them.

An important feature of the editing tool will be provenance maintenance. For any new feature created, the reference information for the base map used to derive the new feature will be stored and associated with the new feature.

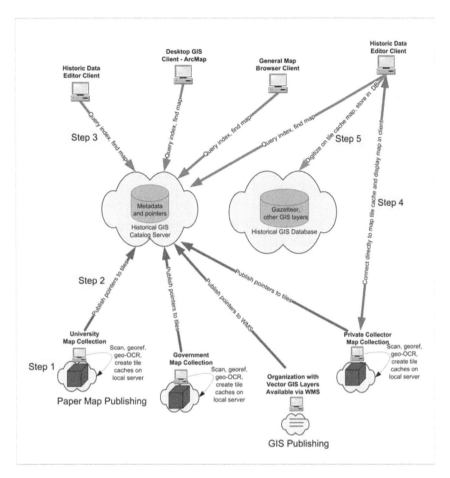

Figure 5.   High level views of stages in collaboration.

Step 1 – Librarian at an organization with paper map collection scans map, (optionally) georeferences, (optionally) extracts place name features, creates tile cache, and puts cache on web accessible space. An online tool for georeferencing scanned maps will be provided.

Step 2 – Librarian notifies the Catalog Server of the existence of the tile cache by URL, fills in basic metadata regarding source, date, georeferencing technique used, etc.

Step 3 – Researcher looking for historic maps to support work in a given region queries the Catalog server using the system's web client and finds a list of matching datasets.

Step 4 – Researcher turns on desired tile cache and a request is made directly to the tile cache server.

Step 5 – The researcher uses the tile cache as a backdrop and digitizes new place name features to add to an historic layer.

5.3   *Proposed collaboration workflows*

There are several ways in which historical map images can be made available to the AfricaMap system for use in building new GIS datasets. As shown in Figure 5, the source data providers are on the bottom, and the map viewing and editing clients are on the top.

Figure 6 illustrates the proposed paper map publishing data flow in the next release of AfricaMap.

In addition to the manual extraction of place name features from historic maps, we also propose the development of an innovative means of gazetteer population: a specialized version of OCR (Optical Character Recognition) that is adapted to support the extraction of place names and locations (lat long) from scanned, georeferenced historic maps. We are calling this

273

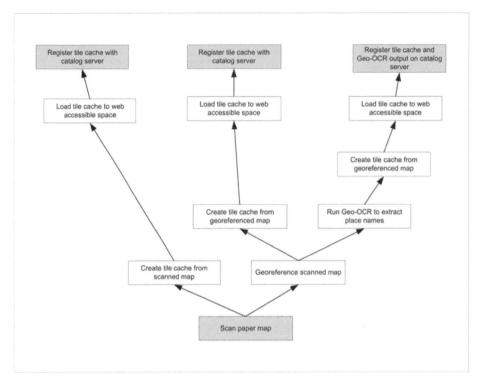

Figure 6.   Paper map publishing data flows.

technology Geo-OCR or Map-OCR. Such an approach, in a collaborative environment such as this one, does not have to be perfect to be useful. Even if it starts with a relatively low accuracy rate, it will provide an automated "guess" as a jump start; iterative corrections as more reference maps are made available through the system will gradually improve the accuracy, thus making significant contributions to an ever-growing historical gazetteer.

Geo-OCR combined with a gazetteer could even provide the beginnings of an automated georeferencing system. The Holy Grail would be a completely automated georeferencing process for scanned maps uploaded to the system followed by a fairly complete place name and feature extraction. It is just a matter of time before this problem is solved. Our estimate is that it will take at least 5 years for a generic, Open Source solution to be developed that will work for most paper maps in most languages, that is, unless Google gets serious about it first.

6   CONCLUSIONS

Today a vast amount of human cultural knowledge is available online, providing more and more materials and sources for research an all discipline. In most cases these and sources materials must be accessed on an individual basis by going to the provider's web site and searching the holdings using the particular site's search interface. Not only are these searching methods often limited, but also there is no viable way of addressing these data comparatively within one system or across multiple systems. In short, unique challenges exist for accessing and effectively comparing and integrating these data. GIS (Geographic Information System) technologies offer a means of starting to address some of the issues of accessibility and comparability. Maps provide a common language for understanding the world.

The mapping revolution made possible by the web provides new means for integrating, analyzing, presenting, and sharing this information. We see web-based GIS as a promising technology for expanding the range of search techniques available to users, making possible more precise searches of materials otherwise isolated by discipline category; this in turn and thus inspires and spurs individuals and institutions to offer their data to the web. GIS provides a uniform geospatial cataloguing format, and because it allows users to view multiple layers comparatively, it encourages a multi-disciplinary approach to understanding events and the related artifacts. GIS technology, used to view historical, environmental, socio-cultural, economic, and political spatial data layers in conjunction with multi-media data (images, video, audio, 3d models) will allow for the exploration and imagining of the earth's past and future in new perspectives.

AfricaMap begins to address also the problem of interdisciplinary geospatial data availability for Africa. The system provides some of the highest resolution mapping publicly available for the African continent, alongside an array of other data. AfricaMap is currently good at display, but it does not allow users to add comments or new materials. As potentially useful as AfricaMap is in its present state, to break new ground in improving access to materials and facilitate collaborative research, it needs to be enhanced in several ways.

Based on feedback from users, the improvements should include enabling users to upload, georeference, time-enable, tag, and control access to their materials. Also important will be a method by which users can comment on materials and email or syndicate their comments. To take advantage and motivate the creation of enhanced metadata, we propose creating a new kind of faceted search and visualization environment that supports visualization of search results by map, time graph, and tag cloud. The authors believe that the lessons learned from AfricaMap, once implemented as enhancements to the system, could help jump-start the next level of online geospatial collaboration.

ACKNOWLEDGEMENTS

The authors thank Professors Suzanne Blier and Peter Bol, co-principle investigators of the AfricaMap project, for their vision and guidance in conceptualizing, developing and promoting AfricaMap. The authors also thank Dr. Frank Chang for his review comments on this manuscript. The AfricaMap project team at CGA includes the authors and David Siegel, Bo Zhao, Julia Finkelstein, Molly Groome, and Giovanni Zambotti.

The AfricaMap project was funded by the following funds and organizations of Harvard University:

- The Provost's Fund for Innovative Technology
- The William F. Milton Fund
- The Center for Geographic Analysis
- The W.E.B Du Bois Institute
- The Department of African and African-American Studies
- The Committee on African Studies

Collaborating Organizations in data, infrastructure and system development include the following:

- Organizations within Harvard University
  - The Center for Geographic Analysis
  - Harvard Map Collection
  - Harvard Geospatial Library
  - Harvard-MIT Data Center
  - The Dataverse Network Project
  - The Institute for Quantitative Social Science
- External contributors
  - East View Cartographic
  - Avencia
  - MetaCarta

For a complete list of the AfricaMap project steering committee members, please see the "About" tab on http://africamap.harvard.edu

REFERENCES

Blood, Rebecca. (2000) Weblogs: A History and Perspective. In Rebecca's Pocket. 07th September 2000. 25th October 2006. [Online] Available from: http://www.rebeccablood.net/essays/weblog_history. html

BOINC. *Open Source software for volunteer computing and grid computing*, [Online] Available from: http://boinc.berkeley.edu/

Giles, J. (2005) Special Report Internet Encyclopedias Go Head to Head. Nature 438, 900–901.

Global Roads Workshop. (2008) *A Strategy for Developing an Improved Global Roads Data Set*. 2. Columbia University. [Online] Available from: http://www.ciesin.columbia.edu/confluence/download/attachments/3407907/GlobalRoads_StrategyPaper_20feb08.pdf

Goodchild, M. (2007) Citizens as Sensors: The World of Volunteered Geography: 1, 2. Santa Barbara: Position Paper for the Workshop on Volunteered Geographic Information. [Online] Available from: http://www.ncgia.ucsb.edu/projects/vgi/docs/position/Goodchild_VGI2007.pdf

Harley, J.B. & Woodward, D. (1987) The History of Cartography: Cartography in prehistoric, ancient, and medieval Europe and the Mediterranean: 112. Chicago: Humana Press.

Hill, L.L. (2006) Georeferencing: The Geographic Associations of Information: 5. Cambridge, Massachusettes: The MIT Press.

Hobbes, Z. (2010) *Hobbes Internet Timeline 10*. [Online] Available from: http://www.zakon.org/robert/internet/timeline/

King, G. (2007). An Introduction to the Dataverse Network as an Infrastructure for Data Sharing. 195. Sociological Methods and Research, Volume 36, No. 2. [Online] Available from: http://gking. harvard.edu/files/dvn.pdf

Mannak, M. (2008) Development: Africa Still Hampered by Lack of Geographical Data. Inter-Press Service News Agency [Online] Available from: http://ipsnews.net/news.asp?idnews=43685

Miller, C. (2006) A Beast in the Field: The Google Maps Mashup as GIS/2: 188. Cartographica volume 41, issue 3.

Nielsen, J. (2008) Nielsen's Law of Internet Bandwidth. useit.com. [Online] Available from: http://www. useit.com/alertbox/ 980405.html

Ogodo, O. (2009) Africa Facing Climate Data Shortage. 11th November 2009. Science and Development Network, [Online] Available from: http://www.scidev.net/en/news/africa-facing-climate-data-shortage.html

Orwant, J. (2009) Comment in Language Log Blog. [Online] Available from: http://languagelog.ldc. upenn.edu/nll/?p=1701 #comment-41758

Petras, V. (2004) Statistical Analysis of Geographic and Language clues in the MARC record. Technical Report for Going Places in the CatelogL Improved Geographical Access Project. School for Information Management and Systems, University of California Berkeley. [Online] Available from: http://metadata.sims.berkeley.edu/ papers/Marcplaces.pdf

Scharl, A. & Tochtermann, K. (eds.) (2007) The Geospatial Web. London: Springer-Verlag.

Schutsberg, A. (2007) Google KML Search: What Does it Mean for Geospatial Professionals? [Online] Available from: http://www.directionsmag.com/article.php?article_id=2409

Tobler, W.R. (1970) A Computer Movie Simulating Urban Growth in the Detroit Region. Economic Geography, 46, 244–40.

Wegener, M. & Fotheringham, S. (2000) Spatial Models and GIS: New and Potential Models, London: CRC.

Wikipedia, (2010) *Moore's Law* [Online] Available from: http://en.wikipedia.org/wiki/Moore%27 s_law.

Wikipedia, (2010) *Ramsden Theodolite* [Online] Available from: http://en.wikipedia.org/wiki/Ramsden_theodolite

Wilson, T. & Burggraf, D. (ed.), (2009) KML Standard Development Best Practices: vii, Open Geospatial Consortium.

Zuckerman, E. (2008). *my heart's in accra:* 1/18/08 Blog Posting. [Online] Available from: http://www.ethanzuckerman.com/blog/2008/01/18/kenya-mapping-the-dark-and-the-light/

*Advances in Web-based GIS, Mapping Services
and Applications – Li, Dragićević & Veenendaal (eds)*
© *2011 Taylor & Francis Group, London, ISBN 978-0-415-80483-7*

A geospatial Web application to map observations and opinions in environmental planning

Claus Rinner
Department of Geography, Ryerson University, Toronto, Ontario, Canada

Jyothi Kumari
*Ecologie fonctionnelle et PHYSique de l'Environnement (EPHYSE),
Institut National de la Recherche Agronomique, Villenave d'Ornon, Bordeaux, France*

Sepehr Mavedati
Department of Geography, University of Toronto, Toronto, Ontario, Canada

ABSTRACT: The geospatial Web enables virtually everyone to contribute to the growing collection of geographically referenced information on the World-Wide Web. In this chapter, we present a Google Maps-based tool that enables Web users to contribute two types of information: annotations and their reference locations. We further differentiate annotations into observations and opinions regarding specific places. The potential of this approach for integrating local knowledge into environmental planning was assessed by conducting an online map-based discussion of organic farming among expert stakeholders in the Kawarthas area in Central Ontario, Canada. The discussion contents shed light on the participants' perceptions of the organic food market. Moreover, the experiment demonstrated how a map-based discussion forum can be useful for obtaining public input on planning and policy issues.

Keywords: Argumentation mapping, decision support, geospatial web, local knowledge, organic farming

1 INTRODUCTION

Web 2.0 technology enables virtually everyone to contribute to the growing information base of the World-Wide Web. Since only a few years, mapping platforms such as Google Maps have enabled citizens to share their geographically referenced information and create an abundant geospatial Web, or GeoWeb (Lake et al. 2004, Haklay et al. 2008). Although the professional applications of this technology are still sparse, it could provide a means to engage stakeholders and the public in societal issues such as climate change adaptation and response (Sieber 2007).

For the most part, recent GeoWeb advancements were made possible by new technologies and products available to users and developers. In particular, Web mapping services such as Microsoft Live Maps, Yahoo Maps, and Google Maps have provided a reliable foundation for building GeoWeb applications. These services provide a transparent and interactive user interface with preloaded maps in combination with open application programming interfaces (APIs) and data formats, enabling the upload of user-generated, geographically referenced content.

GeoWeb and Web 2.0 technologies have mutually contributed to each other's progress by making it possible to create geographic "mashups", i.e. Web applications combining two or more data sources to provide a transparent service to end-users. Geographic mashups have put location-based information and services at the centre of media attention (e.g. Black 2009) and contributed significantly to engaging the public in the opportunities of the Web 2.0.

In spite of its progress, the GeoWeb still lags behind in terms of primary characteristics of the Web 2.0, such as communication and collaboration. Geographic mashups are mostly limited to merging and displaying geographic data with little evidence of user-generated content that would support spatial analysis and geocollaboration. A notable exception is presented by Roth et al. (2008). It can be expected that improvements in related Geomatics technologies will soon broaden the scope of serious GeoWeb applications. Open-source GIS tools have become much more powerful and easier to use, which eliminates the significant cost of access to commercial GIS programs. GPS and other positioning techniques have also become more popular as a component of mobile devices, making it much easier to generate geographically referenced data during daily activities.

A major problem that once faced GIS developers was the lack of geographic databases. Fortunately, many relational databases systems now have added support for geospatial data. For instance, PostgreSQL, MySQL, Oracle, and Microsoft SQL Server all have introduced spatial extensions that allow developers to work with spatial data natively and support spatial queries.

All the aforementioned advancements have made it possible to develop sophisticated geographic applications more efficiently. Such applications could, in turn, support end-users to shift from content consumers to content producers, and by doing so, close the gap between GeoWeb applications and the rest of the Web 2.0.

This chapter is organised into six sections. Section 2 provides a summary of research on map-based communication and deliberation within Geomatics and reviews public participation in environmental planning. In Section 3, we describe the architecture and current development stage of the Argoomap tool. This tool was used in the case study of organic farming in the Kawarthas, detailed in Section 4. The discussion of results in Section 5 is followed by conclusions and an outlook on future research in Section 6.

2 RESEARCH BACKGROUND

2.1 *Map-Based Communication and Deliberation*

Argumentation mapping (Rinner 1999, 2001) combines online discussion forums with online mapping to encourage discussion participants to provide explicit geographic references for their contributions. Such references will enable map-based access to read the current state of a discussion as well as more advanced spatial queries and analyses of discussion contributions, such as finding contributions that refer to a specific area or finding all authors of any replies to contributions within a specific distance from a given location. The general goal of using argumentation maps is to facilitate communication and deliberation in spatial decision-making, as it occurs in many planning procedures.

In a series of case studies, the usability and usefulness of argumentation mapping for geocollaboration, public participation, and community engagement was assessed. Using the first-generation "Argumap" tool, Sidlar & Rinner (2007, 2009) developed simple metrics for geographically referenced participation, and Rinner & Bird (2009) conducted a contents analysis of map-based discussion messages. Using the re-developed "Argoomap" tool, Rinner et al. (2008) analyzed the geospatial relations between reference locations of arguments. Across these case studies, the researchers encountered large variations in the levels of participation and engagement, depending on the type of participants (student volunteers vs. general public) and timing of the user tests (during academic term vs. summer vacation time). Participants provided extensive feedback on different versions of argumentation mapping tools that were subsequently modified, in particular to offer a simplified user interface that mimics widely used online mapping tools such as Google Maps. Other tools that focus on mapping opinions and comments include MapChat (Hall & Leahy 2008) and GeoDF (Tang & Coleman 2008). The name "MapChat" was also used for a prototype tool for collaborative event planning that combines social networking applications with online mapping (Churchill et al. 2008).

Table 1. Decision-making phases (Simon, 1965*, 1977**; Sabherwal and Grover, 2007+) and types of user contributions on the GeoWeb.

Phase	Explanation[+]	Type of contribution	Example
Intelligence*	"The environment is searched for conditions calling for decisions"	Observations	Field observations of invasive species indicating possible need for protective action
Design*	"Possible courses of action are invented, developed, and analyzed"	Contents	Ideas, suggestions, and proposals on how to protect native species
Choice*	"An alternative course of action is selected"	Opinions	Argumentation and voting to support, or object to, specific conservation policies
Review**	"... assessing past choices"	Comments	Evaluation of conservation success through re-iteration of intelligence, and possibly other phases of decision-making

In recent years, the emergence of the Web 2.0 and user-friendly online mapping techniques has created public interest in contributing information through Web-enabled geospatial tools ("volunteered geographic information", Goodchild, 2007). This trend has allowed researchers to instill public participation in planning processes (Seeger, 2008). For example, the publicly available Google Maps API has been used by researchers to develop public participation tools for gathering location-specific information for land-use planning (Mueller et al. 2008), emergency planning (Tanasescu et al. 2006), and urban planning (Rinner et al. 2008).

When exploring Web 2.0 and GeoWeb applications, different types of user input can be distinguished. In many instances, users are enabled to upload stand-alone media (e.g. text, photographs, or videos), while in other cases, users contribute observations or opinions in response to existing information, or are invited to rate existing information or vote upon a given question. With reference to Simon's (1965, 1977) framework for rational decision-making, Table 1 attempts to structure types of user contributions to different phases of decision-making. In the exploratory intelligence phase, field *observations* from users may inform the search for conditions that call for a formal decision-making process. In that case, a design phase would follow, during which users could propose alternative solutions to the decision problem in the form of comprehensive *contents*. In the subsequent choice phase, user *opinions* responding to, and expressing a preference among, the proposed solutions would be gathered and considered in choosing an alternative. Finally, if a review of a past decision occurs, evaluative *comments* would assist with assessing the success or failure of the chosen solution.

2.2 *Participation in Environmental Planning*

Environmental planning requires the integration and synthesis of scattered information from numerous sources and coupling of this information for problem solving and decision-making (Fedra 1993, Elmes et al. 2004). The emphasis is on increasing the use of local geospatial knowledge through local stakeholders' participation (Steinmann et al. 2006). Ensuring public (or stakeholder) participation in the environmental decision-making process is advantageous to both planning officials and stakeholders. This is because any issue of local importance can be resolved at an early stage of planning and a better decision can be made. Furthermore, stakeholders' indigenous knowledge aids planners in interpreting local environmental and social data.

While stakeholder participation is important for the success of participatory decision-making, various conventional methods of public participation such as meetings or work-shops have been found to be ineffective in addressing the needs of stakeholders, because participation in such cases are affected by space-time constraints, and bureaucratic procedures (Carver et al. 2001). Parallel advancements in GIS technology, the Internet and the World Wide Web (WWW) have improved the possibilities for supporting public participation through "e-Participation" (Kearns & Bend 2002, Macintosh 2004, Hansen & Reinau 2006) or asynchronous distributed discussions (Beaudin 1999, Rinner 2001 2006). To date, a variety of Web-based GIS having various levels of technological sophistication have been used for environmental planning and participatory decision-making process (Kingston et al. 2000, Lee et al. 2000, Cetin & Dieker 2003, Brown 2007). Web-based participatory GIS (PGIS) is becoming a promising approach to incorporate local knowledge in spatial decision-making processes (Craig et al. 2002). The objective of PGIS is to integrate spatial technologies with stakeholder and expert knowledge for collective decision-making and problem solving.

The main challenge in developing an efficient PGIS, however, is that local spatial knowledge is often narrative, qualitative, and contradictory, hence, its incorporation in a GIS is often very difficult (Elwood 2006, Elmes et al. 2004). To address this issue, attempts have been made by some researchers to make use of freely available and popular online map services (e.g. Google Maps) to integrate geographically-referenced discussion forums (i.e. qualitative data) and GIS components (Tanasescu et al. 2006, Mueller et al. 2008, Rinner et al. 2008).

Other challenges faced by Web-based PGIS concern the "digital divide" (Peng 2001) and public understanding of spatial decision problems (Kingston 2000, Carver 2001). Addressing these challenges is an important step for the success of a PGIS as an efficient decision support tool. The challenges related to technological skills of the stakeholders lead to the need for more research regarding the improvement of the usability of existing PGIS by making it simple to use on one hand, and an efficient data collection, distribution, and display platform, on the other.

3 A GEOWEB-BASED DISCUSSION FORUM

The Argoomap tool (Rinner et al. 2008) consists of three major software components: the server-side database and application logic, and the client interface. On the server side, the MySQL database is used for storing and retrieving data (including locations, discussion structure and contents, and user data). The PHP programming language was used for data processing on the server and providing the client functionality. The MySQL database and PHP language are two popular choices for today's Web developers for a variety of reasons; they are both open source and therefore are of lower cost, in addition to having large active developer communities, which contribute to the support and documentation of these tools. On the client side, the Google Maps API is used to provide the core mapping functionality. The choice of the Google Maps API for geographic functionality was also driven by the same factors. Although Yahoo and Microsoft provide similar APIs, the Google Maps API was considered the most popular choice for Web-based mapping development.

Since the Google Maps API is a critical component of the Argoomap tool, it will be helpful to include a brief description of its architecture. The API provides a great way for programmers to build Web mapping application on top of Google Maps. It provides "objects" that developers can use and customize in their Web applications. These objects can be categorized and described, as shown in Table 2.

The Argoomap tool makes use of all Google Maps API object categories. In addition to core and map control objects, Argoomap uses GMarker overlay to display referenced locations on the map and GInfowindow overlay to display the discussion contents of each marker. Interaction with users has been accomplished by using GEvents that respond to different actions

Table 2. Google Maps API object categories (Google 2009) and their description.

Object category	Description
Core and base objects, such as GMap, GLatLng, and GLatLngBox	These objects provide the core functionality for Google Maps. GMap is the primary object that defines the map presented to the user. GLatLng and GLatLngBox are used to represent points or rectangles of geographic locations.
Map controls objects	These are the objects that can be added to GMap to support switching between map types (street map, satellite view, hybrid) and navigating the map through zooming and panning. The controls are customizable allowing developers to modify their appearance and behaviour.
Overlay objects, such as GMarker, GPolyline, and GPolygon	These objects are used to add geographically referenced data to the map, to create a mashup. Point features are represented by GMarker, lines by GPolyline, and areas by GPolygon.
Events	Events are what makes Google Maps application interactive, enabling the map to respond to user actions using the mouse or keyboard. Developers can define GEventListeners for different events and define what action to initiate. Events can be defined on any object that exists in the application, such as map controls, markers, or any other object that could be interacted with.
Services	These objects provide other functionalities that do not fall within the previous categories, including helpers to parse data, (reverse) geocoding, traffic overlays, KML and GeoRSS overlays, directions, Google Earth integration, etc. This category is being updated regularly, adding new features and services to facilitate geographic operations.

based on the application's current state. Argoomap also utilizes parsing services to retrieve information from its database and display the initial state of the map-based discussion.

Figure 1 illustrates various states of Argoomap application. The initial view of Argoomap provides users with a full screen map that contains markers indicating the locations that have already been provided by users (Fig. 1, main area). A click on any of the existing markers opens a window containing all the discussion threads that reference that geographic location (Fig. 1a). Clicking on any of the discussion threads will open the post body and highlights all the markers referenced by this post (Fig. 1b). Users can then reply to the post, which, by default, will include all the currently highlighted markers, but they also have the option to add other references or remove existing references from their reply (Fig. 1c). Another way to add a new contribution is to click on an empty area of the map to create a new thread with one or more reference markers.

The previous version of the Argoomap tool has gone through major improvements for this project. An important enhancement was the introduction of user management. This feature also grants the ability to restrict user access to different actions, if desired. There are currently five levels defined: no login required; login required for creating a new discussion thread; login required for replying to other posts; login required for reading the posts; and finally, login required for viewing the map. The administrator of the Web application has the ability to change this level in a configuration file. If any access restrictions are enforced, users will be prompted to login or register upon initiating a corresponding action.

Another improvement was the implementation of a side panel for the discussion forum, which makes browsing the existing posts easier. The side panel offers three alternative views (Fig. 2): a thread view, displaying a tree structure of the discussion (Fig. 2a); a user-based view, which groups each user's posts together (Fig. 2b); and finally, a date view, which sorts the posts based on their submission date (Fig. 2c). The main objective of this feature was to improve the usability for end users allowing them to access the discussion easily and effectively.

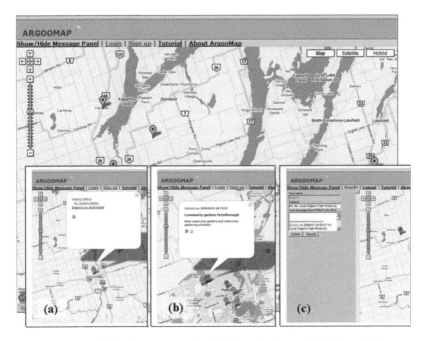

Figure 1. Various states of the Argoomap tool: initial view (background); (a) list of all messages referencing a selected marker; (b) display of a single discussion message and its highlighted reference markers; and (c) input form for replying to an existing discussion message.

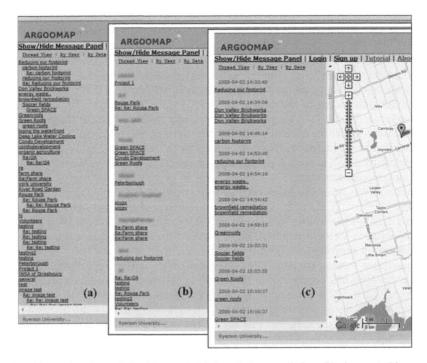

Figure 2. Alternative views of the side panel: (a) thread-structured view; (b) view sorted by user; and (c) view sorted by date of submission.

4 CASE STUDY OF ORGANIC FARMING IN THE KAWARTHAS

4.1 *Outline of Methodology*

A study was conducted to evaluate the usability of the Argoomap tool as a Web-based GIS, as well as its effectiveness in community engagement. Such a topic was chosen wherein the local-knowledge levels of the stakeholders greatly vary and the issues that could be discussed are multi-faceted. This would aid in inferring usability and prospects of Argoomap as a Web-GIS planning tool for various other issues. Because organic food production is a rapidly growing field for entrepreneurship, it was chosen as the topic of discussion to gather the stakeholders' and experts' local knowledge. The study was conducted in the Kawarthas (Kawartha Lakes and Peterborough region), an important agricultural belt of Ontario. This study was carried out in collaboration with the Kawartha Heritage Conservancy (KHC), a non-profit organization based in Peterborough, Canada. The KHC has a Farmlands program where they work with farmers and local organizations to support innovative agriculture practices and promote the sale of organic and locally grown food.

The Argoomap tool was used to engage stakeholders in a discussion of the problems and prospects of organic farming. A selected group of stakeholders was invited to participate in a discussion related to the prospects and problems of organic farming in the Kawarthas. Potential participants were identified and selected after a thorough internet search of websites related to organic farming in Kawartha and Peterborough. Approximately 70 email requests for participation were sent based on the review of farmers' forums and social networking sites, non-governmental organizations in agriculture, related government organizations, and universities. Twelve individuals registered to participate in the study.

The discussion phase started on July 7, 2008 for a period of one week. The link to Argoomap was activated and the discussion during this period was monitored by the researcher. One starter thread (message) entitled "What are the problems and prospects of organic farming in Kawartha and Peterborough region" was posted by the researcher to facilitate the discussion. A reminder E-mail was sent to all the participants on July 10, 2008 to encourage more contributions. Another E-mail was sent on July 14, 2008 to inform the participants about an extension of the discussion period by one more week. Although increased participation was anticipated, the majority of the messages were posted in the first week and a more engaged discussion could not be obtained even after extending the time for discussion for one more week. During the course of the discussion, the researcher posted a reply to a discussion thread posted by a participant to demonstrate the use of the "reply" function of Argoomap and to encourage the participants to respond to the comments posted by others. The discussion was officially closed on July 21. However, on July 22, two discussion threads were submitted by a participant and these contributions were included in the data analysis.

Information from the Web server log file and questionnaire-based survey was used to assess the usability of the Argoomap. The data from Web server log file that were used for analysis included the number of times the participants accessed the tool, the time of the day during which the participants accessed the tool, the number of times the participants started a discussion thread or replied to a message, and the amount of time spent on the Argoomap page. The contents of contributions by participants were used to subjectively assess the quality of the discussion in understanding the role of Argoomap in engaging community through a common forum.

4.2 *Case Study Results*

During the two-week discussion period, Argoomap was accessed 24 times by the twelve participants. The chronology of contributions is shown in Figure 3. It should be noted that there was no contribution on the first two days (July 7 and July 8, 2008). On the third day

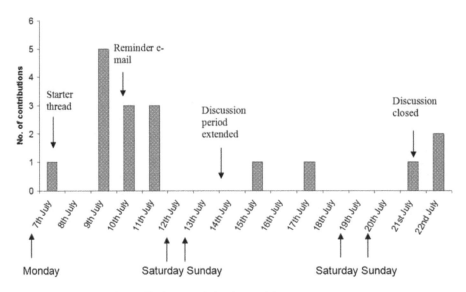

Figure 3.   Chronology of contributions made by the participants.

(July 9, 2008), five messages were contributed by three participants. However, neither the reminder email nor the message confirming an extension of the discussion period generated significant input. Instead, the discussion tapered off after its fifth day.

The discussion was focused on the organic food market and local organic food production. In addition, some participants contributed opinions about agriculture in Ontario in general. Topics such as Ontario's farmland preservation, organic food consumption, prospects of urban agriculture and community gardening appeared in the discussions. Box 1 documents the participants' original messages contributed during the discussion phase.

The performance of the Argoomap was evaluated using a questionnaire survey. The survey consisted of three parts:

- Participants' skills in using a GIS and the Internet
- Understanding of the discussion topic by participants
- Ease with which various functions in the Argoomap could be used

Seven out of twelve registered participants gave their feedback by completing the question-naire survey form, and two participants provided descriptive feedback through E-mail. The survey results from the seven participants are shown in Table 3. Five participants had a good GIS background. All participants were familiar with Web-based discussion forums. Five participants agreed that their familiarity with Google Maps was helpful in becoming familiar with the Argoomap, whereas one disagreed. Another participant did not comment on this question. Five respondents had a general understanding of organic farming and felt comfortable posting comments on the Argoomap. They agreed that the discussion topic was relevant to the current environmental issues, yet disagreed that their interest in organic farming increased by participating in this study. Only one participant agreed that the Argoomap is an effective platform for remotely gathering opinions whereas the other six did not provide any response.

The ease of use of the Argoomap was evaluated with respect to various functions of the tool. The results of this evaluation are shown in Table 4. In general, the participants found the Argoomap interface menu easy to understand. Only one participant found it "somewhat difficult". Except for one participant, the respondents found the functions such as registra-tion, creation of user names and log-in names either "very easy" (4) or "easy" (2).

Map navigation through the zoom and pan tools did not pose any problems and, in general, was found to be "easy" by four participants and "very easy" by three of the participants.

Organic Farming What are the problems and prospects of organic farming in Kawartha and Peterborough? [starter thread by researcher]

Importance of local production I think the first major issue to promote organic farming is to reduce the import \""cheaper\"" organic produce from countries and sell it at cheaper rates. This way, the farmers here are not encouraged to do organic farming. They concentrate on field crops like wheat and corn which have higher market, especially due to the recent oil price boom"

> Re: Importance of local production "Pick Ontario Freshness" marketing campaign by Ontario govt. is a good step to show support for farmers producing Local Food. [reply from researcher]

> > Re: Re: Importance of local production Probably in the current context of oil price hike, things may change

Need for a new By-law Ontario govt should create new by-laws to set a minimum level of food stocked by Ontario\'s grocery chains so that local produce have higher demands.

Organic at the Peterborough Farmers Market There is a very limited amount of certified organic produce for sale at this market. This is disappointing since I would certainly purchase local grown organic produce if it was available. I currently have to get my organic produce from the larger supermarkets in town. I am not sure if these supermarkets buy from local organic farmers. On the other hand, this market does have vendors that sell good certified organic local meat products (beef, chicken and pork). The prices are relatively reasonable compared to non-organic products available. My family stocks up once a month on these meat products."

Organic in Peterborough area For fresh vegetables in season, you can also try By the Bushel delivery service. Not certified though farmer vendors are carefully selected. [reply through a new discussion thread]

> Re: organic in Peterborough area I think the govt should invest more money into advertising organic food. This can be easily done at public places such as public transportation facilities etc. In Toronto TTC trains one can see many forms of advertisements regarding health, education etc. Why not organic food?

Local Organic Farm Products Several local farms have started selling direct to consumers from their properties. These include those who are certified organic. Kawartha Choice marketing, promoting locally grown food, has a website that lists farmgate sales farms and these say whether they are organic or not.

> Re: Local Organic Farm Products This is a great way of promoting of a website like this will help in reducing the time delays in getting certification, I think Current Issues? What do you think in your opinion are the barriers that discourage farmers to grow organic in your area?

Issues are not only \"Organic\" but general agricultural issues too We should understand that organic farming is a subset of general agricultural practice or occupation. Therefore, one needs to think about the current agricultural issues too.
For example, the Ontario Govt\'s farmland Preservation plan prevents the farmers from selling their lands even though they get attractive prices from buyers. This I think should be made less stringent. Land is one of the \"assets\" of a farmer and in times of need, because of this by-law, a farmer cannot meet his urgent financial needs. This makes him remain in the vicious cycle only relying on crop profits to meet his financial needs. These types of stringent laws drastically reduces the entrepreneurship and the risk taking ability of of the framing community preventing a farmer from venturing into new activities like organic farming.

Organic Food Consumption. The organic food consumption pattern has been changed in Ontario.

Organic farming in Ottawa We encourage farmers to grow organic produce by setting up farmers market right in the downtown Ottawa. I think farmers get good returns for their produce and encourage them to grow organic.

Organics vs. local ledger What are the metrix to value local factory farmed chicken vs. organic chicken from NY?.

Community gardens Urban agriculture should be a major priority for city planners. There are many gardens around the city of Toronto such as garden of friends located here.

Community gardens Various garden spots.

*A new discussion thread is identified by **bold** text, whereas others are replies.*

Box 1. Original contributions to the discussion.

Table 3. Participants' computer skills and understanding of the topic.

Participants' computer and internet skills	Strongly agree	Agree	Disagree	Strongly disagree	No comment
I have a good GIS background	3	2	2	–	–
I have good Internet skills and I am familiar with Web-based discussion forums	6	1	–	–	–
I am familiar with Google maps and hence, it was easy to learn and use the Argoomap	2	3	1	–	1
Understanding of the discussion topic					
I have a understanding of the organic agriculture issue	2	3	1	–	1
I felt comfortable giving my opinions by posting comments	1	4	1	–	1
The discussion was relevant to the current environmental issues	–	5	–	–	2
My interest in organic agriculture increased by participating in this study	–	–	5	–	2
The Argoomap is an effective platform for remotely gathering opinions	–	1	–	–	6

Table 4. Subjective ease of using various functions of the Argoomap tool.

Functions	Very easy	Easy	Somewhat difficult	Difficult	Very difficult	Did not use the function
Reading/understanding menu on the Argoomap interface	2	4	1	–	–	–
Creation of a username and login	4	2	–	–	1	–
Map navigation: Zooming and pan (moving around the map)	3	4	–	–	–	–
Ease of reading various discussion threads	2	3	2	–	–	–
Reply to a discussion thread	–	3	2	–	–	2
Starting a new discussion thread	–	4	1	–	–	2
Learning to use the Argoomap	1	4	2	–	–	–
Understanding the Argoomap tutorial	1	3	2	–	–	1
Ease with which one can switch between a task and the tutorial?	–	1	–	1	2	3

Reading the discussion threads was "somewhat difficult" for two of the participants whereas the remaining five found it either "easy" (3) or "very easy" (2). Only three participants found replying to a discussion thread "easy" while two found it "somewhat difficult". The other two did not use this function at all.

The ease with which one is able to start a new discussion in the Argoomap got similar response to that of replying to a message. Four participants found it "easy" to start a new thread whereas two found it "somewhat difficult" and the other two participants did not use this function at all. It was interesting to find that five of participants did not face any difficulty in learning to use the Argoomap. Four participants reported it to be "easy" whereas one found it "very easy". The other two participants found the learning experience "somewhat difficult". The instructions on how to use the Argoomap provided through the tutorial were considered to be "very easy" to understand by one of the participants whereas it was "easy" for three of the participants. Of the remaining three participants, two found it somewhat difficult whereas one did not use the tutorial at all.

The participants encountered problems with switching between the task and the tutorial windows. Because the tutorial window opens in the same window-frame as the task, it was reported to be "very difficult" to use by two of the participants and "difficult" for one participant. Only one person found this option "easy". The overall reaction of the evaluators regarding Argoomap's efficacy as a Web-GIS tool were "good" (by four participants) and "satisfactory" (three participants).

5 DISCUSSION

The nature of the contributions during the discussion phase (Box 1) implied that the participants had a general understanding of organic farming and were interested in exploring organic food market issues. The discussion was focused on topics such as availability of organic produce, locally grown food, "buy local" vs. "buy organic" concepts, farm gate sales, and popularization of organic food market. All these topics are some of the ongoing issues in the global food market. Many consumers view buying local (through farmers markets or community-supported agriculture) as a promising alternative to an unsustainable global food production system that has adverse effects on the environment and local economy.

In this discussion, it was interesting to note that the participants raised the topic of "local vs. organic" farm products by asking how to "value local factory farmed chicken vs. organic chicken from NY". Understanding the difference between locally grown produce (conventional farming using fertilizer and pesticides), local organic produce (organic farming method with no synthetic fertilizer and pesticides), and imported organic food is important for consumers because consumers get confused with the concept of "buy local" vs. "buy organic". The concept of "buy local" is to invest in the local food economy and food grown by local farmers. Local food, generally sold through farmers markets and community-supported agriculture has the advantage of being fresh (LaSalle 2008). However, local does not necessarily reduce the environmental impacts of fertilizers and pesticides unless it is organic. On the other hand, imported organic produce is criticized for its high food miles (fossil fuel consumption through long distance transportation) and carbon footprint (through processing and packaging). Thus, in the absence of locally grown organic produce, consumers are left with a difficult choice.

The issue of "Organic in Peterborough area" was mentioned by one of the participants who, even though interested in locally grown organic produce, had to buy imported organic produce from supermarkets. This suggests the need for encouraging local farmers to adopt organic methods of farming to meet consumer demand. The participants also discussed the regional availability of community food co-ops (cooperative grassroots organisations) and support services for farm gate sales. Some participants suggested the importance and need for government support for the popularization of local organic farm products. All these discussion topics shed light on consumers' perceptions of the organic food market.

Although the topics discussed by the participants indicated their interest and awareness of the organic farming market, the quality of the discussion itself could not be considered as a fully engaged discussion. It was expected that the participants would raise questions about the current issues in organic farming in the Kawarthas and provide their opinion about these, as well as respond to queries raised by other participants. In order for them to be actively involved in the discussion, the participants were invited to access the Argoomap frequently to see the new messages posted by others. However, with the exception of one, each participant accessed the Argoomap only once during a period of two weeks. Most of the participants, posted comments only once and did not involve themselves actively in the discussion. There were only six replies (four first-order replies and two second-order replies), and these replies were from only one participant who was actively engaged in the discussion.

The reason for this limited involvement may be the small number of twelve participants. This may be attributed to the lack of incentives provided to the participants for their time. The only benefits the participants gained were to learn about organic farming issues in the Kawarthas and current research in GeoWeb technology. An increased participation could perhaps have been obtained had some financial compensation been provided. The study could have also benefited, were it conducted as part of an official planning process.

The timing and length of the discussion phase is also likely a factor for the low rate of participation and lack of an engaged discussion. Due to time constraints, the discussion was open for two weeks only. A larger participant group and a longer time frame might have further revealed information about the Argoomap as a participatory GIS in community engagement. Furthermore, an in-depth discussion of the core issues of organic farming such as the current cost of organic certification, labour, weed management, problem soils for organic farming operations, or nearby organic processing facilities might have arisen, had any farmers participated in this study. However, the study was conducted during the summer growing season and therefore involvement from farmers was not achieved. One farmer, who had, at first, agreed to participate, could not do so because of time constraints. This identifies the necessity of allowing more time for discussion so that an engaged and effective discussion can take place.

Information from the Web server log file and questionnaire survey was used to evaluate the usability of the tool in terms of the ease of learning the Argoomap tool, the ease with which various functions of Argoomap can be used, and the participants' overall reaction to the tool's interface.

In order to be efficient and user-friendly, a tool should allow quick learning without spending a lot of time on training. In this study, no workshop or training was provided to familiarize the participants with Argoomap. The participants learned to use the tool by reading the tutorial provided within the Argoomap interface. The fact the participants were able to post a comment after reading the procedures given in the tutorial clearly indicates the "ease of learning" this tool.

The participants understood the tool and appreciated the concept of argumentation mapping. Every participant who contributed to the discussion created a new discussion thread. With one exception, all participants created discussion threads at the place about which they posted their comments (one participant created the placemarks at the place of his residence). The fact that the participants correctly created a discussion thread linked to a place of interest suggests that they understood the underlying concept. Moreover, multiple messages linked to a single placemark were submitted by one of the participants (Fig. 1b), which was surprising since instruction on linking multiple messages to a single placemark was not included in the tutorial.

The assertion that the Argoomap is an easy to learn software application has also been confirmed by the survey in which five participants rated the learning experience to use the Argoomap as "easy". The participants' reaction to the Argoomap interface was either "good" or "satisfactory." The survey results show that the respondents were familiar with Web-based discussion forums and five survey respondents agreed that their familiarity with Google Maps helped them to learn to use the Argoomap tool. Two participants who did not complete the survey form provided their comments through E-mail. They described the Argoomap graphical user interface as "user-friendly".

It is difficult to ascertain why the participants chose not to comment on the effectiveness of the Argoomap tool as a platform for remotely gathering opinions. Only one participant agreed that the Argoomap is an effective platform for remotely gathering opinions whereas the other six did not provide any response. It can be speculated that the participants were exposed to a Web-based GIS tool for the first time, and therefore did not have knowledge of the effectiveness of the other GIS tools; thus, they did not have a benchmark with which to compare the Argoomap, and so chose not to comment on its effectiveness.

Some participants liked the concept of the tool and they were able to appreciate the potential applications of the Argoomap for various planning purposes such as community gardening, locations for farmers markets, farmland preservation, and educational farms. The variety of themes that arose during the case study demonstrated the utility of the Argoomap; that it could be used for various kinds of planning processes for collecting the observations and opinions of local citizens.

6 CONCLUSIONS AND OUTLOOK

In this chapter, we presented a revised version of a GeoWeb tool for deliberation in spatial decision-making, along with a case study of public participation in environmental planning. Among the different types of volunteered geographic information enabled by the GeoWeb, the Argoomap tool specifically supports the expression of opinions and comments as part of the choice and review phases of rational decision-making.

The revision of the Argoomap tool was guided by previous case studies, in which users with computer skills could not easily handle the custom map navigation that was akin to commercial GIS user interfaces. We therefore chose the Google Maps API as the development platform in order to exploit the familiarity of a significant proportion of potential users with the mapping tools. However, the simplification of the user interface reduced the availability of analytical tools to monitor and assess the current state of a map-based discussion. This contrasts with other work including Simão et al. (2009), and Boroushaki & Malczewski (2010), who extend map-based discussion support with multi-criteria analysis to achieve full analytic-deliberative decision support (Jankowski & Nyerges 2003).

A different avenue of future research and development is the introduction of advanced discussion support functionality such as picture and video upload, querying, and automatic geo-referencing. We envision the Argoomap tool linked more closely with existing social networking platforms such as facebook, so that users could be notified about discussion activities in their favourite online environments. GIS functionality such as route calculation could also be useful to participants. In addition, we are exploring map-based discussion support on mobile devices. Finally, research on the Semantic Web may be of relevance to further development of argumentation mapping. For example, the meaning of discussion contributions needs to be clear, and a measure of trust in authors be provided, in order to enable serious uses such as in government decision-making.

ACKNOWLEDGEMENTS

This research was partially funded by the GEOIDE Network of Centres of Excellence (Project PIV-41). We wish to thank the discussion participants for their contributions and the Kawartha Heritage Conservancy for their support of this project. An earlier version of this manuscript was revised by Blake Walker.

REFERENCES

Beaudin, B.P. (1999) Keeping online asynchronous discussions on topic. *Journal of Asynchronous Learning Networks*, 3 (2).

Black, D. (2009) Web spurs map-making renaissance. [Online] The Toronto Star, 07 February 2009. Available from: http://www.thestar.com/News/GTA/article/583888

Boroushaki, S. & Malczewski, J. (accepted). ParticipatoryGIS: A WebGIS-based Collaborative GIS and Multicriteria Decision Analysis. URISA Journal, articles accepted for publication.

Brown, S.C. (2007) The brown county online GIS: An example of a multi-agency collaborative mapping system. *Extension Journal*, 54 (5).

Carver, S. (2001) Public Participation Using Web-Based GIS. *Environment and Planning*, B28 (6), 803–804.

Carver, S., Evans, A., Kingston, R. & Turton, I. (2001) Public Participation, GIS and Cyberdemocracy: Evaluating Online Spatial Decision Support Systems. *Environment and Planning,* B28 (6), 907–921.

Cetin, M. & Diker, K. (2003) Assessing drainage problem areas by GIS: a case study in the eastern Mediterranean region of Turkey. *Irrigation and Drainage*, 52 (4), 343–353.

Churchill, E., Goodman, E.S. & O'Sullivan, J. (2008) Mapchat: conversing in place. In proceedings of Conference on Human Factors in Computing Systems, 5–10th, April 2008, Florence, Italy.

Elmes, G., Challig, H., Karigomba, W., McCusker, B. & Weiner, D. (2004) Local Knowledge Doesn't Grow on Trees, In: Fisher, Peter F., (eds.) *Community-Integrated Geographic Information Systems and Rural Community Self- Definition, in Advances in Spatial Data Handling (Proceedings of 11th International Symposium on Spatial Data Handling)*, Berlin, Springer. pp. 29–40.

Elwood, S. (2006) *Participatory GIS and Community Planning: Restructuring Technologies*, Social Processes, and Future Research in PPGIS. In: Dragicevic, S. & Balram, S. (Eds.) Collaborative GIS. Idea Group Publishing, pp. 66–84.

Fedra, K. (1993) Integrated Environmental Information and Decision-Support Systems. *Proceedings of the IFIP TC5/WG5.11 Working Conference on Computer Support for Environmental Impact Assessment.* 16, pp. 269–288.

Goodchild, M.F. (2007) Citizens as voluntary sensors: Spatial data infrastructure in the World of Web 2.0. *International Journal of Spatial Data Infrastructures Research*, (2), 24–32.

Google (2009) *Google Maps API Reference.* [Online] Available from: http://code.google.com/apis/maps/documentation/reference.html

Haklay, M., Singleton, A. & Parker, C. (2008) Web Mapping 2.0: The Neogeography of the GeoWeb. *Geography Compass*, 2 (6), 2011–2039.

Hall, G.B. & Leahy, M.G. (2008). Design and Implementation of a Map-Centred Synchronous Collaboration Tool Using Open Source Components: The MapChat Project. In Hall GB, Leahy MG (eds): Open Source Approaches in Spatial Data Handling, pp. 221–246. Springer, Berlin, Germany.

Hansen, H.S. & Reinau, K.H. (2006) *The Citizens in E-Participation.* In Lecture Notes in Computer Science 4084, Berlin, Germany, Springer-Verlag. pp. 70–82.

Jankowski, P. & Nyerges, T. (2003) Toward a framework for research on geographic information-supported participatory decision-making. *URISA Journal*, 15 (APA I), 9–17.

Kearns, I. & Bend, J. (2002) *E-Participation in Local Government.* Published by Institute for Public Policy Research, London, England.

Lake, R., Burggraf, D.S. & Trninic, M. (2004) *Geography mark-up language (GML) – Foundation for the Geo-Web.* Hoboken, NJ, USA, John Wiley and Sons.

Lee, B.L., Ninomiya, S., Kim, Y., Kiura, T. & Laurenson, M. (2000) *Map Broker: A Prototype of Multi-tiered Web Interface Employing CORBA for Simple GIS Applications in Agriculture.* AFITA 2000 Second Asian Conference of the Asian Federation for Information Technology in Agriculture, Suwon, Korea.

LaSalle, T. (2008) Local vs. Organic: Exploring the Controversy. *Earth Matters*, 15 (2).

Macintosh, A. (2004) Characterizing e-participation in policy-making. *Proceedings of the 37th Annual Hawaii International Conference on system science.* 5–8th Jan 2004. Int. Teledemocracy Centre, Napier Univ., Edinburgh, UK.

Mueller, T.G., Hamilton, N.J., Mueller, J.E., Lee, B.D., Pike, A.C., Karathanasis, A.D., Nieman, T.J, Carey, D.I. & Zourarakis, D. (2008) Google Maps as an Aid for Land Use Planning Decisions. *Joint Annual Meeting (5–9th Oct 2008).* Houston Texas.

Peng, Z. (2001) Internet GIS for Public Participation. *Environment and Planning*, B28 (6), 889–905.

Rinner, C. (1999) Argumentation Maps – GIS-Based Discussion Support for Online Planning. Ph.D. Thesis, University of Bonn, Germany. GMD Research Series Nr. 22/1999, Sankt Augustin, Germany. pp. 128.

Rinner, C. (2001) Argumentation Maps – GIS-based Discussion Support for Online Planning. *Environment and Planning*, B28 (6), 847–863.

Rinner, C. (2006) *Argumentation Mapping in Collaborative Spatial Decision Making.* In: Dragicevic, S. & Balram, S. (eds.) Collaborative GIS. Idea Group Publishing. pp. 85–102.

Rinner, C. & Bird, M. (2009) Evaluating Community Engagement through Argumentation Maps - A Public Participation GIS Case Study. *Environment and Planning*, B36 (4), 588–601.

Rinner, C., Keβler, C. & Andrulis, S. (2008) The Use of Web 2.0 Concepts to Support Deliberation in Spatial Decision-Making. *Computers Environment and Urban Systems*, 32 (5), 386–395.

Roth, R.E., Robinson, A.C., Stryker, M., MacEachren, A.M., Lengerich, E.J. & Koua, E. (2008) Web-based Geovisualization and Geocollaboration: *Applications to Public Health. Extended abstract and presentation at the 2008 Joint Statistical Meetings in Denver, CO.* 3–7th August 2008.

Sabherwal, R. & Grover, V. (2007) Computer Support for Strategic Decision-Making Processes: Review and Analysis. *Decision Sciences*, 20 (1), 54–76.

Seeger, C.J. (2008) The role of facilitated volunteered geographic information in the landscape planning and site design process. *GeoJournal*, 72 (3, 4), 199–213.

Sidlar, C. & Rinner, C. (2007) Analyzing the Usability of an Argumentation Map as a Participatory Spatial Decision Support Tool. *URISA Journal*, 19 (1), 47–55.

Sidlar, C. & Rinner, C. (2009) Utility Assessment of a Map-Based Online Geo-Collaboration Tool. *Journal of Environmental Management*, 90 (6), 2020–2026.

Sieber, R. (2007) Geoweb for social change. Position paper submitted to Workshop on Volunteered Geographic Information, 13, 14th December 2007, Santa Barbara, USA. [Online] Available from: http://www.ncgia.ucsb.edu/projects/vgi/supp.html

Simão, A.C.R., Densham, P.J. & Haklay, M. (2009) Web-based GIS for Collaborative Planning and Public Participation: an application to the strategic planning of wind farm sites. *Journal of Environmental Management*, 90 (6), 2027–2040.

Simon, H.A. (1965) The shape of automation. New York: Harper and Row.

Simon, H.A. (1977) The new science of management decision. Englewood Cliffs, NJ : Prentice-Hall.

Steinmann, R., Krek, A. & Blaschke, T. (2006) *Can Online Map-Based Applications Improve Citizen Participation?* In: Lecture Notes in Computer Science, Springer Verlag, TED Conference on eGovernment, Bozen, Italy.

Tanasescu, V., Gugliotta, A., Domingue, J., Villarias, L.G., Davies, R., Rowlatt, M. & Richardson, M. (2006) *A Semantic Web Services GIS Based Emergency Management Application. Lecture Notes in Computer Science*. The Semantic Web - ISWC 2006 .Volume 4273.

Verutes, G., Jankowski, P. & Bercker, J. (under review) Discourse Maps - *A Usability and Evaluation Study of a Participatory Geographic Information System*. URISA Journal, articles under review.

Advances in Web-based GIS, Mapping Services and Applications – Li, Dragićević & Veenendaal (eds)
© *2011 Taylor & Francis Group, London, ISBN 978-0-415-80483-7*

Web-based collaboration and decision making in GIS-built virtual environments

Christian Stock, Ian D. Bishop, Haohui Chen, Marcos Nino-Ruiz & Peter Wang
Department of Geomatics, The University of Melbourne, Parkville, Australia

ABSTRACT: The collaborative virtual environment framework SIEVE allows users to automatically build virtual environments and explore them collaboratively in real-time to aid decision making. SIEVE is currently being used in several application areas around landscape visualization and management and security and emergency response. Specific application areas include climate change, future land use exploration, land use productivity analysis and marine security response scenarios. This paper focuses on extensions to SIEVE based on collaborative data sharing web technologies. SIEVE Builder Web allows users to access remote SDI data via a web-mapping service to create and download 3D environments. Another component currently in development allows the import of ancillary data into SIEVE by creating a data mashup. To integrate online data and shared computing facilities we are building a web-based framework to integrate multiple applications to complement SIEVE. Finally, we allow users to exchange spatially referenced photographs remotely within SIEVE Viewer.

Keywords: Decision making, SIEVE, web-based collaboration, virtual environments

1 INTRODUCTION

3D visualisations of Spatial Data Infrastructures (SDI) are increasingly becoming popular due to advances in visualisation hardware, one popular application field being land management (e.g. Dockerty et al. 2005, Sheppard 2005, Stock et al. 2007). Recently, such visualisation systems also include a collaborative component, allowing users to meet in 3D spaces via the Internet (e.g. Churchill et al. 2001, Pettit et al. 2006, Guralnick et al. 2007).

There are a number of challenges to be overcome, such as the time and resource intensive task of building 3D models, the integration of scientific models and the realism of the visualisations. We have developed a system called SIEVE (Spatial Information Exploration and Visualization Environment) which addresses some of the issues by integrating Geographical Information Systems (GIS) and Virtual Reality. In particular, SIEVE aims at being able to build 3D models from any SDI fully automatically, being able to integrate environmental process models into the 3D models and lastly to provide a Collaborative Virtual Environment (CVE).

A first working prototype of SIEVE has been demonstrated in other papers (e.g. Stock et al. 2008). Recently (especially over the last two years), new web based technologies have found their way into mainstream applications, including cloud computing, mobile computing, distributed data sharing and live data streaming (RSS feeds). We have analyzed these new web technologies and identified uses to further enhance collaborative virtual environments. This paper focuses on these new developments around the concept of data sharing using the Internet. These developments are separate of each other, but have the common theme of extending the existing functionality of SIEVE through web based technologies.

As cloud computing is becoming mainstream, it has become clear that SIEVE can take advantage of data sets that are being shared over networks. For instance, any organization can expose their SDI via a web mapping service over the Internet. This enables users that do not have access to spatial data to use 3rd party services to build their own 3D landscape models for use with SIEVE. In the first section of this paper we discuss SIEVE Builder Net which allows users to build 3D landscape models fully automatically on remote data servers. This technology will allow spatial data providers to not only host traditional 2D mapping data, but also 3D environments that are fully automatically built from 2D mapping data without any human intervention.

As computing moves onto distributed computer networks it can be of advantage to have a system that allocates different computing resources to different tasks. Additionally, software applications can be highly specialized and it may not make much sense to develop a single application that handles all possible functionality, but instead leveraging the power of multiple applications may be a more viable and efficient option. We are developing a remote computing system which can integrate different data sources and farm different tasks to different machines. Depending on what a user wants to visualize, the system will initiate appropriate algorithms and also provide different visualization options. This extends our work with SIEVE to integrate other existing visualization systems (such as 3D Nature's Virtual Nature Studio) and is discussed further in the second section of this research.

One application of cloud computing is live data mashups, which allow the integration of multiple data sources in one web service. Such data may not necessarily be coming from an SDI, but could be supplementary real-time data such as weather or traffic. We are developing a system that allows the integration of multiple data sources and then takes the integrated information into the visualization. This allows us to visualize the current weather or traffic condition in an area. Visualizing such up-to-date information can significantly improve decision making in CVEs, especially if decisions have to be made in real-time. This technology is discussed in section three of this paper.

In the final section of the paper we demonstrate remote file exchange capabilities of SIEVE. This allows users to exchange real life photographs using the CVE. These photographs are spatially referenced and can be shared amongst all collaborating users. This technology allows users to capture their own data and share it with other users over a CVE in real-time.

SIEVE is currently being used in two domains, land management and security and emergency response. In the land management domain we are exploring applications in the field of climate change, land management on the individual farm level and precision agriculture. We are working together with the Victorian State Government (through the Departments of Primary Industries and Sustainability and Environments and local Catchment Management Authorities) and the University of New England. In the domain of security and emergency response we are working together with the Defence Science and Technology Organization and the Defence Imagery Geospatial Organization. Our integration of web based technologies with CVEs (in this case SIEVE) will provide both application areas with improved options for collaboration and decision making.

2 SIEVE SUMMARY

The idea behind the SIEVE system is to provide functionality for automatic 3D model building from a given SDI, to allow integration of scientific data, such as environmental process models, and to provide a Collaborative Virtual Environment (CVE). The architecture behind SIEVE has been discussed extensively in other papers (e.g. Stock et al. 2008)—we provide a short summary in this section.

SIEVE consists of three main components; SIEVE Builder, SIEVE Viewer and SIEVE Direct (Fig. 1). SIEVE Viewer is the visualisation component and CVE front end of the system. SIEVE Builder and SIEVE Direct integrate the CVE with the SDI by providing a base dataset for the visualization and enabling direct manipulation of data objects through either the SDI or the CVE.

SIEVE Builder is an ESRI ArcObjects module built in VB.Net which accesses an SDI and automatically converts the 2D spatial data into a 3D landscape model. The SIEVE Builder interface allows users to select an area and with the press of a button start the building process.

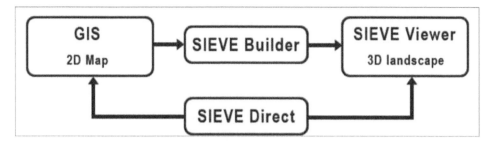

Figure 1.   Connecting GIS and visualization: SIEVE Builder converts 2D mapping data from the GIS into a 3D landscape model that can be viewed in SIEVE Viewer. SIEVE Direct links SIEVE Viewer to the GIS and allows live data manipulations in both directions.

SIEVE Builder needs a DTM and a matching aerial photo to build a 3D base terrain. To place surface objects, such as houses and trees, SIEVE Builder uses point layers to place 3D objects at mapped locations. SIEVE Builder can also place vegetation based on polygon layers.

There are two different varieties of SIEVE Builder. SIEVE Builder Desktop is built as an extension to ESRI ArcMap and allows users to build 3D models locally (O'Connor et al. 2005). SIEVE Builder Net, a more recent development, is built on the ESRI ArcServer platform and allows users to build 3D models remotely via a web mapping interface. Remote 3D model building via SIEVE Builder Net is discussed in detail in the next section.

SIEVE Viewer is built using Garagegames' Torque Game Engine and forms the CVE part of the system. Via SIEVE Viewer users can load 3D landscape models generated by SIEVE Builder and explore them jointly via the Internet (Stock et al. 2008). SIEVE Viewer also allows viewing of underground models, such as ground water tables, to allow users to see the connections of surface and underground data (Stock et al. 2008). Finally, SIEVE Viewer is also used as an augmented reality system (Chen et al. 2006) which allows users to take hypothetical models into the real world or update the SDI with recent visible changes to match the real world.

In some applications it is of advantage to be able to manipulate the 3D world directly from the GIS and vice versa. SIEVE Direct, another ESRI ArcMap extension, allows such functionality (Chen et al. 2006). Via SIEVE Direct users can, for example, change the land cover in specific paddocks from the GIS and the changes are viewable instantaneously in SIEVE Viewer. SIEVE Direct can also be used to edit the 2D data base (e.g. tree locations) directly from the 3D environment.

3   REMOTE MODEL BUILDING

The purpose of SIEVE Builder is to allow users to rapidly build 3D landscape models for exploration in SIEVE Viewer. However, not every user will have access to the necessary spatial data. Therefore, we developed SIEVE Builder Net which allows users to create 3D models via a web mapping interface.

The idea behind SIEVE Builder Net is that a spatial data agency can host their SDI via a web mapping service, which users of the SIEVE system can access and use to automatically build 3D landscapes via the push of a single button for any selected area. After the conversion has finished, users will be able to download the 3D model for viewing in SIEVE Viewer. We are currently trialing the system with the Victorian Department of Sustainability and Environment which is hosting a statewide SDI.

SIEVE Builder Web (Fig. 2) is an ESRI ArcGIS Server solution and consists of a module written in Visual Basic .Net and ESRI ArcObjects. The SDI needs to be exposed via ArcGIS Server. The map to be hosted with ArcGIS Server can either be a single map or can be a more complex infrastructure feeding into ArcGIS Server via ESRI ArcSDE. Once a map has been published via ArcGIS Server it can be made available via a web mapping interface.

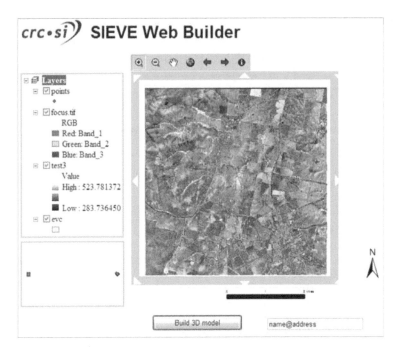

Figure 2. SIEVE Builder Net interface: Users can select an area in the main map view and build a 3D model of the selected area by selecting the build button. When the conversion process has been completed an email will be sent to the specified address.

The ArcGIS Server web mapping interface is using Microsoft ASP.Net and ESRI provides several mapping controls to host maps through this interface. These controls can easily be connected to an ArcGIS Server map. SIEVE Builder Net extends a simple web mapping interface by adding code for the 3D model conversion.

All data sets are hard coded into the SIEVE Builder back end as the setup depends on the SDI. Once this is set up it does not have to be reconfigured unless the SDI changes. Unlike SIEVE Builder Desktop, which allows users to select their own data sets and gives several options to the end user, SIEVE Builder Net only features a map view for area selection and a 'build' button. There are no options available to the end user as the data building process is done remotely and entirely depends on the spatial data sitting behind the web mapping application. This leaves little control to the end user, but at the same time makes producing 3D models from a 3rd party SDI very easy and accessible.

As a minimum requirement, the hosted map needs to contain a Digital Terrain Model (DTM) and aerial photography. SIEVE Builder Net will clip both the DTM and aerial photography to the extent of the current view in the map view. Both data sets are resampled to 256×256 and 4096×4096 pixels respectively. The resampled data sets are then packed into the TGE native TER format. This is similar to SIEVE Builder Desktop and described in more detail in O'Connor et al. (2005).

Like SIEVE Builder Desktop, SIEVE Builder Net can also convert point layers into locations for 3D objects, such as houses and wind pumps. Objects simply have to be linked to locations via the point layer as an attribute and will automatically be placed in the landscape. SIEVE Builder Net can also populate vegetation polygon layers with 3D objects. For this purpose, we are using the Ecological Vegetation Classes (EVC) of Victoria as polygon attributes. Based on the EVC classes we can populate the virtual landscapes with 3D objects of different species based on natural occurrence. This is still work in progress and explained in more detail in Bishop et al. (in press).

One prerequisite to allow automatic integration of 3D objects is availability. To build a geotypical landscape anywhere in Victoria, we need 3D objects for each species, ideally multiple ones for different growth stages. Similarly, there is a need for 3D objects of typical man made structures. We are currently building a 3D object library together with the Departments of Primary Industries and Sustainability and Environment that will allow different organizations to build an ever growing data set of 3D objects that can be accessed by SIEVE Builder. A prototype which allows automatic extraction of 3D objects from a centrally hosted 3D object library and subsequent population of 3D environments with these 3D objects has been built and tested. Further work to refine this model will be done over the next year.

The final piece to fully automatic 3D object integration is a database that maps EVC classes to species occurrence and frequencies. The Department of Sustainability and Environment has such a database and it just a matter of connecting it with the 3D object database and SIEVE Builder. Figure 3 shows how all system components are interlinked with each other.

Each of the three components (SDI, EVC database and object library) typically sit on different servers and all communicate with the SIEVE Builder Web Interface.

End users can request an area through the web mapping interface which then accesses the SDI. The SDI will then query the EVC database and the 3D object database to populate the 3D base terrain with appropriate 3D objects. The final outputs of this process are a TER file and a MIS file which are both native to SIEVE Viewer and a BMP file containing the ground texture.

Once the model conversion process has been finished (this typically take 10 minutes) the 3 files are compressed into a zip file and placed onto a file server. The end user is then notified via email that the file is available and given a download location. The user can download the file and view it with a local SIEVE Viewer installation.

Another important issue is that SIEVE Viewer requires 3D objects to be available on the local user's machine. This means somehow the user's machine needs to be able to download required 3D objects from the file server, or out of the 3D object library. Users will never need the whole 3D object library, at the same time, users will not want to download objects they already have on their local machine (for example, from previous scenario downloads).

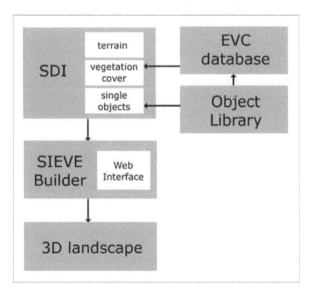

Figure 3.   SIEVE builder net architecture: The SDI is linked with the EVC database and the object library to generate 3D vegetation cover and single objects which together with the terrain form a 3D landscape model. Users of SIEVE can queries the SDI through the web interface and start an automatic process for building a 3D landscape.

This is currently implemented by putting all requested 3D objects into a compressed file which is automatically downloaded to the end users computer and also automatically extracted into the appropriate files. The final solution only downloads 3D objects on a need basis.

There are multiple visualization packages that are aimed at visualizing spatial data (e.g. Leica Virtual Explorer and Google Earth). Our SIEVE packages differs in that it a) includes a highly realistic virtual environment, b) allows collaboration between different users, c) allows visualization of underground layers, and d) a link to an object library which allows fully automatic population of the 3D landscape.

4 DISTRIBUTED COMPUTING

The SIEVE system has a range of features, including visualization of spatial mapping data and environmental process models and remote collaboration in 3D environments. However there are other software tools that can provide complimentary or additional functionality.

Currently, most of the integration of environmental process models into SIEVE has to be done manually. This is due to the fact that model output usually differs depending on context, therefore manual input is required to interpret the model outputs. As another example, users of the visualization may want to view a more photo realistic view of a location in the landscape. 3D Nature's Virtual Nature Studio (VNS) is a visualization package which allows high realism—at the cost of interactivity. By combining SIEVE Viewer with VNS users can combine a high degree of interactivity with high visual realism.

To allow integration of multiple systems, we are building a web-based framework to integrate multiple applications to complement SIEVE. For example, using this framework a user can from within SIEVE Viewer start a VNS rendering process to get a more realistic view of a point of interest. The framework, which is currently under development, will include tighter integration to process modeling software and allow communication between applications via high-level protocols, such as the IEEE standard High Level Architecture—HLA (IEEE 2007). Users will have access to a front-end which allows selection of suitable applications based on the task at hand.

The system will utilize distributed computer technologies based on grid computing and web services, which are the pillar of the Service Oriented Architecture (SOA) paradigm. Not only is grid computing poised to become the essential part for most e-science collaboration platforms (Riedel et al. 2008), but the SOA paradigm has become the framework of choice to design complex, enterprise level solutions for most organizations. The SOA paradigm is based on loosely-coupled modules that are orchestrated together by means of standard communication protocols, Web Service Description Language (WSDL), Simple Object Access Protocol (SOAP) and Universal Description Discovery and Integration (UDDI; W3C 2004).

This Web Services technology communicates between its parties regardless of platform and language implementations, by using standard eXtensible Markup Language (XML) schemas, which provide well-formed data packages and conformity to consensus standards, thus allowing automatic information extraction and verification. When using this web services framework on a grid middleware platform, many computers (potentially thousands) share data, applications and computing capacity to achieve a desirable outcome, a process hidden to the end-user who only interacts with a single entity.

In particular, in the area of geospatial visualisation, the Open Geospatial Consortium (OGC) has extended the aforementioned XML schemas to specify Geography XML (GML) and CityGML (OGC 2008). These schemas define a common semantic information model to represent 3D urban and geographical models, going beyond purely graphical models, and adding semantic and topological aspects to it, and by the same token, enabling lossless information exchange between spatial systems and users.

Following the same line of thought, in order to integrate different computer generated simulations, the HLA enforces a similar collaboration paradigm upon simulations.

HLA dictates the implementation of a Real Time Infrastructure to standardise communication and dataflow between engaging simulations.

4.1 *Related work*

A reusable framework to integrate distributed services for collaboration has been proposed by Luo et al. (2007). In this framework a web services "bus protocol" integrated self-made and third party collaborative tools, following a "mash-up approach" to meet specific platform needs, including security and management. Similar approaches using grid technologies have been successfully implemented (Foster 2006, Riedel et al. 2006). For instance, The Earth system grid (Kendall et al. 2008) not only enables grid sharing of analysis and climate modelling, but also real time distributed visualisation of simulation output. The Large Hadron Collider at CERN is engaging in one of the largest data crunching experiments to date (Clery 2006).

The same grid services technology has been widely applied in designing distributed virtual environments for geospatial data (Zhang et al. 2007). The CSIRO's Solid Earth and Environment Grid also aims to address the issue of transparent access to resources through OGC Web Services architecture (OGC 2006), thus facilitating the management of Australia's natural and mineral resources (Wybon 2006). Likewise, a 3D-GIS architecture platform based upon CityGML Web services standards was proposed by Wang & Bian (2007). This architecture implemented a similar three-tiered layered framework proposed here, emphasizing 3D geo-data applications and OGC's web features services. Analogous technology implementations were carried out by Wang et al. (2007), where a web-based framework was used to target specific urban data sharing dynamic planning.

In the area of HLA web based distributed simulation, Boukerche et al. (2008) and Möller & Dahlin (2006) proposed frameworks that used web services as its communication protocol to control HLA compliant simulations. These platforms successfully implemented flexible and extendible applications to interface with simulations in real-time. When simulation models are integrated in a distributed platform, it has the potential to create sound and relevant assessments of complex environmental impacts, as well as offering to a community of institutions the flexibility to assembly individual modules and modeling paradigms (Rivington et al. 2007, Warren et al. 2008).

These approaches have made substantial contributions to the analysis of climate change impact, and have been widely used in the exploration of predicted change in land and natural resource management (Lee et al. 2008, Ghadirian & Bishop 2008).

On the whole, we strive to define a modular, real-time collaborative framework based upon user and tool-wrapping interfaces that are compliant not only with our exploratory virtual environment (SIEVE), but also with commercial off-the-shelf tools, Web Grid/Services and the High Level Architecture (HLA) standard guidelines.

4.2 *Architecture description*

The overall guiding principles for this framework rely on modules that offer services through standard communication protocols between each other, while maintaining a layered architecture that organizes and orchestrates functionality among services (Fig. 4). In this manner, our architecture ensures both the sharing of spatial information and collaborative work upon it.

The Visualization Layer Services are composed of modules that offer the end visualization outcome, which depends on performance/quality of detail required to visualize the same data provided by the data layer. This layer includes:

– Web Client Services: These are third party web applications that range from the common Internet browser to more sophisticated readers of web 2.0 content like mobile phones (e.g. Google maps mobile API).

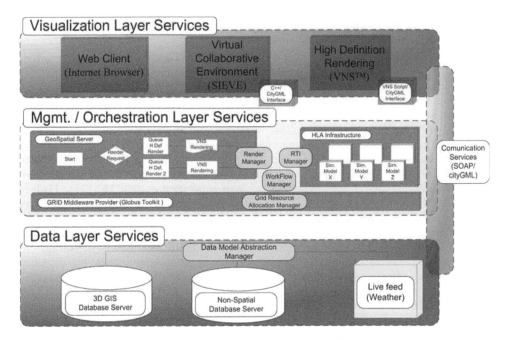

Figure 4. Architecture diagram with a modular SOA platform built upon aggregation of modules according to their services offered.

– Virtual Collaborative Environment Services: Real-time, immersive environments that are suitable for virtual collaboration. SIEVE Viewer falls into this category.
– High Definition Rendering Services: Third party tools for photo-realistic visualization needs (e.g. 3D Nature's Visual Nature Studio).

The Visualization Layer has been partly completed and tested, including the SIEVE and VNS modules).

The Management/Orchestration Layer Services depend on process services that link and sequence services according to existing and potentially new visualization requirements. These automated services further delegate specialized functions such as management, security, batch processing and similar features. This layer includes:

– Workflow Manager: Responsible for managing a sequence of operations/processes to achieve a specific organizational goal (like rendering a sequence of images), orchestrating the interaction of both human and machine actors that may intervene in the process.
– Real Time Infrastructure (RTI) Manager: Responsible for managing the Real Time Infrastructure that implements HLA compliant API and rules. It enforces the standards that any engaging simulation should adhere to. Consequently, it will coordinate data feed/ exchange and operations between simulation federates running on the framework's execution platform.
– Render Manager: Responsible for scheduling batch or simultaneous rendering tasks to available GPU machines (e.g. rendering farms).
– Grid Middleware Manager: Responsible for enabling grid technology, thus sharing resources across multiple machines, while masking this implementation to the other layers which only interact with a single "virtual" entity.

The Management/Orchestration Layer is only partly implemented, specifically the Render Manager has been implemented and tested while the other parts are work in progress.

Data Layer Services manage the data sources which can be composited to feed spatial and non-spatial information requirements that the orchestration layer needs to fulfill its lifecycle, thus abstracting the need for a particular data source, whether this source is a GIS Database, a RSS live feed or other machine available sources such as anonymous FTP repositories.

The GIS Database Server and the 3D object Library have been implemented and tested while the life feed module is partly implemented and can be used for demo purposes.

The Communication Services encapsulate CityGML information using Web Services protocols (Web Service Description Language -WSDL, Simple Object Access Protocol -SOAP, and Universal Description Discovery and Integration -UDDI). Data is transferred from all layers through wrapper interfaces that are implemented by standard contracts on each module. Every "contract" is a code implementation of the cityGML standard written to comply with the corresponding component. For instance, the wrapper interface of SIEVE is written in TorqueScript, a piece of code that obtains the current location and camera orientation, creates a cityGML object, wraps it and sends it through http to the Geoserver. This data object will be unwrapped by standard J2EEE xml parser tools which in turn wraps the message (with additional data as explained previously) and sends it to the VNS Interface.

The Data Layer Services has been partly implemented and tested (see below).

4.3 Link between SIEVE and VNS

As a proof-of-concept in leveraging on this collaborative software, we integrated our current platform of high definition rendering, using the commercial software 3D Nature's VNS, with the open source Geoserver (OGC 2006) and our Virtual Collaborative Environment (SIEVE). A typical flow of information through this architecture is shown in Figure 5. The following discussion illustrates an example of how this system could be used (the work on the SIEVE-VNS link having been completed).

A user is navigating a virtual landscape in SIEVE [1]. The user may want to obtain one or more high resolution images of a particular viewpoint. The requests to render those views are dispatched asynchronously (without obstructing the natural workflow of the collaborative session) through the wrapper interface of SIEVE [2], as SOAP messages to the Orchestration layer. The Workflow Manager acknowledges this request and start a new process [3a], gathering available/optional data from other sources (for instance, the weather time stamp from the Internet [1.3b], but it could be any HLA simulation layer output as well), also delegating best allocation of resources to another process service, in this case the Render Manager [4]. The requests are processed according to VNS rendering machines' availability, including ancillary info linking appropriate resources of view locations and bearings, Level of Detail, shape files, 3D Objects, etc., [5]. The Render Manager then queues and/or dispatches render requests to the High Definition wrapper interface [6]. When the render process is completed [7], the Render manager will notify the Workflow Manager [8], which will in turn publish

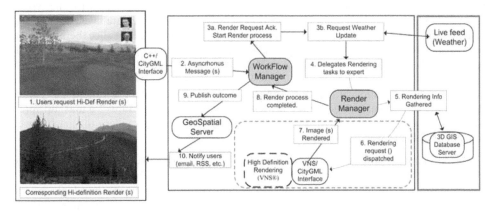

Figure 5. Framework's workflow example of a high-definition rendering request.

corresponding links on the Geoserver web page [9], also notifying users of the outcome of the task through email, RSS, etc., [10]. The user can then download the final rendered image from the Geoserver.

In summary, this framework orchestrates the use of heterogeneous software tools which collectively support distributed spatial analysis and complex environmental decision-making processes. It is also extensible by adhering to service contracts that are defined collectively by regulatory organization like OGF and IEEE. For instance, due to the modularity of the architecture, it would be possible to add another high definition render of choice, or extending modular Render Farm's capabilities, without considerable overhead or software refactoring.

Equally important, we aim to reflect on the technical plane the ideal of managing environmental resources with a broad perspective, one that takes into account all, and often conflicting, interests in different spatial and temporal scales. At the same time, this framework attempts to mirror the following perception: systems dealing with complex environmental concerns should not be dependent on a specific software or economic/scientific paradigm (Pahl-Wostl 2007). Moreover, when this loose-coupling architecture is enabled, it permits a better uncertainty analysis, where a holistic notion of the system can be obtained (Warren et al. 2008).

4.4 *Future work and application*

On the whole, this framework aims to lay the foundation for a more flexible manner to use Collaborative Environments when visualizing environmental models. One of our case studies will be a demo dairy farm in Western Australia. This case study will draw on multiple data sets, including spatial data to model the existing conditions, environmental process models to model predicted futures affected by a range of climate change scenarios, a 3D object library and possibly a live weather data feed. Our main visualization platform will be SIEVE, but end users may also request visualizations from VNS and Google Earth.

5 LIVE DATA FEED

Feeding real-time ancillary data into a 3D visualization is especially beneficial in security and emergency response applications where such data can have an impact on decisions people make. We are planning to import the current weather conditions, existing traffic conditions (cars, boats, airplanes) and incident reports. Presenting this kind of data in a real-time 3D environment will give additional insights and decision options within a complex scenario.

The two challenges are: how we are going to feed such data into our simulation and how we are going to visualize such data? The data feed will be integrated with the system discussed in the distributed computing section. It will be part of the data layer services and feed into the visualization via the communication services. For each data slice there will need to be a translator to make the data accessible to the visualization layer. In detail, the ancillary data will be imported into ESRI ArcGIS Server and sent to SIEVE Viewer via SIEVE Direct.

Mashups are web applications that combine different types of data or functionality from two or more sources into a single integrated application. The architecture of a mashup application (see Fig. 6) is comprised of three participants: API/content providers, the mashup web site, and the client's web browser (Merrill 2006).

- The API/content providers provide the content which can be mashed up by different users. Providers often publish their content through web-protocols to facilitate data retrieval.
- The mashup web site is where the mashup is hosted and where the mashup logic resides. Mashups can be implemented similarly to traditional Web applications using server-side dynamic content generation technologies like Java servlets, CGI, PHP or ASP.

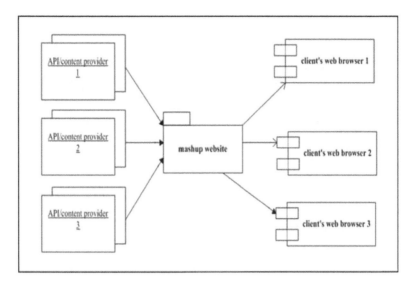

Figure 6. Mashup application: The content providers are feeding data into a mashup website which can be viewed with different browsers.

– The client's web browser. This is where the result is displayed and where user interaction takes place.

At present mashup applications are invariably embedded in web-based applications, and usually delivered in text content or in 2D environments such as maps or graphics. Although such applications may convey information effectively, new applications could be employed in more fields, such as virtual training, if they are extended into 3D collaborative virtual environments.

Our overall system concept is shown in Figure 7. Real-time data are transferred from different sources to the real-time database engine as a first step. These data may have dissimilar types but usually are in a standard data format in order to facilitate data processing. The Real-time integration engine will assemble different data types into one common format. The Visualization Modeling Engine will transfer abstract data to a visual representation which will easily communicate the data. Using this system we will feed real-time data into our visualization. Finally, different users at different client machines may share the final collaborative virtual environment (via a UDP/TPC connection). The three components constitute a Mashup Editor, with functions for data storage, integration, and visualization modeling. The ArcGIS Geodatabase is used for spatial data management and integration.

5.1 *Live-data visualization*

In this research, real-time data are divided into 2 datasets according to their source and providers—public and secure.

Public data are open to the general public. They are mostly accessible through websites, and include datasets like current weather condition, traffic information as well as georeferenced incident reports (e.g. Victorian bushfires and Mexican swine flu—to mention two recent examples).

Secure data is restricted in access, possibly due to proprietary or security concerns. This data can include real-time vehicle movement covering cars, ships, and aircraft. Such vehicles are often equipped with location devices and their movement can be monitored. Another kind of data could be classified information at distinct localities used by defence organizations.

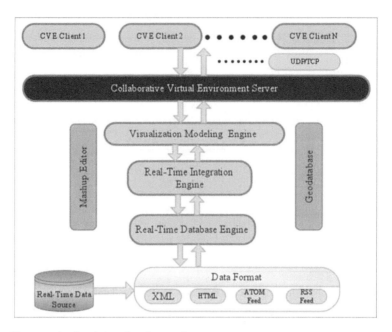

Figure 7. Framework of real-time data integration.

A key issue is that each data set needs to be visualized in an appropriate manner. Table 1 shows example real-time data and visualization data key parameters as well as the connection between them.

Weather condition information is often provided in the form of an XML file, including location of the weather station (longitude and latitude), current time (date/hour/minute), wind parameters (direction and speed), sun position (time of sunrise and sunset), precipitation conditions (type and quantity) and condition codes (indicating weather conditions such as sunny, cloudy, rainstorm or blizzard). SIEVE's weather and atmospheric conditions are represented by datablocks called 'SkyObject', 'Sun' and 'Precipitation'. When a new live weather feed is translated by the Visualization Modeling Engine, variables for each datablock will need to be updated. For example, to visualize wind speed, we can adjust the speed and direction of the cloud layer in the SkyObject. To visualize rain we can set the appropriate PrecipitationType and adjust the splashSize to simulate the severity of the rain in the Precipitation datablock.

Real-time traffic data are often represented by three parameters: road name, start and end point and current traffic jam level. The visualization result will demonstrate the 3D traffic data flow in real-time. Firstly, the local road network is imported into SIEVE Viewer, then the location and traffic conditions in live data are translated by the visualization modeling engine into locations in SIEVE Viewer coordinates and traffic flow speed. 3D cars are placed along the road networks and will drive at given speed across the landscape. The speed and car density will be determined by the current traffic conditions.

Incident reports will need to be handled on a case by case basis. An incident report will always include location and time, but other parameters depend on what kind of incident is being reported. Additionally, each incident will need to be visualized in a different way. An explosion can be represented with a loud noise and flames. However, non-visual information (such as areas of danger where an incident is predicted to have the potential to occur but it has not occurred yet) will also need to be represented in some form.

The live data link is still under development, however a real-time weather stream, real-time ship traffic in Sydney harbor and tide information has mostly been integrated and tested in the SIEVE environment at this point in time.

304

Table 1. Visualization data conversion.

Category	Real-time data	Visualization	SIEVE parameters
Weather	Location and time	Non-visual references	
Condition	Wind	Moving clouds	Speed and direction of cloud layer
	Sun	Position in sky	Sun location
	Precipitation	Rain effect	Particle size and frequency
Traffic	Road location points	Car location	Location and direction
Condition	Traffic level	Car density/speed	3D car model density and speed

5.2 Sydney harbor case study

A security response scenario at Sydney harbor has been selected as a case study to test real-time data integration and visualization technology. This project is based on a hypothetical terrorism incident scenario that describes a chain of events at the Captain Cook dock in Sydney Harbor. This scenario will be used in collaborative manner to simulate decision making in a testing environment where decisions will have to be made quickly, but events cannot be predicted as they happen 'live' and under different frame conditions (e.g. different states of weather and traffic).

For this scenario, a virtual Sydney Harbor model has been created, featuring terrain, water, vegetation, buildings, and vehicles. As a second step we are now integrating a real-time data engine into SIEVE Viewer. The system will connect to online data sources selected by users, which will largely vary with time and will be updated frequently. Finally, some vehicle movements will need to be scripted to simulate actions that occur during the scenario's run time. Alternatively, users may be able to control vessels and initiate actions that need timely response. As a third option, users can be located in the real world to report 'imaginary' incidents to the simulation in a manner that allows users in the training system to respond live.

Integration of real-time weather in the CVE may provide users with a more realistic experience and may lead them to respond differently to different weather condition in an emergency. Traffic information may inspire forward planning so users can quickly select alternative routes when facing traffic congestion. Changing incident locations and statuses gives users additional decision options as different measures will need to be taken to identify the most urgent emergencies, which then need to be attended to—without losing sight of the 'big picture'.

6 REMOTE FILE EXCHANGE

In a collaborative environment, it can be of advantage if users have the capability to exchange images (or other) files. For example, if a user has a photograph of the real world it can be beneficial to be able to share the photo with collaborating users. We have implemented such functionality within SIEVE Viewer.

In order to enable SIEVE Viewer to handle the exchange of files, we have developed uploading and downloading functionalities within SIEVE Viewer. SIEVE Viewer can connect to an external FTP Server through the Internet. To enable uploading and downloading capabilities within SIEVE Viewer we used cURL. cURL is free and open software which can run under a wide variety of operating system. It supports a large quantity of communications protocols.

Apart from exchanging photographs over an FTP server, we also want to record attributes of the photograph, such as location and orientation. To record extra parameters, we created a File Index document, which sits on the FTP Server. The File Index document records an ID, the file name, the location and orientation, a description, the name of the person who shared the photograph, and the submission date.

To be able to share different sets of photographs with different users, photographs are linked to individual scenarios. It is assumed that if a set of users is sharing a scenario, the users will want to be able to share photographs that match this scenario. This means that on the FTP Server there is a separate directory for each scenario and each scenario also has its own File Index document. As an example, an advisor could share three different scenarios with three different farmers. Using the scenario setup, the advisor has access to all three scenarios and all photographs. However, farmer A only has access to scenario A and the photographs belonging to scenario A. He cannot, however, access photographs from farmer B and C as he does not have access to scenario B and C. The remote photograph exchange workflow is shown in Figure 8.

The uploading process works as follows: After a user has taken photographs on a site, he uploads the photographs onto his computer and runs SIEVE Viewer. The user needs to navigate the avatar within SIEVE Viewer to the virtual location where the photographs were taken. To start the uploading process the user can bring up a dialog box (Fig. 9). In this dialog box the user can write a description for the photograph, and when finished click on the upload button to navigate the hard drive and choose a photograph to be uploaded.

After choosing the photos and clicking the upload button, SIEVE Viewer will send an upload command to the FTP Server, and FTP Server will start to receive the data. After transferring the photograph, SIEVE Viewer will update the local File Index with a new record of the file, adding name, description and submission date, and also the location and orientation of the avatar which is regarded as the location and orientation at which the photo was taken. SIEVE Viewer will send this File Index to the FTP Server to keep it synchronized. FTP Server will then send back the updated File Index document to all local SIEVE Viewer clients to keep the File Index document synchronized between all client machines. Finally, users can show a 2D map on which each photograph's location is marked with a number (Fig. 10).

The downloading process works as follows: Another user B who is collaboratively working with user A can now view the photos. User B will start SIEVE Viewer in his computer. He can then open the 2D map view which shows all photographs spatially referenced as numbers.

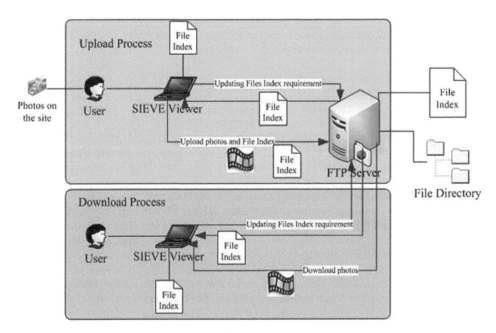

Figure 8.   Photograph exchange between different users via an FTP Server.

306

Figure 9. Map view showing the photograph uploading interface. The locations of existing photographs are shown as black numbers. Photographs can be selected for viewing from the view drop-down box.

Figure 10. Photograph view showing a photograph with name of user who uploaded the photograph, date of upload and description.

When the 2D map view is opened, SIEVE Viewer will send a request to the FTP Server to update the local File Index—in order to ensure user B can view all photographs collected on the FTP Server. He can choose which photo he wants to view by choosing the number of a photograph from a drop down list.

After choosing the photo to be viewed, SIEVE Viewer will detect whether the photo exists on the local computer. If the photo does not exist, SIEVE Viewer will send a download request to the FTP Server, which in response will send the file back to the SIEVE Viewer. A new window will pop up to show the 2D photograph. At the same time the avatar will be relocated to the location where the photo was taken, and the orientation will be updated similarly.

It may be a bit cumbersome to have to use SIEVE for uploading photographs from the field. We have built a prototype of an application for the Apple iPhone to enable users to take and upload photographs directly to the FTP Server. The iPhone helps to determine position and orientation as well. In the future, we may also allow users to exchange other files, such as reports or scientific 2D data graphs.

The photo sharing capability has specifically been developed to make knowledge transfer between farmers, scientist and advisors easier. This way, farmers can take photographs of problem spots of their land and ask advisors questions about these spots. Advisors can then query scientific models to find possible answers. Photographs may also help to improve scientific models as local knowledge can sometimes reveal shortcomings of a particular model.

7 CONCLUSIONS

The integration of GIS and CVEs opens new opportunities for communicating spatial problems. SIEVE is a system that combines SDI and CVEs in the fields of land management and security and emergency response. We have extended the SIEVE platform to incorporate the concept of cloud computing and distributed data. Users can access remote SDIs, stream live supplementary data, such as weather and exchange photographs within the collaborative environment. We are designing a system that can integrate multiple data sources within a local network and the Internet, distribute computing needs onto different servers and provide a selection of visualization options.

We are currently building a few applications around SIEVE. One application is using SIEVE as a remote advisory and knowledge transform system in the field of precision agriculture. Farmers and advisors can meet in the virtual space and analyze and discuss agricultural productivity and consult scientific measurements (such as soil conductivity) to improve the productivity of the land. Another case study is a demo dairy farm in Western Victoria which allows us to test various climate change models and develop possible future scenarios. A third application will be a security response scenario at Sydney Harbor involving a number of maritime vessels and security threats.

In each scenario we will, to a certain degree, test the usefulness of CVEs and live data streaming and the impact these technologies have on decision making. Ultimately, we still have to show that SIEVE does indeed help to communicate spatial problems and offers robust tools to aid better decision making.

For us, the vision to plug into a government SDI and build realistic 3D models with the aid of a 3D object library, stream live supplementary data into the visualization and, at the same time, also offer a look into the science behind the visualized scenarios is an exciting one.

REFERENCES

Bishop, I.D., Chan, P., Chan, T., Lau, A., Pettit, C.J., Stock, C. & Syed, D. (in press) Object libraries: The next step in spatial data infrastructure. In: Bruce, D. (eds.), *Proceedings of SSC 2009 Spatial Intelligence, Spatial Diversity: the national biennial Conference of the Spatial Sciences Institute, Adelaide, September 2009.* Adelaide: Spatial Sciences Institute.

Chen, T., Stock, C., Bishop, I.D. & O'Connor, A.N. (2006) Prototyping an in-field collaborative environment for landscape decision support by linking GIS with a game engine. In: Li, D. & Xia, L. (eds.), *Proceedings of SPIE Volume: 6418 Geoinformatics 2006: GNSS and Integrated Geospatial Applications, Wuhan, October 2006.*

Churchill, E.F., Snowdon, D.N. & Munro, A.J. (2001) *Collaborative Virtual Environments: Digital Places and Spaces for Interaction.* Springer.

Clery, D. (2006) Infrastructure: Can Grid Computing Help Us Work Together? *Science,* 313, 433–34.

Dockerty, T., Lovett, A., Sünnenberg, G., Appleton, K. & Parry, M. (2005) Visualising the potential impacts of climate change on rural landscapes. *Computers, Environment and Urban Systems*, 9 (3), 297–320.

Foster, I. (2006) Globus Toolkit Version 4: Software for Service-Oriented Systems. *Journal of Computer Science and Technology,* 21 (4), 513–520.

Ghadirian, P. & Bishop, I.D. (2008) Integration of augmented reality and GIS: A new approach to realistic landscape visualization. *Landscape and Urban Planning*, 86, 226–232.

Guralnick, R.P., Hill, A.W. & Lane, M. (2007) Towards a collaborative, global infrastructure for biodiversity assessment. *Ecol Lett.,* 10 (8), 663–672.

IEEE Computer Society. (2000) *IEEE Standard for Modeling and Simulation (M&S) High Level Architecture (HLA) –Framework and Rules.* IEEE Computer Society.

Kendall, W., Glatter, M., Huang, J., Hoffman, F. & Bernholdt, D. (2008) Web Enabled Collaborative Climate Visualization in the Earth System Grid. In *International symposium on Collaborative Technologies and Systems. Irvine, May 2008*, 212–220.

Lee, C.H., Huanga, S.L. & Chanb, S. (2008) Biophysical and system approaches for simulating land-use change. *Landscape and Urban Planning*, 86, 187–203.

Luo, T., Song, J., Chen, S., Liu, Y., Xu, Y., Du, C. & Liu, W. (2007) A Services Oriented Framework for Integrated and Customizable Collaborative Environment. In *IEEE International Conference on Information Reuse and Integration, August 2007, Las Vegas.* pp. 385–93.

Merrill, D. (2006) Mashups: *The new breed of Web app.* [Online] *Retrieved March 15, 2009,* Available from: *http://www.ibm.com/developerworks/xml/library/x-mashups.html*

Möller, B. & Dahlin, C. (2006) A First Look at the HLA Evolved Web Service API. In *Proceedings of 2006 Euro Simulation Interoperability Workshop, Simulation Interoperability Standards Organization, June 2006 Stckholm.*

O'Connor, A., Stock, C. & Bishop, I. (2005) SIEVE: An Online Collaborative Environment for Visualising Environmental Model Outputs. In: Zerger, A. & Argent, R.M. (eds.) *International Congress on Modelling and Simulation, December 2005,* Melbourne. pp. 3078–84.

Open Geospatial Consortium. (2006) *Web Map Service Implementation Specification 1.1.0. Open GIS project document.* Document reference: OGC 06–042: 5.

Open Geospatial Consortium. (2008) *OpenGIS City Geography Markup Language (CityGML) Encoding Standard version 1.0.0.* Document reference: OGC 08–007r1.

Pahl-Wostl, C. (2007) The implications of complexity for integrated resources management. *Environmental Modelling & Software*, 22, 561–569.

Pettit, C.J., Cartwright, W. & Berry, M. (2006) Geographical visualization: A participatory planning support tool for imagining landscape futures. *Applied GIS,* 2 (3), 22.1–22.17.

Riedel, M., Eickermann, T. Frings, W., Dominiczak, S., Mallmann, D., Düssel, T., Streit, A., Gibbon, A., Wolf, F. & Lippert, T. (2007) Design and Evaluation of a Collaborative Online Visualization and Steering Framework Implementation for Computational Grid. In *Proceedings of the 8th IEEE/ACM International Conference on Grid Computing, Austin, September 2007.* pp. 169–176.

Riedel, M., Frings, W., Habbinga, S., Eickermann, T., Mallmann, D., Wolf, F. & Lippert, T. (2008) Extending the Collaborative Online Visualization and Steering Framework for Computational Grids with Attribute-based Authorization. In *9th IEEE/ACM International Conference* on *Grid Computing, Tsukuba, September 2008.* pp. 104–111.

Rivington, M., Mathews, K., Bellocchi, G., Buchan, K., Stockle, C. & Donatelli, M. (2007) An integrated assessment approach to conduct analyses of climate change impacts on whole-farm systems. *Environmental Modelling & Software*, 22, 202–210.

Sheppard, S.R.J. (2005) Landscape visualisation and climate change: the potential for influencing perceptions and behaviour. *Environmental Science & Policy*, 8 (6), 637–654.

Stock, C., Bishop, I.D. & Green, R. (2007) Exploring landscape changes using an envisioning system in rural community workshops. *Landscape and Urban Planning* 79: 229–39.

Stock, C., Bishop, I.D., O'Connor, A.N., Chen, T., Pettit, C.J. & Aurambout, J-P. (2008) SIEVE: collaborative decision-making in an immersive online environment. *Cartography and Geographic Information Science*, 35, 133–144.

Wang, H., Hamilton, A., Counsell, J. & Tah, J. (2007) A web based framework for urban data sharing and dynamic Integration. *Architecture, City, and Environment*, 2 (4).

Wang, Y. & Bian, F. (2007) A Extended Web Feature Service Based Web 3D GIS Architecture. In *International Conference on Wireless Communications, Networking and Mobile Computing, September 2007,* Shanghai. pp. 5947–5950.

Warren, R., de la Nava, S., Arnell, N.W. & Bane, M. (2008) Development and illustrative outputs of the Community Integrated Assessment System (CIAS), a multi-institutional modular integrated assessment approach for modelling climate change. *Environmental Modelling & Software*, 23, 592–610.

World Wide Web Consortium, Web services Architecture, (2004) [Online] Available from: *http://www.w3.org/TR/2004/NOTE-ws-arch-20040211/#whatis*

Wybon, L. (2006) Decomposing solid earth and environmental sciences to enable the vision of SEE Grid. In *Proceedings of the Solid Earth and Environmental Grid III Conference, November 2006 Canberra.*

Zhang, J., Gong, J., Lin, H., Wang, G., Huang, J., Zhu, J., Xu, B. & Teng, J. (2007) Design and development of Distributed Virtual Geographic Environment system based on web services. *Information Sciences*, 177, 2968–3980.

*Advances in Web-based GIS, Mapping Services
and Applications – Li, Dragićević & Veenendaal (eds)*
© 2011 Taylor & Francis Group, London, ISBN 978-0-415-80483-7

Development and challenges of using web-based GIS for health applications

Sheng Gao
Department of Geodesy and Geomatics Engineering, UNB, Fredericton, New Brunswick, Canada

Darka Mioc
National Space Institute, Technical University of Denmark, Denmark

Xiaolun Yi
Service New Brunswick, Fredericton, New Brunswick, Canada

Harold Boley
Institute for Information Technology, NRC, Fredericton, NB, Canada

François Anton
Department of Informatics and Mathematical Modelling, Technical University of Denmark, Denmark

ABSTRACT: Web-based GIS is increasingly used in health applications. It has the potential
to provide critical information in a timely manner, support health care policy development,
and educate decision makers and the general public. This paper describes the trends and
recent development of health applications using a Web-based GIS. Recent progress on the
database storage and geospatial Web Services has advanced the use of Web-based GIS for
health applications, with various proprietary software, open source software, and Application Programming Interfaces (APIs) available. Current challenges in applying Web-based GIS
for health, such as data heterogeneity, data privacy and confidentiality, powerful processing
abilities, and appropriate data representation to users are also discussed. The continuous
development of Web-based GIS for health applications will further enhance disease surveillance, health care planning, and public health participation.

Keywords: Health applications, open source software, privacy, web-based GIS

1   INTRODUCTION

Health data are concerned with people's health experiences. Many health care providers, such
as emergency departments, hospitals, clinics, and care facilities are responsible for the health
security of people. Health data cover a wide range of content, including inpatient, outpatient, survey, laboratory, facility, demographic, socio-economic, and environmental information. Their collection can occur through disease surveillance, clinical care delivery, public and
private sector services, environmental monitoring, cohort research findings, and questionnaire surveys.

In the health data collection process, spatial information such as zip codes, postal codes,
or addresses of patients and health care facilities is usually recorded. At different locations
on the Earth, variability in natural earth processes, environmental quality, ecological issues,
and human activities is likely to affect human health. Health geography applies the spatial
perspective in health studies. Boulos et al. (2001) divided health geography into the geography of diseases and the geography of health care systems based on the two intertwined

concepts: health (individual and community health matters) and health care (clinical issues, service planning and management issues). The geography of disease deals with the detection, modeling, and exploration of disease outbreaks, disease risk factor analysis, and etiology hypotheses. The geography of health care systems records details about and abilities of health care providers, and supports health facility planning, management, and delivery for balancing the needs and demands in health care access.

Geographic Information Systems (GIS) have emerged as a powerful technology for health geography. With the use of GIS, health practitioners can visualize health information in relation to demographics, meteorological conditions, administrative boundaries, distance from patient to hospitals/clinics, disease vectors (farm animals, migratory birds, water wells), and much more. A GIS is highly suitable for analyzing epidemiological data, and revealing trends and interrelationships, which would be difficult to discover in tabular formats (WHO 2010). Certain dependencies and relationships between variables, that may not have been previously considered, can be revealed.

The emergence of Web-based GIS further pushes GIS functionalities to the Internet. Web-based GIS harness the power of the Web with basic desktop GIS functionalities (e.g. generating maps, viewing maps, interacting with maps). Web-based GIS have been applied in a wide range of areas in health, such as risk assessment for child safety (Xu et al. 2005), Severe Acute Respiratory Syndrome (SARS) epidemic visualization (Lu 2004), public health surveillance (Kamadjeu & Tolentino 2006a), health care emergency planning and response (Roth et al. 2009), and environment health decisions (Hochstein & Szczur 2006). These cases illustrated the advantages of Web-based GIS technologies for a community-of-practice in response to the growing demand of geospatial information in the health decision making process for medical, social, economic, and environmental benefits.

This chapter summarizes the use and development of health applications using Web-based GIS and identifies the challenges, which need to be addressed in the further improvement of Web-based GIS for health. Section 2 discusses the trends of applying Web-based GIS for health applications. Section 3 provides the description of Web-based GIS applications for disease surveillance, health care, and public participation. Section 4 presents the technologies in the design and implementation of health applications using Web-based GIS. Section 5 discusses the challenges, which need to be addressed in the development of health applications using Web-based GIS. Section 6 concludes this chapter and provides recommendations for future research.

2 TRENDS OF WEB-BASED GIS FOR HEALTH

The Internet dramatically changes the way to distribute information to public health professionals, research groups, and the general public. Traditionally, maps were kept on paper files. Digital maps can be stored in medias (e.g. CD/DVD) for delivery. Internet technology makes online Web mapping possible through Web-based GIS.

Web-based GIS allow one to access the data and the necessary functionalities through Web browsers. Applying Web-based GIS concepts to health applications can reduce the expensive cost on investigation and purchase of advanced hardware and GIS software at health organizations. In addition, it can evidently increase the number of users and be achieved with minimal costs (Maclachlan et al. 2007). As detection at early states is important for health officials to take effective counter-measures to control the spread of disease, Web-based GIS can support quick access to distributed health data for visualization, analysis, planning, and hypotheses. Since the response of a Web-based GIS can be obtained in real-time, it is effective for supporting early detection and appropriate and timely responses for emerging health threats.

Since the use of Web-based GIS for health applications in the early 2000s, several changes have taken place:

– *Towards user-generated maps*. Health maps can be pre-generated then published through Web-based GIS applications. In this case, the cartographic representation and mapping variables are pre-defined. These maps either already exist or are rendered. The pre-generated

312

strategy has been applied for Web-based GIS applications to allow quick interaction with users. To meet the increasing need from users in visualizing and analyzing health information, user-generated maps from Web-based GIS applications are highly demanded. Regarding user-generated maps, the cartographic representation and map variables can be set by users interactively. This on-demand mapping in Web-based GIS applications has been enabled using methods such as SVG (Scalable Vector Graphics), Java Applet, and Web Services.

– *Towards distributed multiple source access.* Health data are distributed by different health organizations. Multiple data sources are usually required in Web-based GIS applications for health decision making. Moving from one source to multiple source investigation and analysis is highly demanded in health applications. However, different data sources often use different generalization models to describe objects, and inconsistency may occur during integration. Standards are needed for health applications using a Web-based GIS to support data and service interoperability. Therefore, national or international data and service standards for health applications using a Web-based GIS will have a great importance.

– *Towards diversification in implementation.* At the beginning stage of applying Web-based GIS for health applications, implementation requires the purchase of proprietary software or long development process. Much work has been done recently on the implementation of Web-based GIS applications for health using free software and open source software, and the Web 2.0 mashup technology further facilitates the development of Web-based GIS applications for health.

– *Towards collaborative environment.* Web-based GIS applications provide the potential for people at different locations to collaborate in health decision making. The collaborative environment is essential in the preparation, response, and recovery stages of disease control. Web-based GIS applications that allow people from different fields (governmental departments, hospitals, rescue centers, the public) to exchange information actively and quickly would be more suitable for real-time disease outbreaks.

– *Towards the push mode.* Web-based GIS applications are typically designed to wait for a user input (pull information). For public health and public safety, this typical method of information delivery is not adequate. When a disease outbreak or emergency occurs, Web-based GIS applications need to "push" the information to appropriate people through active channels such as telephone, pop-up, e-mail, alarm, and GeoRSS.

3 WEB-BASED GIS APPLICATIONS FOR HEALTH

Web-based GIS can analyze health parameters, provide critical information in a timely manner, support health care policy development, monitor disease events, examine health facility distributions, and educate decision makers and the general public. The data used in these applications cover the health, environmental, and socio-economic sources. Common data include hospital and emergency room admissions, ambulance data, patient locations at the time of incidents, cumulative ambient concentrations obtained from air-monitoring and weather stations, hospital and emergency room admissions, questionnaire survey and interview data, hospital staff data, remote sensing imagery (used to extract land cover), groundwater-surface water hydrologic fluxes and water quality data, demographic statistics, and economic vectors.

3.1 *Web-based GIS and disease surveillance*

Monitoring disease rates and related health vulnerabilities assist the investigation of disease outbreaks over space and time. Web-based GIS can illustrate health events at multiple scales, showing the geographic distribution and variation of health information. Time information is also often incorporated in Web-based GIS to explore both the spatial and temporal trends in health surveillance. To alleviate confidential concerns for health applications using Web-based GIS, strategies such as the aggregation of low frequency events over geographic areas and time, and blocking the display of details of subpopulation information like race and gender can be applied (Grigg et al. 2006). A common practice is to develop Web-based GIS applications with interactive browsing and querying functions, showing spatio-temporal health phenomena (Conte et al. 2005, Tsoi 2007, Wang et al. 2008). Utilizing Web-based GIS to develop atlases for different kinds of diseases can support governments in disease prevention. MacEachren et al. (2008) designed a cancer atlas for Pennsylvania, USA, by integrating a dynamic geovisualization, interactive atlas, and Web-map services. This atlas allows the generation of maps and graphics from available cancer data, and enables flexible exploration by linking maps, graphics, and tables.

Data warehouses and On-Line Analytical Processing (OLAP) are being utilized in Web-based GIS for efficient analysis and knowledge discovery. Baptiste et al. (2005) described a Web-based GIS application using data warehouses, enabling the access of spatio-temporal trends of epidemiology for decision support. Spatial OLAP (SOLAP) can support interactive navigation and analysis on different levels of data for decision making (Bedard et al. 2001, Rivest et al. 2005). In health surveillance, spatial data warehouses can consider multiple dimensions such as spatial level, age group, time, and disease type. With Spatial OLAP tools, OLAP operators such as slice, dice, roll up, drill down, and pivot can be applied in both thematic and spatial dimensions. Bernier et al. (2009) showed a Web-based GIS application utilizing Spatial OLAP to explore the relation of health, socio-economic and environmental issues. The roll-up and drill-down operations on thematic, spatial, and temporal dimensions in Spatial OLAP are beneficial to support health decision making.

Besides basic visualization and query functionalities offered in most Web-based GIS applications for health, the provision of processing and modeling abilities in these applications has begun to emerge. Zeilhofer et al. (2009) explored modules in spatial modeling for health and environment related phenomena in a Web-based GIS application to study the habit suitability of mosquito vectors. Gao et al. (2009a) proposed a framework of geospatial infrastructure for health, with the support of online processing abilities through Web Processing Services. With these processing services, Web-based GIS applications can be built based on such infrastructure to support various functionalities. Applying spatial statistics with Web-based GIS, one can detect spatial clusters and spatio-temporal clusters, which can inform people about the excess or unusual disease occurrence in real-time. Reinhardt et al. (2008) developed a Web-based GIS application which enables the detection of spatio-temporal clusters supported by the free software SaTScan.

3.2 *Web-based GIS and health care*

Web-based GIS provide the abilities to show geographic distribution and change of health care facilities, examine the needed variation of health care access, and explore improvements on health care delivery. It can assist stakeholders and policy-makers in effectively distributing health care resources to overcome geographic inequalities in accessing health care among different population groups. Examples of Web-based GIS for health care includes identifying population segments vulnerable to differential geographic access to critical medical treatment, providing optimal routes for emergency responses, assessing resource allocations, monitoring health facility utilization patterns, and planning intervention strategies. Phosuwan et al. (2009) described a Web-based GIS application for evaluating population registration and its geographic relation with participating voluntary health care providers using Google maps and the Keyhole Markup Language (KML). A Web-based GIS application for assessing timely access to trauma centers in the U.S. was developed to address the resource concern using the speed and location of helicopters and ambulances, and the number and location of trauma centers in all regions (Skinner 2010).

Spatial Decision Support Systems (SDSS) are being designed for health care applications in exploring alteratives and identifying potential solutions for specific problems in an interactive environment (McLafferty 2003). Schuurman et al. (2008) presented a Web-based GIS application that integrates SDSS to support policy makers in visualizing multiple health resource allocations within pre-specified scenarios.

3.3 *Web-based GIS and public health participation*

Web-based GIS have brought a new way for the general public to report, disseminate, visualize, and analyze health data, without the need of extensive training. It enables the development of Web portals which allow public access, awareness, and participation in health decision making. For example, Li & Gong (2008) developed a Web-based GIS application which allows the public to report and browse reported West Nile incidents. With the utilization of Web maps, it is easy to explain the geographic variation of health exposure. People can be informed about the environmental hazards around themselves and prepare themselves for disease outbreaks. Gao et al. (2008) showed a health portal which integrates a map viewer and discussion forum to support user participation in health decision making by sharing map views and text.

4 WEB-BASED GIS APPLICATION DESIGN AND IMPLEMENTATION FOR HEALTH

New opportunities for health information delivery and sharing have been provided by Web-based GIS. In the development of Web-based GIS applications, the main components include a Web interface tier, a Web server tier, an application server tier, and a data server tier. A common practice on the use of Web-based GIS for health applications involves the data level and the application and service level, as shown in Figure 1. The data level deals with data storage and integration. Security in health data access can be addressed using user account permission and IP address permission. The application and

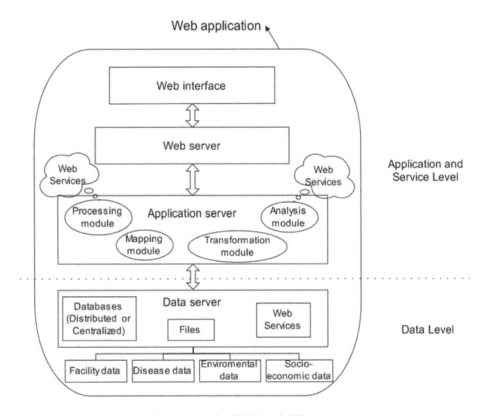

Figure 1.   Health application framework using Web-based GIS.

service level focuses on the provision of value-added information to users through Web interfaces. In the application and service level, secure Web-based GIS applications for health can be provided using methods such as a single point of access, user authentication using directories, security tokens, Secure Socket Layer, HTTP authentication, and Web Services Security.

4.1 *Data storage architecture*

In the data level, a large volume of health-related data such as facility data, disease data, environmental data, and socio-economic data are required for Web-based GIS applications. These data could come from distributed or centralized databases, files, or Web Services. Health data integration from different systems requires standard data formats and message types (e.g. Health Level 7—HL7), vocabulary (e.g. Systematized Nomenclature of Medicine—SNOMED), and communication techniques (e.g. Web Services) (Arzt 2006). Tsui et al. (2003) described a real-time public health surveillance system using Web-based GIS in which clinical data collected by health care providers are transferred to a database through HL7 messages. Zeng et al. (2005) showed a case study of a bioportal system, which gathers all data from different departments through HL7 messages and then integrates them into a bioportal data store consumed by a Web-based GIS.

Database storage is often optimized for Web-based GIS applications. Typically, there are two types of applications: On-Line Transaction Processing (OLTP) systems and On-Line Analytical Processing (OLAP) systems. OLTP systems follow routine and repetitive transactions. Data consistency, integrity, and up to date issues are main concerns in such systems. OLTP systems are suitable in Web-based GIS applications that provide up-to-date hospital patient incidents, emergency department access, and vaccine inventory information. On the other hand, OLAP are best suited for health decision support systems that require a large volume of historical data and need complex query and analysis for fast decisions. Data warehouses, which model data from multi-dimensional views, are usually chosen as the data storage for OLAP systems. The use of Spatial OLAP and spatial data warehouses in Web-based GIS applications can effectively reduce the response time in the exploration of historic health data.

Health authorities considering Web-based GIS technologies should ensure there is a robust database management system and compliance with industry standards in order to guarantee interoperability. A variety of open source database software (e.g. PostGIS-PostgreSQL, MySQL) and proprietary software (Oracle, SQL Server) have been used to store geo-referenced health data in Web-based GIS applications (Conte et al. 2005b, Kamadjeu & Tolentino 2006b, Sanchez et al. 2007). Most database software support OLTP, OLAP, and Spatial OLAP, and provide spatial extensions following the specifications of Open Geospatial Consortium (OGC).

4.2 *Standalone Web-based GIS applications for health*

Standalone Web-based GIS applications for health generally use closely coupled design and are often developed in isolation from one another. The application sever tier of these standalone Web-based GIS applications includes all the abilities of processing, transformation, analysis, and mapping of health data to provide value-added information. The results of processing, transformation, and analysis can generate the mapping values, which would be represented to users through Web maps. Statistical methods (e.g. standard morbidity ratio, crude morbidity ratio) and spatial analysis functions (e.g. proximity analysis, network analysis) can be used to explore health data in specific geospatial boundaries and time periods.

Web-based GIS applications provide a ubiquitous way for user interaction through graphical user interfaces (Web pages) in browsers. Depending on the processing demand on the client side, Web-based GIS applications could be thin or thick-client applications.

Thick-client applications commonly need to install plug-ins for the browser. A large number of standalone Web-based GIS applications have been developed for the generation of health maps dynamically online, with a thin/thick client or hybrid architectures. For instance, Inoue et al. (2003) developed a thin-client, Web-based GIS application to dynamically generate and display infectious disease surveillance data through maps and charts. Blanton et al. (2006) integrated federal, state and local data and developed map tools for rabies surveillance with a Web-based GIS thin client architecture. Some other applications have employed thick client, Web-based GIS approaches to visualize health information through Java Applets or Scalable Vector Graphics (SVG). Qian et al. (2004) provided a thick client, Web-based GIS approach to visualize global SARS information using a Java Applet. Kamadjeu & Tolentino (2006b) implemented a Web-based public health information system to generate district-level country immunization coverage maps and graphs with SVG.

4.3 *Web-based GIS applications using Geospatial Web Services and mashups for health*

4.3.1 *Geospatial Web Services for health applications*

To overcome the integration issues in the design of health applications using Web-based GIS, Service Oriented Architecture (SOA) and Web Services have began to provide new solutions in application development (Boulos & Honda 2006, Gao et al. 2008a). Web Services support machine-to-machine intercommunication functionalities through clearly defined interfaces, independent of development environment and platforms. Geospatial Web Services are a kind of Web Services that provides geospatial functionalities. Therefore, all the functionalities such as the processing, transformation, analysis, and mapping of health data can be published as geospatial Web Services, to provide building blocks of Web-based GIS applications for health. Complex Web-based GIS applications for health can be generated by chaining several geospatial Web Services.

Many standard organizations are working on the construction of basic standards and application specifications for Geospatial Web Services. The Open Geospatial Consortium (OGC) initiated the Open Web Service (OWS) program based on the SOA and Web Services and has proposed or adopted dozens of Geospatial Web Service specifications, such as the Web Map Service (WMS), Web Map Context (WMC), Geographic Markup Language (GML), Web Feature Service (WFS), Web Coverage Service (WCS), Keyhole Markup Language (KML), and Web Processing Service (WPS).

Proprietary software (e.g. ESRI ArcGIS Server, MapInfo and MapXtreme) and open source software (e.g. GeoServer, MapServer) have been adding their support in publishing data or functionalities through OGC standard services. Boulos & Honda (2006) proposed to publish health maps through open source Web-based GIS software that supports OGC specifications. OGC Services, such as WMS, WFS, and WPS have been increasingly applied in Web-based GIS applications for health (Gao et al. 2008, MacEachren et al. 2008, Gao et al. 2009a).

Some Web-based GIS applications for health were built on some free for use tools like Google Map API, and ESRI JavaScript or Flex API, with the consumption of data from Google map services and ESRI REST services respectively (Boulos 2005, Atristain 2009). Plenty of open source tools can also be used to generate thick client Web-based GIS applications (e.g. OpenMap, Geotools) or thin client Web-based GIS applications (e.g. OpenLayers) for health (Bardina & Thirumalainambi 2005, Gao et al. 2009b). Sanchez et al. (2007) stated the advantages of using open source software and open specifications for building Web-based GIS applications for health: 1) suitability for addressing social-cultural-political issues due to its openness development and spirit of cooperation; 2) low cost for the Web-based distributed health spatial-information infrastructure; 3) successful usage in many applications and sectors; 4) simplicity to install, learn, and deploy. Since open source software and open specifications lack user support, the authors also pointed out their disadvantages: 1) need of strong quantitative and qualitative evaluation for large enterprise-wide health application; 2) need of evaluation and comparative studies for the open source or proprietary software alternatives; 3) need of research on performance, scalability, and security; 4) lack of awareness of geospatial open specification in health organizations; 5) need of IT teams with GIS knowledge in health organizations.

In Canada, one of four priority areas in the Canadian Geospatial Data Infrastructure (CGDI) is public health, and the CGDI endeavors to share geospatial information for tracking and monitoring population health (CGDI 2010a). Since 2006, more than 20 projects have been funded by GeoConnections for public health in the federal, provincial, local, or enterprise level, and many of them result in geospatial Web services (CGDI 2010b).

4.3.2 *Geospatial mashups for health applications*

Current Web 2.0 technologies revolutionize the Web to further facilitate data sharing and collaboration between users. Web 2.0 mashups refer to "Web sites or services that weave data from different sources into a new data source or services" (Boulos et al. 2008). The mashup technology changes standalone application development patterns. It supports fast application development and lowers the programming skills in the development of Web-based GIS applications for health. Many great map APIs are available for mashups, such as Google Map API, Yahoo API, and OpenStreetMap API, and many Web 2.0 tools emerge for mashup development such as iGoogle, Pageflakes, Dapper, Open Kapow, Yahoo! Pipes, and IBM mashup center. With the mashup tools and APIs, it is possible to integrate and wire some widgets/gadets together in the design and implementation of Web interfaces. The widget/gadgets actually perform the business logic process with Representational State Transfer (REST)/Simple Object Access Protocol (SOAP) Web Services, Web pages, and (Geo) RSS feeds, as shown in Figure 2. With the mashup technology, many Web-based GIS applications for health were developed. Boulos et al. (2008) presented a series of geo-mashup examples in the health field such as infectious disease surveillance and molecular epidemiology. Li & Gong (2008) developed dead bird reporting and public notice board applications using Google Map API, Yahoo API, Flickr API, Flash, and GeoRSS. Cheung et al. (2008) demonstrated a mashup example on filtering tabular cancer data sets from the US national cancer institute Websites through Yahoo! pipes and displaying the results using Google maps.

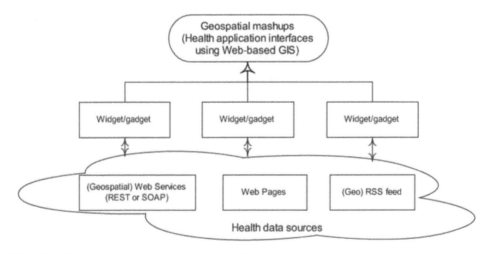

Figure 2. Geospatial mashups for health applications.

5 CHALLENGES AND ONGOING RESEARCH

The experience of disease outbreaks has demonstrated the importance of Web-based GIS technologies for public health. Several challenges still exist in advancing Web-based GIS for health.

5.1 Health data heterogeneity

Public health data tend to be divided into silos: hospitals, physicians, financial management, etc. This data fragmentation is partially due to federal budgets that allocate separate funding blocks for different providers and services. The data collection process varies amongst different health organizations with different tools and methods. The integration of health data across service systems is a challenge (McLafferty 2003). The heterogeneity problems of health data come from different input formats, different spatial levels (e.g. point, postal code, county), different ways in describing a concept, different naming conventions, different terminologies, different information models, and different data transmission standards. For example, no central repository of health data exists in the United States and there is considerable variation in the formats and location requirements of the data that are reported (National Research Council (U.S.) 2007). The variability in the implementation of health standards (e.g. Health Level 7 standards) also makes it difficult to combine data from multiple health care delivery systems (Lober et al. 2002). The sharing of health data across states or regions is uncommon, as inconsistencies across states regarding their use of geocoding references and statistical and mapping software limit the possibilities to integrate data for multi-state studies (Gregorio et al. 2006). Wiafe & Davenhall (2005) pointed out that successful disease surveillance activities require standardized methodology, appropriate tools for promoting data collection, accurate synthesis of the data, continuity over time, and timely dissemination of the resulting information to health officials and the public.

To overcome data heterogeneity, the semantics-based (usually, ontology-based) approach is being utilized in health data integration (Rey et al. 2006, Schuurman & Leszczynski 2008). Ambiguities in data could be removed with an explicit description using ontologies. Many health standards, such as HL7, the Health Insurance Portability and Accountability Act (HIPAA), and the Unified Medical Language System (UMLS) can serve as ontologies in the exchange and integration of health data. Because different ontologies are used for different databases, matching concepts from different ontologies is an important and difficult task (Lee et al. 2006, Ryan 2006). For geo-referenced health data integration, the utilization of geospatial semantics, which describe the underlying meaning of geospatial objects and their spatial relationships corresponding to the real world, would be beneficial.

In addition, semantic interoperability also poses a challenge in applying geospatial Web Services and applications for health. In the development of mashups for health, Cheung et al. (2008) pointed that solving semantic issues would further support the semantic mashup of health data in a user-friendly and social-friendly fashion. The use of ontologies has been a popular approach to support complex knowledge representation, and semantically enriched machine-readable content is needed for semantic mashups for health (Boulos et al. 2008).

5.2 Health data privacy and confidentiality issues

As the U.S. Department of Health and Human Services (2000) pointed out, one major challenge in this decade is the increase of public access to geo-referenced health data without compromising confidentiality. Privacy and confidentiality issues have been given a lot of attention in health studies. Privacy is to protect personal information not to be disclosed and distributed. The privacy rules consider the rights of privacy in doctor-patient relationships and personal health information from the perspective of public access; confidentiality is the responsibility of health practitioners to hold confidential the patient's information (Ölvingson et al. 2002).

From the guidance of the U.S. Centers for Disease Control and Prevention (CDC) and the U.S. Department of Health and Human Services, three types of health information are identified with different levels of privacy rules, namely protected health information, limited data sets, and de-identified information (Thacker 2003). However, due to the vague and subjective nature of data provision, many health organizations simply remove all the geographic identifiers when releasing data (Pickle et al. 2006). The privacy and confidentiality issues always make it difficult to obtain health data at a fine scale for performing spatial analysis in Web-based GIS applications. It is agreed that there is a consistent trade-off between spatial analysis accuracy and privacy rules (Kwan et al. 2004, Sherman & Fetters 2007).

The popularity of mashups creates a new way to mix geo-referenced health information together, and privacy will be a concern as data can be possibly de-identified from the integration of various kinds of sources. AbdelMalik et al. (2008) carried out a survey on views and requirements of public practitioners about privacy issues and spatial data across UK and Canada. The results showed a high demand for personally identifiable data and geographic information in their health activities in both countries. Therefore, the harmonization of privacy and confidentiality with health data requirements in Web-based GIS applications is an urgent issue.

5.3 *Powerful processing ability issues*

Perez et al. (2009) pointed out that one of the greatest problems in health application development is to be able to collect and analyze data and communicate results for real-time or near real-time decision making. Reinhardt et al. (2008) showed a Web-based GIS application for meningococcal disease surveillance, which connects to an automated cluster detection system on a weekly basis to generate cluster maps for online visualization. However, identifying new emerging risks and spatio-temporal clusters from different health factors in real-time is important for health surveillance. The functionalities of Web-based GIS applications for health mostly provide spatial data and static thematic maps on disease risks and other health-related phenomena, and the applications that serve spatial modeling functionalities are limited (Zeilhofer et al. 2009). For example, Bayesian hierarchical models have become more prevalent for spatio-temporal exploration and for understanding the relationships between environmental pollutants and different health hazards, and they are often computationally expensive (LeSage et al. 2009). Due to the computational challenges, the modeling methods combining many health factors for accurate inference and uncertainty assessment have not yet been applied in Web-based GIS applications for health in real-time.

5.4 *Appropriate representation of health information to users*

People usually prefer to use spatial boundaries that they are familiar with, such as administrative boundaries to convey information, in which they can compare disease distributions, health facility locations, and health intervention in different regions. The results of health applications using Web-based GIS could be presented to users through thematic maps, showing health phenomena at different levels. However, maps can easily mislead (Hanchette 1998), and poorly designed maps can inadvertently mis-communicate information (Monmonier 1991). Considerable information is missing from such maps, such as the statistic methods used and health source metadata. As the representation of information is essential for appropriate interpretation, consideration needs to be given to the use of Web-based GIS in interpreting health data. The representation of health information should capture health information distribution while minimizing individual identification potential. A HEalth Representation XML (HERXML) schema considering semantic, geometric, and cartographic representation of the statistic results of health activities was designed to explore the appropriate representation of health information via the Internet (Gao et al. 2009a). The complex nature of the data and the heterogeneity in user skill and knowledge both demand consideration when a data depiction is to be designed for facilitating appropriate interpretation (Buckeridge et al. 2002).

Therefore, good health information representation models are required to facilitate information delivery and overcome confusion.

6 CONCLUSIONS AND FUTURE RESEARCH

The development of Web-based GIS technologies facilitates health organizations to share and exchange health information in a collaborative environment. Web-based GIS will further demonstrate its strong abilities in disease surveillance, health care planning, and public health participation. The real-time data processing and mapping functionalities of Web-based GIS applications for health will be more diverse and powerful.

The SOA and mashup technologies discussed in this chapter will be widely used in the fast development of Web-based GIS applications for health. More and more geospatial Web Services will emerge and serve for health applications using Web-based GIS. Therefore, the trust issues of these services will be significant in the development of Web-based GIS applications, especially for handling critical situations such as disease outbreaks. Systematic frameworks and tools for the evaluation of geospatial Web Service quality would be highly valuable in directing service providers on how to improve services and assist users to acquire optimal services.

Health data heterogeneity will continuously be problematic in Web-based GIS applications for health at different levels of data integration. Incorporating semantics (including geospatial semantics) would facilitate Web-based health data sharing and integration. Furthermore, intelligent health applications should be able to represent medical concepts and terminologies, general rules of disease development, and the association of health problem occurrence with environmental or social-economic issues. This will lead to the explication of existing knowledge using ontologies and rules, which have the potential to support the discovery of new information in the early detection of health phenomena.

With the rapid deployment of sensors, such as human body sensors, air pollution sensors, climate sensors, and satellite sensors, huge amounts of these sensor data and hospitalization data can be collected in real-time and be available through (geospatial) Web Services. In addition, vast amounts of historical data exist, such as disease outbreak data, environmental data, and intervention strategy data. As shown in Figure 3, future research can be on the development of Web-based GIS applications with the use of all real-time and historical health data for online data mining (with the support of grid computing and cloud computing) to extract new knowledge for real-time health decision making.

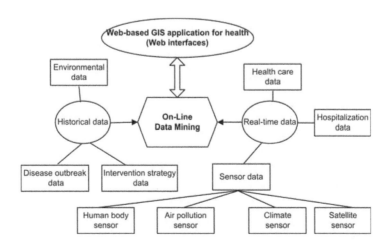

Figure 3.   Online data mining for real-time and historical health data.

Using Web-based GIS technologies can support health practitioners and the general public in health decision making. Public health security needs the participation of all citizens. The Web-based GIS development for health demands the support of citizen empowerment to allow citizens to actively participate in health decision making.

REFERENCES

AbdelMalik, P., Boulos, M.N. & Jones, R. (2008) The perceived impact of location privacy: A web-based survey of public health perspectives and requirements in the UK and Canada. *BMC Public Health*, 8, 156.

Arzt, N.H. (2006) The new alphabet soup: models of application integration. *Journal of Healthcare Information Management: JHIM*, 20 (3) 16–18.

Atristain, B., (2009) *VIPER: Virginia Deploys Web-Based Emergency Management System.* [Online] Available from: http://www.esri.com/news/arcnews/fall09articles/viper.html, [Accessed March 10, 2010].

Baptiste, R., Toubiana, L., Le Mignot, L., Ben Said, M., Mugnier, C., Le Bihan-Benjamin, C., Jaïs, J.P. & Landais, P. (2005) A Web-based GIS for health care decision-support. *AMIA ... Annual Symposium Proceedings / AMIA Symposium. AMIA Symposium.* pp.365–369.

Bardina, J.E. & Thirumalainambi, R. (2005) Distributed web-based expert system for launch operations. *2005 Winter Simulation Conference*, 4–7th December Orlando, USA.

Bédard, Y., Merrett, T. & Han, J. (2001) Fundamentals of spatial data warehousing for geographic knowledge discovery. In Miller, H. & Han, J. (eds.) *Geographic Data Mining and Knowledge Discovery*, Taylor and Francis, London. pp. 53–73.

Bernier, E., Gosselin, P., Badard, T. & Bédard, Y. (2009) Easier surveillance of climate-related health vulnerabilities through a Web-based spatial OLAP application. *International Journal of Health Geographics*, 8 (18).

Blanton, J.D., Manangan, A., Manangan, J., Hanlon, C.A., Slate, D. & Rupprecht, C.E. (2006) Development of a GIS-based, real-time Internet mapping tool for rabies surveillance. *International Journal of Health Geographics*, 5 (47).

Boulos, M.N. (2005) Web GIS in practice III: Creating a simple interactive map of England's Strategic Health Authorities using Google Maps API, Google Earth KML, and MSN Virtual Earth Map Control. *International Journal of Health Geographics*, 4 (22).

Boulos, M.N. & Honda, K. (2006) Web GIS in practice IV: Publishing your health maps and connecting to remote WMS sources using the Open Source UMN MapServer and DM Solutions MapLab. *International Journal of Health Geographics*, 5(6).

Boulos, M.N., Roudsari, A.V. & Carson, E.R. (2001) Health geomatics: An enabling suite of technologies in health and healthcare. *Journal of Biomedical Informatics*, 34 (3), 195–219.

Boulos, M.N., Scotch, M., Cheung, K. & Burden, D. (2008) Web GIS in practice VI: A demo playlist of geo-mashups for public health neogeographers. *International Journal of Health Geographics*, 7 (38).

Buckeridge, D.L., Mason, R., Robertson, A., Frank, J., Glazier, R., Purdon, L., Amrhein, C.G., Chaudhuri, N., Fuller-Thomson, E., Gozdyra, P., Hulchanski, D., Moldofsky, B., Thompson, M. & Wright, R. (2002) Making health data maps: A case study of a community/university research collaboration. *Social Science and Medicine*, 55 (7), 1189–1206.

CGDI, (2010a) *About CGDI.* [Online] Available from: http://www.geoconnections.org/en/aboutcgdi.html, [Accessed 10th March 2010].

CGDI, (2010b) *Public health projects.* http://www.geoconnexions.org/en/communities/publichealth/projects, (Accessed 10th March 2010).

Cheung, K., Yip, K.Y., Townsend, J.P. & Scotch, M. (2008) HCLS 2.0/3.0: Health care and life sciences data mashup using Web 2.0/3.0. *Journal of Biomedical Informatics*, 41(5), 694–705.

Conte, A., Colangeli, P., Ippoliti, C., Paladini, C., Ambrosini, M., Savini, L., Dall'Acqua, F. & Calistri, P. (2005) The use of a Web-based interactive Geographical Information System for the surveillance of bluetongue in Italy. *OIE Revue Scientifique Et Technique*, 24 (3), 857–868.

Gao, S., Mioc, D., Anton, F., Yi, X. & Coleman, D.J. (2008) Online GIS services for mapping and sharing disease information. *International Journal of Health Geographics*, 7 (8).

Gao, S., Mioc, D., Yi, X., Anton, F., Oldfield, E. & Coleman, D.J. (2009a) Towards Web-based representation and processing of health information. *International Journal of Health Geographics*, 8 (3).

Gao, S., Oldfield, E., Mioc, D., Yi, X. & Anton, F. (2009b) *Geospatial Web Services and applications for infectious disease surveillance*. In: Duncan, K. & Brebbia, C.A. (eds.), Disaster Management and Human Health Risk: Reducing Risk, Improving Outcomes, Wessex Institute of Technology, UK. pp. 13–19.

Gregorio, D.I., Samociuk, H., DeChello, L. & Swede, H. (2006) Effects of study area size on geographic characterizations of health events: Prostate cancer incidence in Southern New England, USA, 1994–1998. *International Journal of Health Geographics*, 5 (8).

Grigg, M., Alfred, B., Keller, C. & Steele, J.A. (2006) Implementation of an internet-based geographic information system: The Florida experience. *Journal of Public Health Management and Practice*, 12 (2), 139–145.

Hanchette, C. (1998) GIS implementation of 1997 CDC guidelines for childhood lead screening in North Carolina. *The Third National Conference on GIS in Public Health*, San Diego, CA.

Hochstein, C. & Szczur, M. (2006) TOXMAP: A GIS-based gateway to environmental health resources. *Medical Reference Services Quarterly*, 25 (3), 13–31.

Inoue, M., Hasegawa, S., Suyama, A. & Meshitsuka, S. (2003) Automated graphic image generation system for effective representation of infectious disease surveillance data. *Computer Methods and Programs in Biomedicine*, 72 (3), 251–256.

Kamadjeu, R. & Tolentino, H. (2006a) Open source Scalable Vector Graphics components for enabling GIS in web-based public health surveillance systems. *AMIA ... Annual Symposium Proceedings / AMIA Symposium. AMIA Symposium.* p. 973.

Kamadjeu, R. & Tolentino, H. (2006b) Web-based public health geographic information systems for resources-constrained environment using scalable vector graphics technology: A proof of concept applied to the expanded program on immunization data. *International Journal of Health Geographics*, 5 (24).

Kwan, M., Casas, I. & Schmitz, B.C. (2004) Protection of geoprivacy and accuracy of spatial information: How effective are geographical masks? *Cartographica*, 39 (2), 15–28.

Lee, Y., Supekar, K. & Geller, J. (2006). Ontology integration: Experience with medical terminologies. *Comput. Biol. Med.*, 36 (7–8), 893–919.

LeSage, J., Banerjee, S., Fischer, M.M. & Congdon, P. (2009) Spatial statistics: Methods, models & computation. *Computational Statistics and Data Analysis*, 53 (8), 2781–2785.

Li, S. & Gong, J. (2008) MASHUP: A New Way Of Providing Web Mapping/Gis Services. *ISPRS Congress Beijing 2008*, 3–11th July, Beijing, China.

Lober, W.B., Karras, B.T., Wagner, M.M., Overhage, J.M., Davidson, A.J., Fraser, H., Trigg, L.J., Mandl, K.D., Espino, J.U. & Tsui, F. (2002) Roundtable on bioterrorism detection: Information system-based surveillance. *Journal of the American Medical Informatics Association*, 9 (2), 105–115.

Lu, X. (2004) Web-GIS based SARS epidemic situation visualization. *Proceedings of SPIE - The International Society for Optical Engineering*, 5444, 445–452.

MacEachren, A.M., Crawford, S., Akella, M. & Lengerich, G. (2008) Design and implementation of a model, web-based, GIS-enabled cancer atlas. *Cartographic Journal*, 45 (4), 246–260.

Maclachlan, J.C., Jerrett, M., Abernathy, T., Sears, M. & Bunch, M.J. (2007) Mapping health on the Internet: A new tool for environmental justice and public health research. *Health and Place*, 13 (1), 72–86.

McLafferty, S.L. (2003) GIS and health care. *Annual Review of Public Health*, 24 (1), 25–42.

Monmonier, M.S. (1991) *How to lie with maps*. Chicago: University of Chicago Press.

National Research Council (U.S.), Committee on Research Priorities for Earth Science and Public Health, (2007) Earth Materials and Health: Research Priorities for Earth Science and Public Health. 188.

Ölvingson, C., Hallberg, J., Timpka, T. & Lindqvist, K. (2002) Ethical issues in public health informatics: Implications for system design when sharing geographic information. *Journal of Biomedical Informatics*, 35 (3), 178–185.

Perez, A.M., Zeng, D., Tseng, C., Chen, H., Whedbee, Z., Paton, D. & Thurmond, M.C. (2009) A web-based system for near real-time surveillance and space-time cluster analysis of foot-and-mouth disease and other animal diseases. *Preventive Veterinary Medicine*, 91 (1), 39–45.

Phosuwan, A., Hsu, C.E., Dunn, K., Mansueto, M. & Salisbury, L. (2009) Applying Informatics to Improve Vulnerable Population Registration for Emergency Preparedness in the Gulf Coast Region of Texas. *Texas Public Health Association Journal*, 61 (4), 42–47.

Pickle, L.W., Szczur, M., Lewis, D.R. & Stinchcomb, D.G. (2006) The crossroads of GIS and health information: A workshop on developing a research agenda to improve cancer control. *International Journal of Health Geographics*, 5 (51).

Qian, Z., Zhang, L., Yang, J. & Yang, C. (2004) Global SARS information WebGIS design and development. *International Geoscience and Remote Sensing Symposium (IGARSS)*, 5, 2861–2863.

Reinhardt, M., Elias, J., Albert, J., Frosch, M., Harmsen, D. & Vogel, U. (2008) EpiScanGIS: An online geographic surveillance system for meningococcal disease. *International Journal of Health Geographics*, 7 (33).

Rey, D., Maojo, V., García-Remesal, M., Alonso-Calvo, R., Billhardt, H., Martin-Sánchez, F. & Sousa, A. (2006) ONTOFUSION: Ontology-based integration of genomic and clinical databases. *Computers in Biology and Medicine*, 36 (7–8), 712–730.

Rivest, S., Bédard, Y., Proulx, M., Nadeau, M., Hubert, F. & Pastor, J. (2005) SOLAP technology: Merging business intelligence with geospatial technology for interactive spatio-temporal exploration and analysis of data. *ISPRS Journal of Photogrammetry and Remote Sensing*, 60 (1), 17–33.

Roth, L.H., Criss, K., Stewart, X. & McCann, K. (2009) Preplink: A novel web-based tool for healthcare emergency planning and response. *Biosecurity and Bioterrorism*, 7 (1), 85–92.

Ryan, A. (2006) Towards semantic interoperability in healthcare: ontology mapping from SNOMED-CT to HL7 version 3. *Proceedings of the second Australasian workshop on Advances in ontologies*, 5th December, Hobart, Australia.

Sanchez, R., Anderson, G., Cruz, J. & Hayden, M. (2007) The potential for the use of Open Source Software and Open Specifications in creating Web-based cross-border health spatial information systems. *International Journal of Geographical Information Science*, 21 (10), 1135–1163.

Schuurman, N., Leight, M. & Berube, M. (2008) A Web-based graphical user interface for evidence-based decision making for health care allocations in rural areas. *International Journal of Health Geographics*, 7 (49).

Schuurman, N. & Leszczynski, A. (2008) A method to map heterogeneity between near but non-equivalent semantic attributes in multiple health data registries. *Health Informatics Journal*, 14 (1), 39–57.

Sherman, J.E. & Fetters, T.L. (2007) Confidentiality concerns with mapping survey data in reproductive health research. *Studies in Family Planning*, 38 (4) 309–321.

Skinner, R. (January, 2010) *Integrating Location into Hospital and Healthcare Facility Emergency Management - Part 3, Applied GIS.* Directions Magazine, [Online] Available from: http://www.directionsmag.com/article.php? article_id=3388, [Accessed 10th March 2010].

Thacker, S.B. (2003) HIPAA Privacy Rule and Public Health, Guidance from CDC and the U.S. Department of Health and Human Services. http://www.cdc.gov/mmwr/preview/mmwrhtml/m2e411a1.htm, accessed March 10, 2010.

Tsoi, C.W. (2007) Development of a Cross-Domain Web-based GIS Platform to Support Surveillance and Control of Communicable Diseases. In: Lai, P.C. & Mak, A.S.H. (eds.), *GIS for Health and the Environment*, Berlin Heidelberg, Springer. pp. 44–56.

Tsui, F., Espino, J.U., Dato, V.M., Gesteland, P.H., Hutman, J. & Wagner, M.M. (2003) Technical description of RODS: A real-time public health surveillance system. *Journal of the American Medical Informatics Association*, 10 (5), 399–408.

U.S. Department of Health and Human Services (2000) Healthy People 2010: Understanding and Improving Health. U.S. Government Printing Office, Washington, DC, USA.

Wang, Y., Tao, Z., Cross, P.K., Le, L.H., Steen, P.M., Babcock, G.D., Druschel, C.M. & Hwang, S.-. (2008) Development of a web-based integrated birth defects surveillance system in New York State. *Journal of Public Health Management and Practice*, 14 (6), E1–E10.

WHO - World Health Organization (2010) *GIS and public health mapping.* [Online] Available from: http://www.who.int/ health_mapping/gisandphm/en/index.html, [Accessed 10th March 2010].

Wiafe, S. & Davenhall, B. (2005) *Extending Disease Surveillance with GIS.* [Online] Available from: http://www.esri.com/news /arcuser/0405/disease_surveil1of2.html, [Accessed 10th March 2010].

Xu, J.Z., Wang, C., Hu, Y., Croitoru, A. & Tao, V. (2005) A web-based risk assessment system and service for child safety: The case of mid-town Toronto. *Geomatica*, 59 (2), 149–158.

Zeilhofer, P., Neto, P.S., Maja, W.Y. & Vecchiato, D.A. (2009) A web-based, component-oriented application for spatial modelling of habitat suitability of mosquito vectors. *International Journal of Digital Earth*, 2 (4), 327–342.

Zeng, D., Chen, H., Lynch2, C., Eidson, M. & Gotham, I. (2005) Infectious Disease Informatics and Outbreak Detection. In: Chen, H., Fuller, S.S., Friedman, C. & Hersh, W. (eds), *Medical informatics: knowledge management and data mining in biomedicine*, Springer.

Open standards for geospatial services

*Advances in Web-based GIS, Mapping Services
and Applications – Li, Dragićević & Veenendaal (eds)*
© *2011 Taylor & Francis Group, London, ISBN 978-0-415-80483-7*

OGC standards: Enabling the geospatial web

Carl Reed
Open Geospatial Consortium, Inc., Wayland, USA

ABSTRACT: This chapter should be viewed as an overview of the web services standards work of the Open Geospatial Consortium with a special focus on the role of geospatial standards in the evolution and the future of the geospatial web. The geospatial web paradigm suggests the complete integration of geographic (location) and time information into the very fabric of both the internet and the web. Today, the geospatial web encompasses applications ranging from as simple as geo-tagging a photograph to mobile driving directions to sophisticated spatial data infrastructure portal applications orchestrating complex workflows for complex scientific modeling applications. In all of these applications, location and usually time are required information elements. In most of these applications, standards are the glue that enables the easy and seamless integration of location and time. These standards may be very lightweight, such as GeoRSS, or more sophisticated and rich in functionality such as the OGC Web Feature Service (WFS).

Keywords: Geospatial standards, open standards, spatial data infrastructure, web services, standards development, interoperability

1   THE GEOSPATIAL WEB

What is the geospatial web? Opinions vary and there are numerous books, articles and editorials written on this topic (e.g. Scharl & Tochtermann 2007, Liebhold 2005). However, for this chapter, the geospatial web is not defined as just an array of mash-ups or even the hundreds of Spatial Data Infrastructures that have been successfully deployed. Rather, the geospatial web is about the complete integration and use of location at all levels of the internet and the web. This integration will often be invisible to the user. A major premise of this chapter—and of the work of the Open Geospatial Consortium (OGC)—is that the ubiquitous permeation of location into the infrastructure of the internet and the web is being built on standards.

There are numerous issues facing the realization of the vision of the geospatial web, including those related to geospatial standards. These include:

- Lack of standards that support privacy rules related to location;
- Lack of consistent service level agreements related to the use of standards;
- Lack of policies and agreements on what standards should be used (and when);
- Lack of internationally agreed semantics for various information domains;
- Wild-west philosophy of the evolving geospatial web which leads to multiple, often competing, standards for the same interoperability problem;
- Potential impacts of the "cloud" on the geospatial industry. Cloud computing is Internet-based computing using shared resources.
- Disagreements on which architectural patterns, such as Service Oriented Architectures versus Representational State Transfer, the geospatial web should be built on.

All of these issues are discussed on a regular basis in various forums in the OGC.

2 INTRODUCTION TO THE OGC

Founded in 1994, the Open Geospatial Consortium (OGC) is a global industry consortium with a vision to "Achieve the full societal, economic and scientific benefits of integrating location resources into commercial and institutional processes worldwide". Inherent in this vision is the requirement for geospatial standards and strategies to be an integral part of business processes.

The mission of the OGC is to serve as a global forum for the development and promotion of open standards and techniques in the area of geoprocessing and related information technologies. The ongoing standards work of the OGC is primarily supported by the volunteerism of the OGC membership. Consortium activities are supported by membership fees and, to a lesser extent, development partnerships and publicly funded cooperative programs.

The OGC has 410+ members[1]—geospatial technology software vendors, systems integrators, government agencies and universities—participating in the consensus standards development and maintenance process. The interfaces, encodings, and protocols defined by the OGC are designed to meet community requirements in support of the integration of geospatial content and services into solutions that geo-enable the Web, wireless and location-based services, and mainstream IT. In support of meeting the interoperability requirements of the geospatial web, the OGC also supports a major commitment to collaborate with other standards development organizations that have requirements for using location based content.

2.1 What is an OGC standard?

The primary product of the OGC is a "standard" together with related supporting documents. A standard is a document that details the engineering aspects (and rules) for implementing an interface or encoding that solves a specific geospatial interoperability problem. OGC standards are defined, discussed, tested, and approved by the members using a formal consensus process. The structure for and content of what must be contained in an OGC standard document is guided by the policies defined in OGC document "The Specification Model—A Standard for Modular specifications" (OGC 2009c). For example, every OGC standard shall have statements called *requirements*. A requirement is expression in the content of a standard conveying criterion to be fulfilled if compliance with the standard is to be claimed and from which no deviation is permitted. In terms of the actual text of an OGC standard, requirements use normative language and in particular are commands and use the imperative "shall". The point is that all OGC standards are to be structured in the same way, guided by the policies agreed to the OGC Members (Reed 2010).

Information and downloads of the currently approved OGC Standards can be found at the OGC web site. These standards are freely and publicly available (OGC 2010b).

2.2 How does the OGC standards development process work?

There are two key operational aspects to the OGC standards development and maintenance process. These are the OGC Interoperability Program and the OGC Specification Program.

2.2.1 The OGC interoperability program

The OGC Interoperability Program provides a rapid engineering process to develop, test, demonstrate, and promote the use of OGC standards. The Program organizes and manages **Interoperability Initiatives** that address the needs of industry and government sponsors.

There are different types of interoperability initiatives:

- Test Beds: Collaborative, applied research and development efforts to develop, architect and test candidate standards addressing Sponsor requirements.

---

[1] As of December 2010.

- Pilot Projects: Collaborating communities applying and testing technology providers' interoperable offerings in real world settings.
- Interoperability experiments: Similar to a test bed but proposed, managed, and resourced by the OGC members.

The typical Test Bed activity lasts 6 months and results in the development of numerous reports as well as an interoperability demonstration.

OGC Web Services-7 is the most recent OGC test bed (OGC 2009b). OWS-7 has multiple threads. Each thread focuses on a specific domain of interoperability requirements. For example, The Sensor Fusion and Exploitation (SFE) Thread builds on the OGC Sensor Web Enablement (SWE) framework of standards that has achieved a degree of maturity through previous OWS interoperability initiatives and deployments worldwide. SFE focuses on integrating the SWE interfaces and encodings with workflow and web processing services to perform sensor fusion. SFE will continue the development of Secure SWE architecture, as well as the interoperability of SWE and Common CBRN Sensor Interface (CCSI).

Emphasis for SFE during this phase of the OWS testbed will be the following:

- Motion Video Fusion. Geo-location of motion video for display and processing. Change detection of motion video using Web Processing Service with rules.
- Dynamic Sensor Tracking and Notification. Track sensors and notify users based on a geographic Area of Interest (AOI). The sensor and the user may be moving in space and time.
- CCSI-SWE Best Practice. Building on OWS-6, develop an Engineering Report to be considered by the OGC Technical Committee as a Best Practice.

The Oceans Interoperability initiative is an example of a very successful, recently completed interoperability experiment. The first phase of the Oceans IE was created to investigate the use of OGC Web Feature Services (WFS) and OGC Sensor Observation Services (SOS) for representing and exchanging point data records from fixed in-situ marine platforms. Phase II consolidated consensus for a portion of the Ocean-Observing community on its understanding of various OGC standards, solidified demonstrations for Ocean Science application areas, hardened software implementations, and produced an OGC Best Practices document that can be used to inform the broader ocean-observing community on the use of OGC standards.

2.2.2 *The OGC specification program*

The OGC Specification Program **Technical Committee (TC)** is where the formal OGC standard consensus discussion and approval process occurs. The Technical Committee is comprised of a number of working groups (WGs). Technical Committee **Working Groups** provide an open, collaborative forum for discussions, presentations and recommendations. These forums are critical for discussions of key interoperability issues for specific information communities, discussion and review of standards, and presentations on key technology areas relevant to solving geospatial interoperability issues. Working groups typically work on domain specific interoperability issues.

The primary product of the TC is the processing and adoption of OGC standards (which are often drafted in OGC interoperability test beds). The TC is also responsible for the maintenance and revision of the adopted standards. The Technical Committee is organized to focus on both general and domain-specific standards development. The OGC TC Policies and Procedures provide the rules and governance for the work of the TC.

2.3 *Current OGC standards focus areas*

The OGC membership currently is focusing in these major standards activities:

- Sensor Web Enablement: The SWE initiative is focused on developing and enhancing standards to enable the discovery of sensors and corresponding observations, exchange, and processing of sensor observations, as well as the tasking of sensors and sensor systems. Sensors include both dynamic, such as satellite and UAV, and in-situ, such as weather stations and traffic, sensor systems.

- Earth Systems Science: This broad OGC activity focuses on enhanced collaboration between and among research and scientific communities for requirements related to and the use of OGC and related standards. The goal is to increase the ability of the earth systems science community to share location based information. Currently, the OGC has very active work in the Oceans Science, Hydrology, and Meteorological research and science domains.
- Mass Market Geo: The mission of the OGC Mass Market activity is to broaden the use of location-aware technologies in mainstream consumer and business IT infrastructures. Recent activity and discussions in the group are focusing on the use of OpenSearch as a lightweight protocol for use with OGC standards, a candidate GeoSMS standard, and other lightweight protocols and Application Programmers Interfaces.
- Security and Rights Management: This long term OGC activity has two main goals: to establish an interoperable security framework for OpenGIS Web Services to enable protected geospatial information processing, and to coordinate and mature the development and validation of work being done on digital rights management for the geospatial community.
- 3d Information Management (and related CAD-GIS integration): The 3D Information Management (3DIM) activity is facilitating the creation of new standards and enhancement to existing OGC standards that enable infrastructure owners, builders, emergency responders, community planners, and the traveling public to better manage and navigate complex built environments.

Decisions as to what are the key standards development focus areas are determined by the membership, typically as a result of market trends.

3 THE OGC REFERENCE MODEL AND ABSTRACT SPECIFICATION

In 1994 when the OGC held its first meeting, the participants quickly realized that a key issue facing the geospatial standards community was the lack of a consistent vocabulary upon which standards could be built. For example, there was no clear industry agreement on what was meant by a feature or a curve or even a point. So, for the first two years, most of the energy of the OGC was directed at defining a vocabulary and a related reference model. The result of this activity was the development of what in the OGC is called the Abstract Specification (AS). The AS provides the conceptual foundation and semantics upon which OGC implementation standards can be built.

The first adopted OGC standard was Simple Features (OGC 1997). Simple Features describes the common architecture for simple feature geometry as well as common access and storage mechanisms. The geometry is defined in OGC Abstract Topic volume 1: Feature Geometry and Topic 2: Spatial Referencing by Coordinates. Both of these OGC Abstract Specifications are also ISO standards.

Since 1997, the OGC Members have approved many more OGC standards. These standards themselves reference other international standards from ISO, the Organization for the Advancement of Structured Information Standards (OASIS) and the Internet Engineering Task Force (IETF). As such, the Members determined that what was needed was an overarching reference model that describes the entire OGC standards baseline and the various relationships and dependencies contained in the standards baseline.

The OGC Reference Model (ORM) was first released in 2007 and has since been revised twice (Percivall et al. 2008). The purpose of the ORM is to:

- Provide an overview of OGC Standards Baseline;
- Provide insight into the current state of the work of the OGC;

- Serve as a basis for coordination and understanding of the documents in the OGC Standards Baseline;
- To provide a useful resource for defining architectures for specific applications.

3.1 *OGC document types*

In addition to implementation and abstract standards, the OGC Membership develops and processes other supporting documents types. These include:

Best Practices: A document containing guidance related to the use and/or implementation of an adopted OGC document, usually in a specific information domain or application area. Best Practices Documents are an official position of the OGC and thus represent an endorsement of the content of the paper.

Public Engineering Reports: A document generated by an OGC Interoperability Initiative, such as a Test Bed or Interoperability Experiment, and approved for public release by the OGC Membership. Engineering Reports are not the official position of the OGC and contain a statement to that effect.

Discussion Papers: A document containing discussion of some technology or standard area for release to the public. Discussion Papers are not the official position of the OGC and contain a statement to that effect.

White Papers: A publication released by the OGC to the Public that states a position on a social, political, technical or other subject, often including a high-level explanation of an architecture or framework of a solution.

4   THE OGC STANDARDS BASELINE

OGC standards are typically either interface or encoding standards. Additionally, the members define profiles and application schemas of these standards. An **interface standard** is a standard that describes one or more functional operations necessary to allow the exchange of information between two or more (usually different) systems. An **encoding** standard specifies rules, often as XML schema, for converting geospatial content into an encoding payload that can then be shared between and among systems.

As of late 2010, there were 32 approved OGC standards. A number of these OGC standards have been submitted to ISO and have been approved as International Standards.

This section provides a synopsis of the OGC standards baseline and provides more detail on those OGC standards that are most often implemented as part of a Spatial Data Infrastructure (SDI).

Table 1 provides a list and short description of all of the current OGC approved standards. All of these OGC standards are freely available for download and free use from the OGC website.

The next subsections describe the most commonly implemented OGC standards with examples. Typically, these standards are implemented as part of a portal application or some enterprise workflow.

4.1   *Web Map Service (WMS)*

WMS provides a simple HTTP interface for requesting geo-registered map images from one or more distributed geospatial databases. A WMS request defines the geographic layer(s) and area of interest to be processed. The response to the request is one or more geo-registered map images (returned as JPEG, PNG, etc.) that can be displayed in a browser application. The interface also supports the ability to specify whether the returned images should be transparent so that layers from multiple servers can be combined or not.

Table 1.   Approved OGC standards.

Catalogue 2.0.2* (AKA CSW)	Publish and search collections of descriptive information (metadata) about geospatial data, services and related resources.
Catalogue Services Standard 2.0 Extension Package for ebRIM Application Profile: Earth Observation Products	Describes the mapping of Earth Observation Products defined in the GML Application schema for Earth Observation products to an ebRIM structure within an OGC® Catalogue 2.0.2.
CSW ISO Metadata Application Profile	Specifies the interfaces, bindings, and encodings required to publish and access digital catalogues of metadata for geospatial data, services, and applications that comply with ISO 19115 (metadata).
Coordinate Transformation Service 1.0	Provides a standard interface and mechanism for software to specify and access coordinate transformation services for use on specified geospatial content.
CityGML 1.0*	Encoding Standard for the representation, storage and exchange of virtual 3D city and landscape models. CityGML is implemented as an application schema of the Geography Markup Language version 3.1.1 (GML3).
Filter Encoding 2.0+	Defines an XML encoding for filter expressions (search criteria).
Geographic Objects 1.0*	Provides an open set of common, lightweight, language-independent abstractions for describing, managing, rendering, and manipulating geometric and geographic objects.
Geography Markup Language (GML)*+ 3.2	An XML grammar for expressing geographical features. GML serves as a modeling language for geographic systems as well as an open interchange format for geographic transactions on the internet.
GML in JP 2000 1.0*	Geographic Imagery Encoding Standard defines the means by which GMLis used within JPEG 2000 compressed images.
GeoXACML 1.0	Encoding Standard (GeoXACML) defines a geospatial extension to the OASIS standard "eXtensible Access Control Markup Language (XACML)".
KML 2.2	An XML language focused on geographic visualization, including annotation of maps and images, in earth browser applications.
OGC Location Services (OpenLS) 1.2	Specifies interfaces and operations for a Directory Service, Gateway Service, Geocoder Service, Presentation (Map Portrayal) Service, Routing Service, and a Navigation Service
Observations and Measurements (O & M) 1.0.0+	Defines an abstract model and an XML schema encoding for observations and it provides support for common sampling strategies. Part of OGC Sensor Web standards framework.
Sensor Model Language (SensorML) 1.0.0*	Specifies models and XML encoding that provide a framework within which the geometric, dynamic, and observational characteristics of sensors and sensor systems can be defined. Part of OGC Sensor Web standards framework.
Sensor Observation Service (SOS) 1.0*	An API for managing deployed sensors and retrieving sensor data and specifically "observation" data. Part of OGC Sensor Web standards framework.

(Continued)

332

Table 1.   (*Continued*)

Sensor Planning Service (SPS) 1.0.0*	Defines interfaces for queries that provide information about the capabilities of a sensor and how to task the sensor. Part of OGC Sensor Web standards framework.
Simple Features (SF) 1.2*+	Provides a well-defined and common way for applications to store and access feature data in relational or object-relational databases, so that the data can be used to support other applications through a common feature model, data store and information access interface.
Style Layer Descriptor (SLD) 1.1.0*	Defines an encoding that extends the WMS standard to allow user-defined symbolization and coloring of geographic feature and coverage data.
Symbology Encoding 1.1.0*	Defines an XML language for styling information that can be applied to digital geographic feature and coverage data.
SWE Common 2.0	Encoding Standard defines low level data models for exchanging sensor related data between nodes of the OGC® Sensor Web Enablement (SWE) framework.
Table Join Service 2.0	Defines a simple way to describe and exchange tabular data that contains information about geographic objects.
TransducerML 1.0.0	Application and presentation layer communication protocol for exchanging live streaming or archived data to (i.e. control data) and/or sensor data from any sensor system.
Web Coverage Service 2.0	Defines a standard interface and operations that enables interoperable access to geospatial "coverages". Coverages are content such as satellite imagery or orthophotos.
WCS Transaction operation extension	The WCS Transaction operation allows clients to add, modify, and delete grid coverages that are available from a WCS server.
WCS Processing Extension (WCPS) 1.0	Specifies an additional processing operation that may optionally be implemented by WCS servers.
Web Feature Service (WFS) 2.0+	Defines an interface for specifying requests for retrieving geographic features across the Web using platform-independent calls.
Web Map Context (WMC) 1.1	Defines the creation and use of documents which unambiguously describe the state, or "Context," of a WMS Client application in a manner that is independent of a particular client.
Web Map Service (WMS) 1.3*+	Provides a simple HTTP interface for requesting geo-registered map images from one or more distributed geospatial databases.
Web Map Tile Service 1.0	A WMTS enabled server application can serve map tiles of spatially referenced data using tile images with predefined content, extent, and resolution.
Web Processing Service 1.0.0*	Specifies rules for standardizing how inputs and outputs (requests and responses) for geospatial processing services, such as polygon overlay.
Web Services Common 1.2.0	Specifies parameters and data structures that are common to all OGC Web Service (OWS) Standards.
*	In revision as of December 2010.
+	Is either ISO standard or are in the processing of being approved as an ISO standard.

The following hypothetical URL requests the US National Oceanographic and Atmospheric Administration hurricane image shown in Figure 1 below.

4.2 *Web Feature Service (WFS)*

The WFS standard defines interfaces and operations for data access and manipulation on a set of geographic features stored in one or more distributed geospatial databases. The operations include:

- Retrieve (Query) features based on spatial and non-spatial constraints;
- Create a new feature instance;
- Get a description of the properties of features;
- Delete a feature instance;
- Update a feature instance;
- Lock a feature instance.

A WFS request typically results in a response that includes a GML encoded payload of the features that satisfy the query. However, the response payload can be in other encodings, such as GeoRSS[2] or KML.

The following example selects the geometry and depth from the Hydrography feature set for the area of the Grand Banks:

```
http://a-map-co.com/mapserver.cgi?  VERSION=1.3.0&REQUEST=GetMap&
CRS=CRS:84&BBOX=-97.105,24.913,-78.794,36.358&
WIDTH=560&HEIGHT=350&LAYERS=AVHRR-09-27&STYLES=&
FORMAT=image/png&EXCEPTIONS=INIMAGE
```

Figure 1.   Response from a WMS request.

```
<?xml version="1.0" ?>
<GetFeature
  version="1.1.0"
  service="WFS"
  handle="Query01"
  xmlns="http://www.opengis.net/wfs"
  xmlns:ogc="http://www.opengis.net/ogc"
```

---

[2] While GeoRSS is not specifically an OGC standard, Version One was developed by OGC staff and OGC Members. Further, there is a GML profile of GeoRSS.

```
    xmlns:gml="http://www.opengis.net/gml"
    xmlns:myns="http://www.someserver.com/myns"
    xmlns:xsi="http://www.w3.org/2001/XMLSchema-instance"
    xsi:schemaLocation="http://www.opengis.net/wfs../wfs/1.1.0/WFS.
        xsd">
    <Query typeName = "myns:Hydrography">
      <wfs:PropertyName>myns:geoTemp</wfs:PropertyName>
      <wfs:PropertyName>myns:depth</wfs:PropertyName>
      <ogc:Filter>
        <ogc:Not>
          <ogc:Disjoint>
            <ogc:PropertyName>myns:geoTemp</ogc:PropertyName>
            <gml:Envelope srsName="EPSG:63266405">
              <gml:lowerCorner>-57.9118 46.2023<gml:lowerCorner>
              <gml:upperCorner>-46.6873 51.8145</gml:upperCorner>
            </gml:Envelope>
          </ogc:Disjoint>
        </ogc:Not>
      </ogc:Filter>
    </Query>
  </GetFeature>
```

Note that the Grand Banks are bounded by the box: [-57.9118, 46.2023,-46.6873, 51.8145]. The output of such a request might look as follows:

```
  <?xml version="1.0" ?>
  <wfs:FeatureCollection
    xmlns="http://www.someserver.com/myns"
    xmlns:wfs="http://www.opengis.net/wfs"
    xmlns:gml="http://www.opengis.net/gml"
    xmlns:xsi="http://www.w3.org/2001/XMLSchema-instance"
    xsi:schemaLocation="http://www.someserver.com/mynsHydrography.xsd
    http://www.opengis.net/wfs ../wfs/1.1.0/WFS.xsd">
    <gml:boundedBy>
      <gml:Envelope srsName="http://www.opengis.net/gml/srs/epsg.
          xml#63266405">
      <gml:lowerCorner>10 10</gml:lowerCorner>
      <gml:upperCorner>20 20</gml:upperCorner>
    </gml:Envelope>
    </gml:boundedBy>
    <gml:featureMember>
      <HydrographyHydrography gml:id="HydrographyHydrography.450">
        <geoTemp>
          <gml:PointsrsName="http://www.opengis.net/gml/srs/epsg.
              xml#63266405">
          <gml:pos>10 10</gml:pos>
          </gml:Point>
        </geoTemp>
        <depth>565</depth>
      </HydrographyHydrography>
    </gml:featureMember>
    <gml:featureMember>
      <HydrographyHydrography gml:id="HydrographyHydrography.450">
        <geoTemp>
          <gml:Point   srsName="http://www.opengis.net/gml/srs/epsg.
              xml#63266405">
```

```
        <gml:pos>10 11</gml:pos>
      </gml:Point>
    </geoTemp>
    <depth>566</depth>
  </HydrographyHydrography>
 </gml:featureMember>
 <!--
 .
 . … more HydrographyHydrography instances …
 .
 -->
</wfs:FeatureCollection>
```

4.3   *Web Coverage Service (WCS)*

A Web Coverage Service (WCS) defines a standard interface and operations that enables interoperable access to geospatial *coverages*[3] stored in one or more distributed geospatial and/or imagery databases. The term *grid coverages* typically refers to content such as satellite images, digital aerial photos, digital elevation data, and other phenomena represented by values at each measurement point. A WCS request includes specifying the response payload encoding, such as GeoTIFF, NetCDF, or HDF-EOS.

A client might issue the following GetCoverage operation request with minimum contents—encoded in KVP:

```
http://my.service.org/path/script?
service=WCS
&version=1.1.2
&request=GetCoverage
&identifier=Cov123
&BoundingBox=-71,47,66,51,
urn:ogc:def:crs:OGC:EPSG:4326
&format=image/netcdf
```

This corresponds to the following minimum request encoded in XML:

```
<?xml version="1.0" encoding="UTF-8"?>
<GetCoverage mlns="http://www.opengis.net/wcs/1.1"
xmlns:ows="http://www.opengis.net/ows/1.1"
xmlns:xsi="http://www.w3.org/2001/XMLSchema-instance"
xsi:schemaLocation="http://www.opengis.net/wcs/1.1
../wcsGetCoverage.xsd" service="WCS" version="1.1.2">
  <ows:Identifier>Cov123</ows:Identifier>
    <DomainSubset>
     <ows:BoundingBox crs="urn:ogc:def:crs:EPSG:4326">
      <ows:LowerCorner>-71 47</ows:LowerCorner>
      <ows:UpperCorner>-66 51</ows:UpperCorner>
     </ows:BoundingBox>
    </DomainSubset>
    <Output format="image/netcdf"/>
</GetCoverage>
```

---

[3] Feature that acts as a function to return values from its range for any direct position within its spatio-temporal domain. A Digital Elevation Model is an example of a coverage.

4.4 Geography Markup Language (GML)

The Geography Markup Language (GML) is an XML grammar for expressing geographical features (GML 2007). GML serves as a modeling language for geographic systems as well as an open interchange format for geographic transactions on the internet. As with most XML based grammars, there are two parts to the grammar—the schema that describes the document and the instance document that contains the actual data.

A GML document is described using a GML Schema. This allows users and developers to describe generic geographic data sets that contain points, lines and polygons. However, the developers of GML envision communities working to define community-specific application schemas based on an agreed information or content model. Using application schemas and related content models, users can refer to roads, highways, and bridges instead of points, lines and polygons. If everyone in a community agrees to use the same schemas they can exchange data easily.

The following GML example illustrates the distinction between *features* and *geometry objects*:

```
<abc:Building gml:id="SearsTower">
  <gml:name>Sears Tower</gml:name>
  <abc:height>52</abc:height>
  <abc:position>
    <gml:Point>
      <gml:coordinates>100,200</gml:coordinates>
    </gml:Point>
  </abc:position>
  <app:extent>
    <gml:Polygon>
      <gml:exterior>
        <gml:LinearRing>
          <gml:coordinates>100,200</gml:coordinates>
        </gml:LinearRing>
      </gml:exterior>
    </gml:Polygon>
  </app:extent>
</abc:Building>
<abc:Building gml:id="SearsTower">
  <abc:position xlink:type="Simple" xlink:href="#p21"/>
</abc:Building>
<abc:SurveyMonument gml:id="g234">
  <abc:position>
    <gml:Point gml:id="p21">
      <gml:coordinates>100,200</gml:coordinates>
    </gml:Point>
  </abc:position>
</abc:SurveyMonument>
```

The *Building* feature has several *geometry objects*, sharing one of them (the *Point* with identifier *p21*) with the *SurveyMonument* feature.

4.5 Web Processing Service (WPS)

In 2007 the OGC Members approved the Web Processing Service (WPS) standard. Since then, there have been numerous implementations of the WPS and much experimentation in the GRID community.

WPS defines a standardized interface that facilitates the publishing of geospatial processes, and the discovery of and binding to those processes by clients. Processes include any algorithm, calculation or model that operates on geospatially referenced data. Publishing means making available machine-readable binding information as well as human-readable metadata that allows service discovery and use.

A WPS can be configured to offer any sort of GIS functionality to clients across a network, including access to pre-programmed calculations and/or computation models that operate on spatially referenced data. A WPS may offer calculations as simple as subtracting one set of spatially referenced numbers from another (e.g. determining the difference in influenza cases between two different seasons), or as complicated as a hydrological modeling (Diez et al. 2008).

The data required by the WPS can be delivered across a network, or available at the server. This standard provides mechanisms to identify the spatially referenced data required by the calculation, initiate the calculation, and manage the output from the calculation so that the client can access it. This WPS is targeted at processing both vector and raster data. The standard is designed to allow a service provider to expose a web accessible process, such as polygon intersection, in a way that allows clients to input data and execute the process with no specialized knowledge of the underlying physical process interface or API. The WPS interface standardizes the way processes and their inputs/outputs are described, how a client can request the execution of a process, and how the output from a process is handled. Because WPS offers a generic interface, it can be used to wrap other existing and planned OGC services that focus on providing geospatial processing services.

5 COLLABORATION

No relevant standards development can happen in isolation. Collaboration with other standards organizations and information communities is critical to the viability, maturity, and uptake of any OGC standard. This section provides information on how the OGC collaborates with other standards organizations and information communities.

5.1 *Collaboration with Other Standards Developments Organizations (SDOs)*

Geospatial standards cannot be developed without consideration of and collaboration with other international standards. Further, there are other internet and web standards organizations that have requirements for encoding and communicating location content.

The OGC collaborates with a number of other standards organizations. The primary reasons for these collaborations are to ensure that:

- OGC has input into related standards, such as those pertaining to workflow, grid computing, security, and authentication;
- Location encodings and protocols required by other standards organizations are harmonized with the work of the OGC and ISO TC 211 (geomatics);
- OGC community remains up to date with significant and related standards.

Four standards organizations that the OGC works closely with on a regular basis are now described in the next subsections.

5.1.1 *ISO TC 211*

The OGC has a Class A Liaison relationship with TC 211. TC 211 work aims to establish a structured set of standards for information concerning objects or phenomena that are directly or indirectly associated with a location relative to the Earth. Numerous ISO standards are used as the abstract foundation for work in the OGC. Further, the OGC often submit OGC standards into the ISO process for review and approval as International Standards. Once the OGC standard is also an ISO standard, the document is available from either the OGC or from ISO.

Currently the following OGC standards are also ISO standards. These are:

- Web Map Service (WMS) 1.3
- Simple Features Access (SF) 1.1
- Geography Markup Language (GML) 3.2

The Web Feature Service (WFS) and Filter Encoding (FE) are also ISO standards[4].

5.1.2 *The Internet Engineering Task Force (IETF)*

The Internet Engineering Task Force (IETF) is a large open international community of network designers, operators, vendors, and researchers concerned with the evolution of the internet architecture and the smooth operation of the internet. The primary task of the Geo-PRIV WG is to assess the authorization, integrity and privacy requirements that must be met in order to transfer location information, or authorize the release or representation of such information through an agent as a part of the internet infrastructure.

In addition, the working group selected an already standardized format to recommend for use in representing location per se[5]. A key task has been to enhance various and protocols using the enhanced location encoding to ensure that the security and privacy methods are available to diverse location-aware applications.

In terms of the OGC standards, the GeoPRIV community developed (in concert with this author) a GML application schema for encoding location payloads (Thomson & Reed 2007). This schema was designed to be used with any number of existing, deployed internet standards, such as the Session Initiation Protocol (SIP) and the Presence Information Data Format (PIDF).

5.1.3 *OASIS (Organization for the Advancement of Structured Information Standards)*

OASIS and the OGC have a formal memorandum of understanding. There is extensive synergy between the work of the OGC and the work being done in OASIS. The OASIS community works a variety of standards development activities focused on the requirements of the business and e-government community. The OGC actively participates in a number of OASIS technical committees, such as Emergency Services, Service Oriented Architecture, and Energy Market Information Exchange. Further, the OGC tests and defines best practices for the use of OGC standards with specific OASIS standards, such as BPEL (Business Process Execution Language) and ebXML Registry. Finally, the OGC has defined a geospatial extension for the eXtensible Access Control Markup Language (XACML) (Matheus & Hermann 2008).

5.1.4 *Open Grid Forum (OGF)*

In 2007, the OGC and the Open Grid Forum (OGF) signed a memorandum of understanding to collaborate. The primary goal of the MoU has a suite of open interfaces, encodings and standards for managing and presenting geospatial data and wants to ensure that these standards "work" with the capability for distributed resource management, i.e. grids. The initial goals of the collaboration include:

- Integrate OGC's Web Processing Service (WPS) Standard with a range of "back-end" processing environments to enable large-scale processing. The WPS could also be used as a front-end to interface to multiple grid infrastructures, such as the United Kingdom's National Grid Service. This would be an application driver for both grid and data interoperability issues.

---

[4] October 2010.
[5] The IETF developed a GML application schema for the mandatory payload encoding of location elements for a variety of IETF standards. This application schema was submitted to the OGC and approved as an OGC Best Practice.

- Integration of WPS with workflow management tools.
- Integration of OGC federated catalogs/data repositories with grid data movement tools.

WPS is just a starting point for the collaboration. The real goal is to do more than scientific research; it is to greatly enhance operational mapping such as hurricane forecasting and location-based services.

5.2 *Collaboration with information communities*

The OGC does not have expertise in all domains. An example is the geology information domain. Therefore, the OGC strives to maintain relationships and/or collaborations with other information communities to insure that community's requirements are understood in the OGC and that these information communities can use OGC standards.

As an example, a need was identified in geology community to be able to query and exchange digital geological information between data providers and users. To meet this community objective, an international working group comprised of many national level geology organizations developed a conceptual geoscience data model. Development and approval of the data model required the participants to agree on vocabularies and semantics.

One the initial content model was agreed to, the next step was to map the conceptual data model to an encoding that supports interchange. The GeoSciML application schema provides a framework for application-neutral encoding of geoscience thematic data and related spatial data (GeoSciML 2009 and CSIRO 2010). The GeoSciML schema is based on the OGC Geography Markup Language (also ISO DIS 19136) for representation of features and geometry, and the Open Geospatial Consortium (OGC) Observations and Measurements standard for observational data. Geoscience-specific aspects of the schema are based on a conceptual model for geoscience concepts and include geologic unit, geologic structure, and Earth material from the North America Data Model (NADMC1, 2004), and borehole information from the eXploration and Mining Markup Language (XMML). Development of controlled vocabulary resources for specifying content to realize semantic data interoperability continues.

Intended uses are for data portals publishing data for customers in GeoSciML, for interchanging data between organizations that use different database implementations and software/systems environments, and in particular for use in geoscience web services. Thus, GeoSciML allows applications to utilize globally distributed geoscience data and information.

GeoSciML is *not* a database structure. GeoSciML defines a format for data interchange. Agencies can provide a GeoSciML *interface* onto their existing data base systems, with no restructuring of internal databases required. Figure 2 depicts how a single client can access geological structure content from multiple WFS instances deployed by a number of national Geological Surveys using different vendor platforms.

5.3 *Collaboration with the open source community*

The collaborative relationship between the OGC and the Free and Open Source Software for Geospatial (FOSS4G) community has evolved over the past ten years (FOSS4G 2010). Today, many FOSS4G projects use OGC standards in their applications. More importantly, the geospatial open source community has made significant contributions to numerous OGC standards, including the OGC Web Map Service, Web Map Context, Style Layer Descriptor, and Catalogue standards. Also, as part of the OGC Compliance Testing Program, the FOSS4G community has provided the Reference Implementations for a number of OGC standards. In 2008, the OGC and OSGeo signed a formal Memorandum of Understanding that codifies the relationship and the joint activities that will most benefit the geospatial community (OGC 2009a).

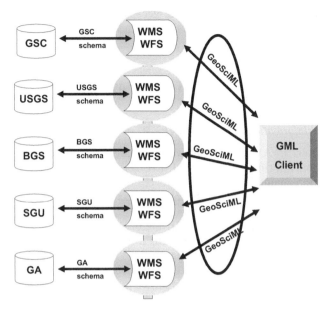

Figure 2.   Architecture of the GeoSciML[6] Test Bed 2.

6   RETURN ON INVESTMENT (ROI) THE VALUE AND BENEFIT OF GEOSPATIAL STANDARDS

There is considerable evidence that standards reduce risk, protect legacy investments, reduce long term maintenance costs, and increase choice and implementation flexibility (NASA 2005).

The NASA Geospatial Standards ROI study documented that projects implementing geospatial interoperability standards had a risk-adjusted ROI, or "Savings to Investment" ratio, of 119.0 percent throughout the five-year project life cycle. Looking over a 10-year project life cycle, a standards based project had a risk-adjusted ROI of 163.0 percent. Therefore, a standard's based project saved 26.2 percent when compared to the project that relied on a proprietary approach. One way to express this result is by saying that for every $100 million spent on projects based on proprietary platforms, the same value could have been achieved with $75 million if the projects had been based on open standards. Other key findings and results from the study included:

- Standards-based projects have lower maintenance & operation (M&O) costs than those that rely exclusively on proprietary products for data exchange. In fact, the majority of Case 2 costs were M&O (89 percent). This cost category is exposed to the greatest risk over time because it is inflexible and not extensible.
- Standards-based projects have greater first-time system planning and development costs, but future projects using the same standards will have significantly reduced planning and development costs. This is because standards and specifications have already been adopted.
- The value generated by the open standards solution was greater than expected. The open solution returned 55 percent more value to its stakeholders than did the proprietary solution. As a result, the open solution would be preferable even if its costs were higher than the proprietary solution.

In short, standards help to form an information culture and information economy that is content-rich and diverse in viewpoint. By clarifying functions, service invocations, and data

---

[6] Slide courtesy of the OGC.

definitions, standards make the distribution of geospatial information understandable—not just for government technologists, managers, and decision support analysts, but for all stakeholders, including industry partners.

7 EXAMPLES OF ENTERPRISE PORTAL APPLICATIONS USING OGC STANDARDS

7.1 *Shared Land Information Platform (SLIP) Western Australia*

In Western Australia (WA), approximately 2000 people per month are now logging on to the Shared Land Information Platform (SLIP 2005) to access more than 20 gigabytes of land information maintained by many government departments. By so doing, users eliminate having to visit and deal with the multiple agencies otherwise needed to gather this information.

Nineteen government agencies and two private organizations are currently connected to the SLIP network, bringing together and making available some 200 vector spatial datasets and over 1000 imagery datasets that in the past resided in isolated systems.

The SLIP initiative stems from the Western Australian Government e-initiative and recommendation of the Functional Review Committee in 2003 to develop a common information framework that would allow agencies to make available and share their information online. Western Australia Land Information System (WALIS) contracted OGC Australia to study needs, requirements, etc., and develop a reference architecture for SLIP based on open standards and specifications, so that SLIP would produce a 'connected government', reduce capital expenditure, avoid duplication of services and make information readily available to government agencies, businesses and citizens from just one place.

SLIP has these objectives:

- Simplify access to the government's valuable collections of land and geographic information;
- Improve the efficiency of government business processes and enable better decision-making;
- Promote better integration across government; and
- Facilitate the development of new applications to meet the changing requirements of the general community and business.

Landgate, Western Australia's land and property information agency, is the lead agency for SLIP's development.

Currently, SLIP utilizes the OGC WMS, WFS, GML and KML standards. The *Global Spatial Data Infrastructure Association* (GSDI) has asked that the *SLIP Enabler* architecture and collaboration model be included within its Cookbook which profiles best-practice systems.

7.2 *Heidelberg-3D uses OGC standards to model city*

A three-dimensional Spatial Data Infrastructure (SDI) for the city of Heidelberg is freely available as part of Project GDI3D (GDI3D 2011). The goal of the project, funded by the Klaus-Tschira-Foundation in Heidelberg, is to develop new technologies based on existing standards for the interoperable processing, visualization and analysis of 3D city and landscape models. The system implements the candidate standard OpenGIS Web 3D Service (OGC W3DS) and implements a number of other OGC standards (Fig. 3):

- Geocoder, part of the OGC OpenLS Utility Service
- OGC Web Processing Service (WPS)
- OGC Sensor Observation Service (SOS)
- OGC OpenLS Directory Service
- OGC Web Map Service (WMS)

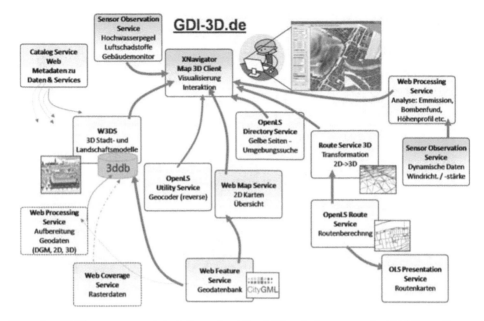

Figure 3. OGC components of the 3-dimensional Spatial Data Infrastructure for Heidelberg. Courtesy OGC: GEOSS AIP Phase 3[7].

- OGC Web Feature Service (WFS)
- OGC Catalogue Service (CS-W)
- OGC CityGML
- OGC Symbology Encoding Specification

The system comes with a free 3D-Client called XNavigator. This client is a Java WebStart-Application. The 3D activity and the use of CityGML is a key component of the German mandated initiative to model noise propagation from transportation systems and to define project to reduce noise to foster a more livable environment.

The 3D model has been generated in very close cooperation with the surveying office of the city of Heidelberg. Additional content is available from other project partners or has been digitized within the project itself. Most of the data has been generated from 2D data automatically. The system includes the complete set of approximately 40,000 buildings of the City of Heidelberg that can be accessed from the web. Available data includes a 5- meter DEM (digital elevation model), land use, aerial photographs, street names and a range of other data such as historic flood areas, parking lots, street and traffic signs, trees and even a detailed model of the Virgin Mary that has been generated from terrestrial laser scan. Important buildings and bridges have been modeled in higher detail and have textures (Fig. 4). All data is preprocessed for efficient streaming and is managed within a database.

7.3 Canada Geospatial Data Infrastructure (CGDI)

The Canadian Geospatial Data Infrastructure (CGDI) provides both a technical framework for building the national SDI in Canada as well as a very well defined governance structure for their SDI. The use of OGC standards is guided by the framework and governance infrastructure. Geoconnections Canada facilitates the development and governance of the CGDI (CGDI 2008, Geoconnections 2010). Within the CGDI there is a group called the Technical

---

[7] *portal.opengeospatial.org/files/?artifact_id=38173.*

343

Figure 4.   OpenLS Route Service enables 3D-Routing with height profile and animation—including routes over bridges. Courtesy Heidelberg 3d[8].

Advisory Panel (TAP). This group evaluates and endorses each standard to be used in the CGDI. Further, they define which version of the standard as well as other pertinent technical information that is used to enhance interoperability.

As an example, consider the OGC Web Map Service (WMS). The TAP initially endorsed WMS in 2001. Since then, the TAP has endorsed more recent versions of the WMS standard, specifically WMS 1.1. Along with the endorsement, the CGDI community has agreed to a CGDI profile that documents that mandatory elementsthat must be implemented for any CGDI compliant WMS implementation. By implementing the mandatory elements based on the CGDI profile, the compliant WMS can be easily plugged into the CGDI infrastructure.

CGDI currently provides users with online access to hundreds of geographic databases repositories throughout the Canada. Local and regional SDIs can be linked into the national SDI. This is all accomplished by using the standards based framework as defined by the governance policies.

The CGDI also standardizes the way information in many of these databases is stored, accessed, and presented online. As with the interface standards, these standards are reviewed and endorsed by GeoConnections—the national partnership program, led by Natural Resources Canada, that promotes the CGDI's use and growth. The use of these standards is consistent with and often applies to geospatial data infrastructures run by other countries. Consequently, data from standards-based Canadian and international sources are compatible.

Any application that implements one or more of the endorsed OGC standards can quickly and easily integrate their geospatial data repositories and related services directly into the CGDI.

8   COMPLIANCE AND INTEROPERABILITY TESTING

ISO-Standards and OGC abstract specifications provide a conceptual or abstract model, while OGC standards provide implementation rules and guidance. However, simply implementing

---

[8] http://www.heidelberg-3d.de/screenshots.en.htm

an OGC standard does not guarantee even a minimum level of interoperability between two different vendor solutions. Therefore, the OCG has for several years offered a compliance testing program (OGC 2010a).

Software products can be "compliant with" or "implementing" OGC standards. If software is "compliant with" an OGC standard, this means that it has passed a formal compliance test. A compliance test determines that a product implementation of a particular OGC Standard fulfills all mandatory elements as specified in the standard and that these elements are operable. This is the first step in ensuring interoperability. At the least, a product that has passed compliance testing for a given OGC standard provides key information to a user or buyer of the software that implements that standard. However, compliance testing will not ensure, or even test, interoperability of software products. Plug fests are required to test whether two or more product implementations of a given standard are interoperable.

The OGC provides a compliance testing program and framework. The purpose of the OGC Compliance Testing Program is to permit vendors and users to take advantage of the standards that OGC has created. The program provides a process for testing compliance of products that implements one or more OGC standards.

When a vendor has completed compliance testing and OGC has confirmed its successful completion, vendors who agree to the terms of the OGC Trademark License Agreement that accompanies this program, and who have paid their trademark license fees, may use OGC's marks (trademarks or certification marks) to indicate to their customers that they have achieved compliance with a given OGC standard.

OGC compliance tests are available for some but not all OGC standards. OGC is implementing its Compliance Testing Program in phases in step with the maturing of its standards baseline.

9   VERY RECENT AND NEAR TERM FUTURE ACTIVITIES

Standards work in the geospatial community is far from done. As more applications implement OGC standards, new interoperability gaps and requirements are identified and submitted to the OGC membership for consideration. Changes in the IT infrastructure and market forces shape the standards work of the OGC. Finally, the ability for information communities to obtain consensus on domain vocabularies and semantics is becoming increasingly important for achieving the ability for global scientific and research communities to share content and processing models.

The following subsections describe recent standards trends shaping the future work of the OGC.

9.1   *Sharing of aeronautical information*

The Aviation Information Management (AIM) activity is a new OGC effort to develop and demonstrate the use of the Aeronautical Information Exchange Model (AIXM) in an OGC Web Services environment. AIXM was developed by the US Federal Aviation Administration (FAA) and Eurocontrol as a global standard for the representation and exchange of aeronautical information. It was designed as a basis for digital aeronautical information exchange and for enabling the transition to a net-centric, global aeronautical management capability. AIXM has been developed using the ISO 19100 modeling framework and has two major components: a conceptual model presented in the form of an UML class model and a data encoding specification which was developed using the OGC Geography Markup Language (GML). Current research is on the use and enhancement of Web Feature Service and Filter Encoding specifications in support of AIXM features and 4-dimensional flight trajectory queries, the definition of a prototype of Aviation client for retrieval and seamless visualization of AIXM, Weather and other aviation-related data, emphasizing time and spatial filtering in order to present just the right information into a given user context anytime, anywhere, and defining an architecture of standards-based mechanism to notify users of changes to user-selected aeronautical information.

9.2 *Hydrology domain working group*

The purpose of the Hydrology DWG is to provide a venue and mechanism for seeking technical and institutional solutions to the challenge of describing and exchanging data describing the state and location of water resources, both above and below the ground surface. The path to adoption will be through OGC papers and standards, advanced to ISO where appropriate, and also through the World Meteorological Organization's (WMO) and it's Commission for Hydrology (CHy) and Information Systems (WIS) activities. While CHy has the recognized mandate to publish and promote standards in this area, OGC contributes to the process with its resources and experience in guiding collaborative development among disparate participants in a rapidly evolving technological environment. It is proposed that the OGC Hydrology DWG will provide a means of developing candidate standards for submission to ISO and for adoption by CHy as appropriate.

9.3 *Meteorology domain working group*

The purpose of the OGC Meteorology DWG is to provide an open forum for work on meteorological data interoperability, and a route to publication through OGC's standards ladder (Discussion paper/Best Practice/Standard, and, if appropriate, to ISO status), thence giving a route for submission to the WMO CBS for adoption. While the WMO CBS has a mandate to promote standards in this area, OGC is able to contribute to this process with its resources and experience in guiding collaborative development among disparate participants in a rapidly evolving technological milieu.

Numerous organisations within the meteorological community, both operational and research, have demonstrated that the OGC standards are effective in meeting their needs. However, meteorology has a number of complexities not originally considered by the geographic community. As a result, many meteorological implementations of OGC standards have chosen different mechanisms to mitigate the complexities and shortcomings in the standards, thus inhibiting interoperability across the community.

Furthermore, the European Commission's INSPIRE Directive also seeks to create an inclusive framework for exchange of geographic, or 'spatial', information throughout Europe. INSPIRE includes meteorology amongst the many 'thematic domains' of geographic information, thus providing a legislative driver to achieve commonality of approach at least within Europe.

10   CONCLUSIONS

Back in the early 1990's GIS data silos and proprietary data formats were normal. Procurement language mandated the ability to read and write these proprietary formats. Competition and market creativity and growth was stifled. A number of government and private sector organizations believed that the ability to more easily share geospatial data was critical to the long-term growth and viability of the industry. From these early discussions, the OGC was born and work began on standards that enabled geospatial interoperability. Today, there is a solid framework of standards from not just the OGC but also from other standards organizations. This framework provides encoding, modeling, and service interface standards. These standards are allowing the evolution of the geospatial web. From low level internet standards to high level client visualization standards, the consistent expression of geospatial content has opened new markets, increased productivity, reduced risk, and enhanced market competition.

Today, there is not a geospatial technology provider or user that does not use some form of geo-enabled web application. The web has opened the use of geospatial content and services to hundreds of millions of users.

There are currently hundreds of operational applications that implement OGC standards and the work continues on evolving and expanding the geospatial web. But the work of the OGC Membership is not done. Work continues on a new generation of standards that reflect market requirements and the ever changing nature of the IT infrastructure.

REFERENCES

CGDI. (2008) Canada Geospatial Data Infrastructure Mandatory Elements for WMS Implementation. GeoConnections, [Online] Available from: http://geoconnections.org/architecture/technical/specifications/wms/CGDI_WMS_profile_v1.1.xls, [Last accessed 26th January 2011].

CSIRO. (2010) Solid Earth and Environment GRID. Interoperability Working Group of the Commission for the Management and Application of Geoscience Information, a commission of the International Union of Geological Sciences, [Online] Available from: https://www.seegrid.csiro.au/twiki/bin/view/CGIModel/InteroperabilityWG

Diez, L., Granell, C. & Gould, M. (2008) Case study: Geospatial Processing Services for Web-bases Hydrology Applications. In *Geospatial Services and Applications for the Internet*. Springer, 2008.

FOSS4G. (2010) Free and Open Source for Geospatial Conference. OSGeo, [Online] Available from: http://foss4 g.org/static/index.html, [Last accessed 26th January 2011].

GDI3D. (2011) Spatial Data Infrastructure for 3D Data. University of Heidelberg, [Online] Available from: http://www.gdi-3d.de/, [Last accessed 26th January 2011].

Geoconnections. (2010) GeoConnections: Mapping the future together online. GeoConnections Canada, [Online] Available from: http://www.geoconnections.org/Welcome.do, [Last accessed 26th January 2011].

GeoSciML. (2009) Why do we need GeoSciML. [Online] Available from: http://www.cgi-iugs.org/tech_collaboration/docs/Why_do_we_need_GeoSciML_v1.doc

GML. (2007) OGC Geography Markup Language 3.2.1. Open Geospatial Consortium, [Online] Available from: http://www.opengeospatial.org/standards/gml

ISO DIS 19136. (2007) Geography Markup Language. International Standards Organization. [Online] Available from: http://www.iso.org/iso/catalogue_detail.htm?csnumber=32554, [Last accessed 30th January 30, 2011].

Liebhold, M. (2005) The Geospatial Web: A Call to Action - What We Still Need to Build for an Insanely Cool Open Geospatial Web. In Schuyler, Erle, Rich Gibson & Jo Walsh, (2005), *Mapping Hacks*, O'Reilly Media, ISBN: 978-0-596-00703-4.

Matheus, A. & Hermann, J. (2008) Geospatial eXtensible Access Control Markup Language (GeoXACML). Open Geospatial Consortium, OGC 07-026r2.

NASA. (2005) *Geospatial Interoperability Return on Investment Study*. National Aeronautics and Space Administration Geospatial Interoperability Office. Booz, Allen, Hamilton, [Online] Available from: http://gio.gsfc.nasa.gov/docs/ROI/Study.pdf

OGC. (1997) OpenGIS® Implementation Specification for Geographic information - Simple feature access - Part 1: Common architecture. Open Geospatial Consortium.

OGC. (2009a) OGC and OSGeo Sign Memorandum of Understanding. Open Geospatial Consortium, [Online] Available from: http://www.opengeospatial.org/pressroom/pressreleases/944

OGC. (2009b) OGC Web Services Testbed 7.0. Open Geospatial Consortium, [Online] Available from: http://www.opengeospatial.org/standards/requests/60, [Last accessed 30th January 2011].

OGC. (2009c) The Specification Model — A Standard for Modular Specifications. Policy Standards Working Group (ed.), Open Geospatial Consortium, [Online] Available from: https://portal.opengeospatial.org/files/?artifact_id=34762.

OGC. (2010a) OGC Compliance Testing. Open Geospatial Consortium, [Online] Available from: http://www.opengeospatial.org/compliance

OGC. (2010b) OGC Implementation Standards documents. Open Geospatial Consortium, [Online] Available from: http://www.opengeospatial.org/standards/is

Percivall, G., Reed, C., Leinenweber, L., Tucker, C. & Cary, T. (2008) The OGC Reference Model (ORM). Open Geospatial Consortium, OGC 06-062.r4, [Online] Available from: http://www.opengeospatial.org/standards/orm

Reed, C. (2010) OGC Technical Committee Policies and Procedures. OGC. [Online] Available from: http://portal.opengeospatial.org/files/?artifact_id=23325

Scharl, Arno & Klaus Tochtermann, (eds.). (2007) *The Geospatial Web: How Geobrowsers, Social Software and the Web 2.0 are Shaping the Network Society*. Advanced Information and Knowledge Processing Series 2007, London: Springer, ISBN 1-84628-826-6.

SLIP. (2005) Shared Land Information Platform – SLIP. Landgate, Western Australia, [Online] Available from: https://www2.landgate.wa.gov.au/slip/portal/home/home.html

Thomson, M. & Reed, C. (2007) GML 3.1.1 - PIDF-LO Shape Application Schema for use by the Internet Engineering Task Force (IETF). Open Geospatial Consortium, OGC 06-142r1.

*Advances in Web-based GIS, Mapping Services
and Applications – Li, Dragićević & Veenendaal (eds)*
© 2011 Taylor & Francis Group, London, ISBN 978-0-415-80483-7

Vector data formats in internet based geoservices

Franz-Josef Behr
*Faculty of Geomatics, Computer Science and Mathematics, Stuttgart University of Applied Sciences,
Stuttgart, Germany*

Kai Holschuh
Marketing and Intercultural Management, Karlshochschule International University, Karlsruhe, Germany

Detlev Wagner
*Faculty of Geomatics, Computer Science and Mathematics, Stuttgart University of Applied Sciences,
Stuttgart, Germany*

Rita Zlotnikova
Blom Romania, I.H. Radulescu, Jud Dambovita, Romania

ABSTRACT: A diverse range of formats and standards are used to exchange and present
2D location awareness information on the internet. The emphasis of this chapter is on the
eXtensible Markup Language (XML), the basis of many (web) geographic information system (GIS) standards and services. The background, structure and elements of the related
languages are introduced, and their relationships to new tools, application programming
interfaces (APIs) and associated areas of application are also described. Throughout the text
examples are used to explain data interchange formats. The illustrations of code are mostly
in XML, the rest in WKT, JSON and GeoJSON. The chapter concludes with a summary of
usefulness of the different standards to data interoperability on the internet.

Keywords: Internet, web GIS, interoperability, data structures, standards, spatial infrastructure, representation, format

1   INTRODUCTION

In recent years, interest in distributed web services has increased within information technology circles. Internet-based geospatial applications are an integral element of the Web 2.0
developments. Web-based geographic information systems (Web GIS) provide the navigational tools on which many distributed web services are based and enhance the appeal and
usefulness of countless others.

The aim of this chapter is to give an overview of the diverse formats and standards used
to exchange and present 2D location awareness information on the Internet. Emphasis is on
the eXtensible Markup Language (XML), the basis of many Web GIS standards. The background, structure and elements of these languages are introduced, and their relationship to
new tools, application programming interfaces (APIs) and associated areas of application
are also described.

While XML plays a leading role in the creation of geospatial web services today, especially
in the establishment of open geospatial standards, such as the Web Map Service (WMS) and
the Web Feature Service (WFS) of the Open Geospatial Consortium (OGC), other related
standards and formats have also been authorized by the World Wide Web Consortium
(W3C, http://w3.org/) and are regularly used in Web GIS. The definitions and applications

of these meta-languages, including XML Schema Definitions (XSD), XML Namespaces, XML Linking Language (XLink), XML Pointer, Extensible Stylesheet Language with XSL Transformations, XML Path Language and XSL Formatting Objects, in combination with Cascading Stylesheets (CSS), are reviewed in this chapter as well. Moreover, since the combination of Extensible Hypertext Markup Language (XHTML) with complementary technologies, such as Scalable Vector Graphics (SVG), has become an essential part of Web GIS as well, it is also covered.

Throughout the text examples are used to explain data interchange formats. The illustrations of programming code are mostly in XML, the rest in WKT, JSON and GeoJSON. The chapter concludes with a summary of usefulness of the different standards to present geographic information on the Internet.

2 EXTENSIBLE MARKUP LANGUAGE

The eXtensible Markup Language (XML Bray et al. 2006, Bray et al. 2008) is fundamental to geoinformatics. It is used to define data objects, structure data, provide metadata, assign attributes and other information, as well as store configuration parameters.

2.1 Introduction to XML elements and objects

Before considering data objects, let us look at the parts that make up an XML element. All types of alphanumerical data can be identified and put together as an *element* which is simply a way to structure information within a pre-defined namespace. Figure 1 shows the basic three-part structure of an XML element: *start tag*, *element content* and *end tag*. Together, the three parts form a "nested" structure. The element name appears in both tags, in the latter preceded by a slash character. These two tags bracket the element content. The content can be either empty, contain text, and / or contain other XML elements.

For the sake of brevity, an empty element may be written in the following abbreviated form:

```
<name />
```

Element content, however, is subject to some restrictions. It may not contain certain characters reserved for the markup language, such as < or &. These must then be coded as *entities*, for example, < instead of <.

In addition to being part of the element content, information may also be provided in the form of attributes in the start tag. Here is an example:

```
<text x="15" y="135">AbcDef</text>
```

Some XML-based languages, such as Scalable Vector Graphics (SVG, explained in section 3.8.2) often include information as attributes in start tags.

Attributes are made up of pairs of keys and values (KVPs) separated by an equal sign. Keys are alphanumeric symbols (letters and numerals) chosen to label aspects of the

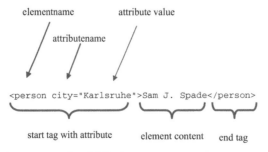

Figure 1. Basic structure of an XML element.

properties of an element, and values are the specific names or numbers the keys represent. Values must be embedded in single or double quotation marks. In Figure 1, `"city"` is the name of an attribute corresponding to the element `"person"` and with the attribute value `"Karlsruhe"`.

Another important feature of XML is that one element may itself contain text and other elements. This characteristic enables elements to support additional and more interesting tasks, but it also increases the complexity of revising and correcting compositions. Hence, it is essential that so-called "nesting" be carried out methodically. Inconsistency in structuring elements not only leads to errors at inception, but inhibits the ability to process the XML instance later.

Elements are principal building blocks of XML data objects. The element that encloses the other elements used to document a data object is called the *root element*. In other words, the code which introduces and finalizes a particular document is actually just one element that surrounds all the other elements in a document.

XML objects *per se* are difficult to define in a concise manner. Comprehension requires familiarity with the nuances of their formation and application. Therefore it is all the more important to gain an understanding of the basic aspects of XML data objects in order to begin to understand their usage in geodata interchange.

In general, when an XML data object is coded according to the conventions used to define elements as indicated above, it is considered "well-formed" and called an *XML document*. More specifically, an XML object consists of:

– a prolog, the so called XML declaration;
– a root element and embedded child elements;
– a linked or embedded Document Type Declaration (DTD) or, alternatively, a linked XML Schema defining the elements, their attributes and hierarchical structure (these are optional and not a prerequisite of an XML object; see section 2.4);
– processing instructions with information on programs used to process the XML object (optional);
– further comments (optional);
– CDATA sections, containing blocks of information, such as JavaScript code, hidden from XML parsers (optional).

The lines of code shown below illustrate the various parts of an XML object. In line 1 the XML declaration (prolog) is given. This indicates the version of XML employed, here 1.0, and the specific character encoding used, in this case the ISO-8859-1 codification of the Latin alphabet. The declaration may also indicate if the document can be considered "stand alone", i.e. is valid without further declarations.

```
 1: <?xml version="1.0" encoding="ISO-8859-1" standalone="yes"?>
 2: <?xml-stylesheet href="point.xsl" type="text/xsl" ?>
 3: <!DOCTYPE point [
 4: <!ELEMENT point (easting, northing, d)>
 5: <!ELEMENT easting (#PCDATA)>
 6: <!ELEMENT northing (#PCDATA)>
 7: <!ELEMENT id (#PCDATA)>
 8: ]>
 9: <point>
10:   <easting>35.0</easting>
11:   <northing>54.0</northing>
12:   <id> 1</id>
13: </point>
```

In lines 3 through 8 the *Document Type Declaration* is defined, indicating how an XML instance of a point element is laid out. In the example, the association to the root element, as defined by the contents of elements easting, northing, and id, is given. Here, these represent geographic coordinates, and identify the element with a unique id. In terms of the syntax used here, they are PCDATA, i.e. character data[1].

Lines 9 through 13 contain the root element, indicating a specific geographic spot. The contents of this element, as specified in line 4, are the child elements containing the actual values of the geographic coordinates and the id.

Instructions on how to process information (*processing instruction*, given in line 2) bring additional programs into play. Such applications can handle tasks involving the style of graphical presentation or to execute an XSL Transformation (XSLT, to be discussed below).

An important part of programming and markup languages are the comments written in source code. These are ignored by a parser. In XML-based languages, a comment may look like the following:

```
<!-- This is a comment -->
```

Any alphanumeric character or other symbol may be employed between the divergent tags (`<--` and `-->`) that frame a comment except the string double-hyphen (`"--"`).

2.2 *XML related languages and concepts*

The term XML stands for a large group of different markup languages and concepts, the vast majority developed or coordinated by the World Wide Web Consortium. Of these, the most important concepts and standards used in the field of geoinformatics are:

- XML Namespaces,
- Document Object Model (DOM; World Wide Web Consortium 2005),
- Document Type Declaration and XML Schema,
- XLink – XML Linking Language and XML Pointer,
- Scalable Vector Graphics (SVG),
- Cascading Stylesheets (CSS) and the Extensible Stylesheet group of languages.

XML is the basis of additional standards like the Geography Markup Language and other specific application extensions used in geoinformatics. XML is also employed in communication interfaces for geo-web services, such as Web Map Service (Open Geospatial Consortium 2004), Web Feature Service (Open Geospatial Consortium 2002) and Coordinate Transformation Service (Open Geospatial Consortium 2001).

2.3 *XML namespaces*

In general, the designation of element names in XML is unrestricted, allowing multiple ambiguous uses of the same name. For instance, the element "title" could be used to identify persons within an organization as well as indicate songs on a CD recording.

XML namespaces (http://www.w3.org/TR/REC-xml-names/) provide the means to specify element and attribute names in a simple and unequivocal manner. Names are connected to namespaces (vocabularies) by a prefix through unique Internationalized Resource Identifier (IRI) references[2]. Thereby, even if they belong to different schemas, through the definition and use of prefixes, elements can have *qualified names* and can be used simultaneously. For example, the namespaces `"tkfd"`, `"gml"` and `"xsi"`, indicated below, are given for the TKFD root element[3] and following elements. The URI reference "http://www.lv-bw.de/tkfd" identifies the namespace tkfd which, in turn, connects to the elements `"TrainStation"`, `"objectType"` and `"id"`. Similarly, the elements `"centerOf"`, `"Point"` and `"pos"` are derived from the namespace `gml` which is typically used for the Geography Markup Language (discussed in section 3).

---

[1] An acronym derived from "parsed character data."
[2] formerly called Unified Resourxce Identifier (URI).
[3] TKFD: Akronym for Thematische Kartenfachdaten (Thematic Cartographic Data, Graf 2004).

```
<?xml version="1.0" encoding="UTF-8"?>
<tkfd:TKFD xmlns:tkfd="http://www.lv-bw.de/tkfd"
  xmlns:gml="http://www.opengis.net/gml"
  xmlns:xsi="http://www.w3.org/2001/XMLSchema-instance"
  xsi:schemaLocation="http://www.lv-bw.de/tkfd">
  <tkfd:TrainStation>
  <tkfd:objectType tkfd:id="EZ00VPK">9201</tkfd:objectType>
    <gml:centerOf>
      <gml:Point>
        <gml:pos>3515955.37 5409276.28</gml:pos>
      </gml:Point>
    </gml:centerOf>
  </tkfd:TrainStation>
</tkfd:TKFD>
```

A namespace without declared namespace prefix can be used as a *default namespace*, associated to all elements without qualifying prefix.

2.4   *Document type declaration and XML schema*

A *Document Type Declaration* (DTD) defines elements according to their attributes and hierarchical structure. DTDs were used in the past in HTML, SVG and GML 1.0, but today this form of element declaration is considered inadequate and is being replaced by the *XML Schema Definitions* (XSD, http://www.w3.org/XML/Schema) which permit significantly more detailed classifications of objects. As compared to DTD however, developing an XML Schema is more complex, principally due to the ability to build on other, previously defined elements.

An XML schema essentially operates on two levels. At the core, the root element of the schema contains namespace declarations and the attribute of the target namespace. This root element is then supported by a structure of geo-objects designated through their own specific elements. Thereby, whether simple and pre-defined or complex and self-defined, data types can be used on the basis of the same well-established concepts as used in object-oriented programming languages.

An XML object, once it is defined by an XML schema, is called an *instance document* or an *XML instance*. An XML instance is usually a file, but can also be a stream of data emitted by a network service or a specific value as indicated in a designated field of a given database. Schema definitions can be found in many kinds of geospatial standards and data models and they play an essential role in all types of services designated by the Open Geospatial Consortium (OGC).

If an XML object is well-formed and complies with the constraints expressed in its DTD or its XML Schema it is considered "valid".

2.4.1   *Data types*

The basic data types used in XML Schema are the same as those used in other programming languages. These data types include string, decimal, those used for date, time, Uniform Resource Locators (URLs), and so on (http://www.w3.org/TR/xmlschema-0/#CreatDt). Elements are then easy to define. For example, an element named Easting can be indicated as having a data type double defined in the namespace xs:

```
<xs:element name="Easting" type="xs:double"/>
```

With *facets* it is possible to create new data types by restricting existing ones. Emphasis is on those aspects which differ from other data types. A new data type called "featureClassIdType", which supports only the integer values 9401, 9402, and 9403, would be documented as follows:

```
<xs:simpleType name="featureClassIdType">
  <xs:restriction base="xs:integer">
    <xs:enumeration value="9401"/>
    <xs:enumeration value="9402"/>
    <xs:enumeration value="9403"/>
  </xs:restriction>
</xs:simpleType>
```

Or, as shown below, geometryType can be designated as a string limited to three values:

```
<xs:simpleType name="geometryType">
  <xs:restriction base="xs:string">
    <xs:enumeration value="Point"/>
    <xs:enumeration value="Line"/>
    <xs:enumeration value="Area"/>
  </xs:restriction>
</xs:simpleType>
```

New complex types may be developed based on previously defined types. The data type "PointType" below, equivalent to the DTD in section 2.1, has two properties, which describe the location of a point. The third element indicates the id used to identify the object:

```
<xs:complexType name="PointType">
  <xs:sequence>
    <xs:element name="Easting" type="xs:double"/>
    <xs:element name="Northing" type="xs:double"/>
    <xs:element name="id" type="xs:nonNegativeInteger"/>
  </xs:sequence>
</xs:complexType>
```

2.4.2   *Deriving types by extension*
In GML it is common that complex data types are defined by extending previously defined data types. Below, the data type "TrainStationType" extends AbstractFeatureType (obtained from the gml namespace) by adding the properties "stationName" and "centerOf".

```
<xs:complexType name="TrainStationType">
  <xs:complexContent>
    <xs:extension base="gml:AbstractFeatureType">
      <xs:sequence>
        <xs:element name="stationName" type="xs:string">
        <xs:element ref="gml:centerOf"/>
      </xs:sequence>
    </xs:extension>
  </xs:complexContent>
</xs:complexType>
```

Moreover, it is possible to define the kinds of derivations to be used in instance documents. Similar to polymorphism in object-oriented programming languages, derivated features can be used instead of the original ones, indicated by the substitutionGroup attribute value. Below, first the element "Station" with the data type "StationType" is declared. Then, according to the attribute "substitutionGroup", this element can be used as a substitute whenever the abstract element "gml:_Feature" is used[4]:

```
<xs:element name="TrainStation" type="tkfd:TrainStationType"
substitutionGroup="gml:_Feature"/>
```

_____

[4]The underlined space in "gml:_Feature" indicates that the element is abstract.

The specification of "BathingPond" below provides a more detailed example of XML Schema. As usual, the first element is made up of two parts: the declaration and the type definition. The declaration includes the name (BathingPond) and data type (BathingPondType). The type definition, also called *content model*, describes the structure of how BathingPond is specified. The value "gml:_Feature" of "substitutionGroup" indicates that occurrences of gml:_Feature can be substituted by BathingPond. Specifically, the data type "tkfd:BathingPondType" is derived from the abstract data type "AbstractFeatureType", which provides general properties for GML features. In this example, "gml:_Feature" is an element of this data type.

```
<xs:element name="BathingPond" type="tkfd:BathingPondType"
substitutionGroup="gml:_Feature"/>
<xs:complexType name="BathingPondType">
  <xs:complexContent>
    <xs:extension base="gml:AbstractFeatureType">
      <xs:sequence>
        <xs:element name="objectType" type="xs:string">
        </xs:element>
        <xs:element name="tkn" type="xs:string">
        </xs:element>
        <xs:element ref="gml:centerOf"/>
      </xs:sequence>
    </xs:extension>
  </xs:complexContent>
</xs:complexType>
```

2.5 *XML linking language and XML pointer language*

XML Linking Language (XLink; http://www.w3.org/XML/Linking) allows elements to be added to XML documents by connecting them to other information resources. XLink does more than attach a clickable hyperlink, as HTML does. In SVG documents, for instance, its href attribute can reference individually defined symbols such as map signatures or coordinates and fix them onto a map as shown in the example:

```
<symbol id="symbol3067" viewBox="0 0 50 50">
  <circle cx="25" cy="25" r="24"/>
</symbol>
<!-- insert the symbol at the position in the map canvas -->
<g transform="translate(145.7, 5.2)
  scale(0.004)"
  <use x="0" y="0" width = "50" height = "50"
    xlink:href="#symbol3067" />
</g>
```

The XLink schema document xlinks.xsd is a fundamental component of GML 3. It allows elements and data types to be defined by indicating references to other declarations where the descriptions of these elements are actually spelled out. When only a specific part of a document needs to be accessed the application of XLink is enhanced by the XML Pointer Language (XPointer; http://www.w3.org/TR/xptr-framework/). In XML Pointer, the symbol "#" initiates the address of the document fragment to be referenced, as shown in the examples provided in section 3.3. Furthermore, it is possible to address single child elements in the document's element hierarchy.

2.6 *Cascading Stylesheets and extensible Stylesheet language*

Cascading Stylesheets (CSS, http://www.w3.org/Style/CSS/) provide a straightforward means of presenting XML documents on output devices such as computer screens. The use of CSS

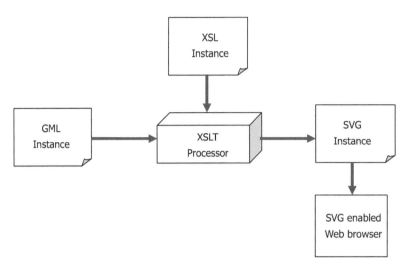

Figure 2.　Transformation of geodata encoded on GML into an SVG instance.

enables document presentation to be separated from content, an essential principle in the work with geospatial data. In SVG, for example, CSS enable the presentation style of all aspects and features of an entire geographic layer to be set at once as well as separately from other layers (Watt & Lilley 2001:151). In the following example style properties like line color (stroke), line widths, etc. for the border of a parcel are defined:

```
<path id="parcel114"
   style="fill:none;opacity:1;stroke:#000000;stroke-width:1px;"
   d="M2294.96 1366.67 l -1.88 -46.52 l -29.49 0.63 l -29.75 1.11
   l 2.33 47.64 l 58.78 -2.86"/>
```

The value of the d attribute comprises the coordinates of the line stroke to draw on the output device.

As shown in Figure 2, XML documents can be transformed into other XML documents (like SVG instances), into data streams or even into other output formats (like PDF) through the application of *The Extensible Stylesheet Language Family* (XSL; http://www.w3.org/ Style/XSL/). This group of languages includes XSL Transformations (XSLT), XML Path Language (XPath) and XSL Formatting Objects (XSL-FO). XSLT is the tool most often used to transform XML documents, because it supports both the structural transformation of XML objects and schema definitions.

Of particular importance to geoinformatics is the ability of specialized programs (XSLT processors) to transform GML instances into SVG instance documents (see Figure 2). XSLT, for example, enables the British Ordinance Survey (Ordnance Survey 2010) to visualize building permits and cadastral data.

XPath (World Wide Web Consortium 2007) is another related concept widely used in the process of transforming XML documents. It allows parts of XML documents to be addressed using a separate syntax. It simply indicates where to look for the document fragment by following the nested hierarchical structure of the XML elements. In the example below the first coordinate value of a point feature is extracted:

```
<xsl:value-of select="substring-
before(gml:centerOf/gml:Point/gml:pos,' ')"/>
```

Here, through the use of a so called *substring-before* function, the first part of the element content (e.g. the first coordinate value) is extracted from the "gml:pos" element.

3 GEOGRAPHY MARKUP LANGUAGE

Geography Markup Language (GML) is the only broadly recognized markup language used for the modeling, transport and storage of geographic information (Lake et al. 2004, Open Geospatial Consortium 2007). The basic definitions for elements, attributes and data types of the most recent version 3 of GML are provided in more than 30 XML schema documents, called basis schemas.

3.1 *Application domain extension*

For many essential tasks that involve GML an application specific schema, or *Application Domain Extension* (ADE), is created. An ADE provides a formal description of all relevant features and relationships used for that particular task or field of work. Such a schema can define separate geographic objects or features (as shown in the BathingPond example below), with their corresponding elements and properties. The use of AFIS-ALKIS-ATKIS-Application schemas[5] published by the Surveying Authorities in Germany is a case in point (Seifert 2005).

3.2 *GML features and properties*

The essential building blocks of GML are *geo-objects* or *features*. Features are made up by the many data types used to specify non-geometric and geometric characteristics. They are meant to represent objects in the real world, such as streets, buildings, rivers or survey markers, all anchored within the specific context of a given application (Open Geospatial Consortium 2002).

The following shows how the instance of the geo-object `"BikePath"` could be coded in GML:

```
<BikePath gml:id="EZ02LLB"> ... </BikePath>
```

The instance is attached to the mandatory attribute `"id"` which, in turn, is a part of the gml namespace. This explicit marker is required for all object instances of a GML document. Further, features become child elements that more closely define object instances. In the case of BikePath, `"objectType"`, `"category"` and `"mapSheetNumber"` can be specified in the following manner:

```
<BikePath gml:id="EZ02LLB">
  <objectType>9102</objectType>
  <category>1470</category>
  <mapSheetNumber>17120</mapSheetNumber>
</BikePath>
```

By convention, names of features begin with a capital letter and names of specific feature traits begin with a small letter. Words that are part of trait names are attached to each other, and the subsequent words (within the continuous string of words) start with a capital letter.

Trait values, as shown in the example above, can be declared within object instances. By applying XLink, however, traits can also be specified in another part of the document or in any other document that can be referenced by a URI.

```
<objectType xlink:href="http://www.anyserver.de/objType.xml#9401" />
```

---

[5]Theses schemas describe the data model for cadastral and topographic data of the Surveying Authorities of the States of the Federal Republic of Germany (AdV, http://adv-online.de). Core of NAS, its data exchange format, is the Geography Markup Language (GML).

Above, the element objectType is an empty element. Its content is referenced by an attribute in the start tag—actual specifications are in fragment 9401 of file "objType.xml" on server "www.anyserver.de".

3.3 *Relationships between features*

Features can have properties which themselves are features. If a bathing pond is near a hiking trail the relationship can be denoted as follows:

```
<BathingPond>
  <objectType> 9401</objectType>
  <nearBy>
    <tkfd:HikingTrail gml:id="trail001" />
  </nearBy>
</BathingPond>
```

As mentioned before, a feature defined elsewhere can be referenced using XLink. Below, the GML instance of BathingPond indicates other sources addressable by an URL:

```
<BathingPond>
  <nearBy xlink:href="#trail002" />
</BathingPond>
```

3.4 *Geometrical properties*

While the previous standards for GML provided only a limited number of simple geometrical properties the current GML 3 standard supports almost all kinds of two- and three-dimensional properties, as well as coordinate reference systems, time properties, dynamic features, topology, spatio-temporal coverages, observations, units of measure and some rules on presentation style derived from SVG (Open Geospatial Consortium 2007:49).

GML features may have several geometric properties, each one embedded in a child element, which specifies the data type responsible for specific geometric properties (e.g. centerLineOf, curveProperty, surfaceProperty). The child element of a property element is one geometric element (e.g. Point, LineString, Polygon, etc.) or an XLink reference to a remote element. The following example shows the geometry of a linear feature specified by three vertices, each defined by a "gml:pos" element:

```
<gml:centerLineOf>
  <gml:LineString>
    <gml:pos>3512126.69 5411713.94</gml:pos>
    <gml:pos>3512219.35 5411712.56</gml:pos>
    <gml:pos>3512748.51 5411740.24</gml:pos>
  </gml:LineString>
</gml:centerLineOf>
```

A second example shows the locational description of a point feature:

```
<gml:centerOf>
  <gml:Point>
    <gml:pos>3512280.93 5410246.16</gml:pos>
  </gml:Point>
</gml:centerOf>
```

As mentioned above, geometrical properties can be referenced using an XLink link to another feature which is identified by its id, such as:

```
<gml:centerLineOf xlink:href="#0002CFV"></gml:centerLineOf>
```

358

The value of this centerLineOf property is the resource returned by traversing the link (Open Geospatial Consortium 2007:25).

3.5 *Spatial reference systems*

Spatial geometry is determined by the system of coordinates used. In GML, the specific system of coordinates chosen must be explicitly designated through a *Coordinate Reference System* (CRS) or *Spatial Reference System* (SRS) (Lake 2004:192). A spatial reference system can in turn be designated by simply using a Universal Resource Identifier (URI) to link the document to a classification system developed by the OGP Geomatics Committee, formerly called European Petrol Survey Group (OGP 2010). Within this system, each spatial reference system is identified by a unique code. For example, "4326" is the code for the World Geodetic System 1984 used in many fields like GPS measurements and Internet mapping APIs:

```
<tkfd:TrainStation gml:id="b00214">
  <stationName>Aguas Calientes</stationName>
  <gml:centerOf>
   <gml:Point
srsName="http://www.opengis.net/gml/srs/epsg.xml#4326">
     <gml:pos>-72.525 -13.155</gml:pos>
   </gml:Point>
  </gml:centerOf>
</tkfd:Station>
```

3.6 *Feature collections*

Features are commonly aggregated into *feature collections*, in a manner similar to the layer concept used in GIS. A feature collection can be empty or contain an unlimited number of feature members. A collection may have spatial properties as well. Furthermore, a collection can itself also be a feature. It is even possible to have feature collections of feature collections!

The following example shows features as members of the feature collection "SportLocations":

```
<tkfd:SportLocations gml:id="ID000001">
  <gml:featureMember>
   <tkfd:BathingPond gml:id="ID000002">
     ...
   </tkfd:BathingPond>
  </gml:featureMember>
  <gml:featureMember>
   <tkfd:BoatRent gml:id="ID000004">
     ...
   </tkfd:BoatRent>
  </gml:featureMember>
 ...
```

3.7 *Assessment*

GML is a highly developed, yet complex set of standards. The complexity is probably why "...the uptake of GML in the mainstream Web community has been slow" (Open Geospatial Consortium 2006b:11). The richness of GML is also its weakness. Usage of GML often results in very large and complex files, which can greatly hinder the ability to transfer data across networks and the further processing. The application specific schema is another problem because, in contrast to KML (see next section), for example, it does not allow data encoded in GML to be shared in an interoperable way directly among arbitrary applications.

3.8   Additional formats

3.8.1   KML

The Keyhole Markup Language (KML) is an XML based grammar, originally created by Keyhole, Inc. Its use has increased significantly since Google purchased the company in 2004, especially since it was incorporated into the Google Earth Browser. The list of applications using KML today is extensive, including Internet mapping APIs, desktop GI systems, and virtual globes. KML received further recognition after it was published as a standard by the OGC (Open Geospatial Consortium 2008).

KML is a relatively concise standard, mixing data and presentation, which makes it easy to use and to learn. It is suitable for representing objects on pre-existing maps; however, it is not rich enough to be able to define a complete, complex map. For that a case more complete languages, like GML, must be used.

KML is effective at programming geographic explanations, but is limited in other ways. It uses only one coordinate reference system (WGS84) and altitude values are defined by EGM96 (Geoid Vertical Date). Furthermore, many KML viewers are two-dimensional and do not render three-dimensional models or provide a means to represent altitude. In addition, some standard aspects, such as animation, are usually not supported. Below is an example of KML code taken from http://www.mygeoposition.com/:

```
<?xml version="1.0" encoding="iso-8859-1"?>
<kml xmlns="http://earth.google.com/kml/2.1">
  <Placemark>
    <name>Stuttgart, Germany</name>
    <description></description>
    <Point>
      <coordinates>9.180769,48.777106,0</coordinates>
    </Point>
  </Placemark>
</kml>
```

3.8.2   Scalable vector graphics

The XML-based markup language *Scalable Vector Graphics* describes two-dimensional vector and mixed vector/raster graphics (World Wide Web Consortium 2009). The image is brought to the screen by so-called *User Agents*. These are either common browsers, separate SVG-viewers, or SVG-enabled Web browsers.

SVG offers a number of advantages over other common graphic formats for Web pages, such as GIF, JPEG and PNG (Watt & Lilley 2001, Ueberschär & Winter 2006):

- it is an open, non-proprietary, standardized language,
- it is directly supported in modern Web browsers, and hand-held devices,
- it supports high-resolution graphics that, among other things, retain the same graphical quality when zooming-in on an object,
- it supports a greater variety and depth of colors,
- it supports animation without increasing file size,
- it supports the Document Object Model, which allows elements written in SVG to be controlled and modified by ECMAScript, or other object-oriented programming languages,
- it facilitates filter effects, such as shading.

The above features are all especially useful when programming maps and other forms of geographic information. Hence, SVG has become the required format for outputs made for the Web Map Service (Kettemann 2005). Some programs automatically generate SVG code; other programs include SVG converters as an extension. In short, the application of SVG is widespread. For instance, SVG graphics are supported by Wikipedia. In the field of geoinformatics, it is used by the Geographical Survey Department of the German State of

Baden-Württemberg (Graf 2004) and by the British Ordnance Survey (Ordnance Survey 2010). SVG plays an essential role in the XSL-based generation process of map tiles for OpenStreetMap (http://wiki.openstreetmap.org/wiki/SVG).

SVG is a standard in many browsers and is used in several APIs to overlay vector data on top of grid tiles. The following example (taken from Zerndl 2009) shows the integration of SVG elements (i.e. polyline, closed polygon, circle, and text) in XHTML. The background image of a topographic map of the Geographical Survey Department of the German State of Bavaria, is derived from its WMS server.

```
<html xmlns="http://www.w3.org/1999/xhtml">
  <head>
    <title>HTML and SVG with XHTML</title>
  </head>
  <body>
    <div style="position:absolute; left:50px; top:50px;">
      <img
src="http://www.geodaten.bayern.de/ogc/getogc.cgi?REQUEST=GetMap
&amp;VERSION=1.1.1&amp;    LAYERS=TK50&amp;    SRS=EPSG:31468&amp;BBOX=
4516489.753,5323730.381,4518264.567,5325505.195&amp;FORMAT=image/png
&amp;WIDTH=500&amp;HEIGHT=500&amp;TRANSPARENT=TRUE&amp; STYLES="/>
    </div>
    <svg xmlns="http://www.w3.org/2000/svg" width="660" height="440"
style="position:absolute; left:50px; top:50px; z-index:100">
      <polygon points="10,10 220,20 220,250 220,296 10,110"
              style="stroke:blue; fill:yellow; fill-opacity:0.5" />
      <circle class="f1st" cx="300" cy="350" r="80"
              style="stroke:blue; stroke-width:2px; fill:grey;
              fill-opacity:0.5"/>
      <polyline points="300,30 350,50 200,200 150, 180 280, 50"
              style="stroke:blue; stroke-width:5px;
              fill:none; opacity:0.5" />
      <text x="50" y="400" style="font-family:Arial;
              font-size:36px;
              stroke:#FF0000;">SVG on top of XHTML</text>
    </svg>
  </body>
</html>
```

The positioning and the styling of the geometric elements are done by CSS which allows the precise overlay of XHTML's div element with the SVG element. The result is shown in Fig. 3.

3.8.3 *GeoRSS*

RSS (usually defined as Really Simple Syndication) is essentially an XML-based grammar used to syndicate news or weblogs (Open Geospatial Consortium 2006).

Normally the RSS feed structure contains a channel (feed), which has a title, a link and a description. The channel includes items (entry), each with a title, a short description and a link to a Web page that contains the complete article.

GeoRSS builds on RSS, to be able to describe the various ways to encode the location of news sites or weblogs (Worldkit 2006). Furthermore, GeoRSS differs from RSS by introducing several namespaces based on XML schema files. Due to its simplicity and versatility GeoRSS has the ability to make systems interoperable. It is used in both popular commercial and for-free APIs used in making maps.

Two versions of GeoRSS currently exist, GeoRSS GML and GeoRSS Simple. GeoRSS GML is intended to be a bonafide GML Application Profile. It is not only the smallest but probably the most "atomic" GML profile (Open Geospatial Consortium 2006:3). According

Figure 3. XML allows the integration of different XML languages, demonstrated in this example by overlaying SVG data to an image derived from a WMS server (Geobasisdaten © Bayerische Vermessungsverwaltung 2010, http://www.geodaten.bayern.de).

to the GML specifications, it supports a greater range of features than GeoRSS Simple. Notable is the ability of GeoRSS GML to support different coordinate reference systems.

Below is an example of how a point feature is coded in GeoRSS GML taken from the GeoRSS Web site (http://georss.org/gml):

```
<entry>
  <title>Crossing Muddy Creek</title>
  <link href="http://www.myisp.com/dbv/2"/>
  <id>http://www.myisp.com/dbv/2</id>
  <updated>2005-08-15T07:02:32Z</updated>
  <content>Check out the salamanders here</content>
  <georss:where>
    <gml:Point>
      <gml:pos>45.256 -110.45</gml:pos>
    </gml:Point>
  </georss:where>
</entry>
```

GeoRSS Simple supports basic geometries (point, line, box, polygon) and covers the typical use cases; the spatial reference system is restricted to WGS84. The following example, taken from the USGS Web site (http://earthquake.usgs.gov/earthquakes/catalogs/1 day-M2.5.xml), demonstrates how point features are coded in GeoRSS Simple. In addition to RSS elements and information on location HTML code is included in a CDATA section.

```
<entry>
  <id>urn:earthquake-usgs-gov:nc:40237628</id>
  <title>M 2.6, Northern California</title>
  <updated>2009-06-04T12:49:49Z</updated>
  <link rel="alternate" type="text/html"
      href="/eqcenter/recenteqsww/Quakes/nc40237628.php"/>
  <link rel="related" type="application/cap+xml"
      href="/eqcenter/catalogs/cap/nc40237628"/>
  <summary type="html"><![CDATA[<img
      src="http://earthquake.usgs.gov/images/globes/40_-120.jpg"
      alt="38.191&#176;N 121.878&#176;W" hspace="20" />
      <p>Thursday, June 4, 2009 12:49:49 UTC
      <br>Thursday, June 4, 2009 05:49:49 AM at epicenter</p>
      <p><strong>Depth</strong>: 20.80 km (12.92 mi)</p>]]>
  </summary>
  <georss:point>38.1910 -121.8778</georss:point>
  <georss:elev>-20800</georss:elev>
  <category label="Age" term="Past hour"/>
</entry>
```

3.8.4 *GPS eXchange format (GPX)*

GPX is an open XML-based format especially designed for the interchange of GPS data. The latest XML schema version 1.1 has been available since August 2004 on Topgrafix's website (http://www.topografix.com/gpx/1/1/gpx.xsd) along with sample files and a validator. GPX is used to describe tracks (records of points along already traveled paths) or waypoints (descriptions of turning points, landmarks or other points along a user's path to a given destination). Each waypoint may also be given a timestamp, in order to record time and place of a visited position for later evaluation.

```
<?xml version="1.0" encoding="UTF-8" standalone="no" ?>
<gpx xmlns="http://www.topografix.com/GPX/1/1"
  creator="MapSource 6.11.6" version="1.1"
  xmlns:xsi=http://www.w3.org/2001/XMLSchema-instance
  xsi:schemaLocation="http://www.topografix.com/GPX/1/1
http://www.topografix.com/GPX/1/1/gpx.xsd">
  <wpt lat="42.998545" lon="6.400465">
    <ele>193.109131</ele>
    <name>VIGIE DE P</name>
    <cmt>04-JUN-09 13:48:46</cmt>
    <desc>04-JUN-09 13:48:46</desc>
    <sym>Flag, Blue</sym>
  </wpt>
  <trk>
    <name>03-JUN-09</name>
    <trkseg>
      <trkpt lat="-40.943039" lon="172.891352">
        <ele>731.219971</ele>
      </trkpt>
      <trkpt lat="-40.942966" lon="172.891563">
        <ele>727.855225</ele>
      </trkpt>
...
```

As is common for most GPS coding, the geographic coordinates are given in decimals based on WGS 84 datum.

The format is supported by a wide range of desktop applications. In addition, mobile phone platforms and handheld GPS devices support this format. An complete list can be found on http://www.topografix.com/gpx_resources.asp.

3.8.5 *JSON and GeoJSON*

Another established format for transferring information is JavaScript Object Notation (JSON, http://json.org/). JSON is applied extensively throughout the Web and uses conventional JavaScript notation to define objects. For example, several of Google's web-based applications and services provide data feeds in JSON.

JSON-coded information can be executed as a JavaScript statement by using the eval() function. With this function, objects defined in a JSON string can be instantiated. A note of caution here, however. A JSON string may contain malicious code intended to hack into or harm programs. Therefore only JSON code obtained from a trusted server should be executed.

A variation of JSON, GeoJSON, is commonly used in the geoinformatics community (Schaub et al. 2008). It is especially useful for combining different geometrical features and shapes. GeoJSON standardizes the way spatial data is represented in JSON by structuring different geographic data according to the Simple Features Specification (Open Geospatial Consortium 2006). The following geocoding of Accra, Ghana is encoded using GeoJSON:

```
{ "name":"Accra, Ghana",
  "Status":
  {"code":200,
   "request":"geocode"},
   "Placemark":[
     {"id":"p1",
      "address":"Accra, Ghana",
      "AddressDetails":
        {"Country":
          {"CountryNameCode":"GH",
           "Locality":
           {"LocalityName":"Accra"}
          },"Accuracy": 4
        },
      "Point":
      {"coordinates":[-0.20738,5.54009,0]
      }
    }
  ]
}
```

3.8.6 *WKT format*

WKT (well known text) is a format for encoding simple features according to the specification of OGC (Open GIS Consortium 2010). It offers a light-weight solution to encode geometrical features like Point, Linestring, Polygon, MultiPoint, MultiLineString, MultiPolygon and GeometryCollections.

In the latest specification, version 1.2.1 (Open Geospatial Consortium 2010), more complex geometries are supported, like PolyhedralSurface, TIN and 3D Points. A simple Polygon shows the basic structure of the encoding:

```
POLYGON ((0 0 0, 0 0 1, 0 1 1, 0 1 0, 0 0 0))
```

Coordinate reference systems can be defined in WKT (Open Geospatial Consortium 2010:73). This is necessary because no default reference system is given, either for GML or GeoRSS GML.

The WKT standard is supported by most major applications and widely used with spatially enabled Data Base Management Systems (DBMS). Due to its compactness and standardization it is used for importing and exporting geodata. Some systems support the

Well-known Binary Representation for Geometry (WKB format, Open Geospatial Consortium 2010:62).

3.8.7 *CSV format*

Comma Separated Values (CSV) format, an implementation of a delimited text file, is a widely supported light-weight format. Because this format is quite simple and supported by almost all spreadsheets and database management systems, it has become a pseudo standard even for GI systems.

The following example shows how CSV is used as part of a two-step coding process. First, the geometrical properties are converted to WKT format. Then, then resulting string is embedded in a CSV encoded string together with information about Road ID, road name, layer name, and geometry type.

```
Road_0001;Aimin_Jie;Road;Polyline;LINESTRING(116.375686
39.931674,116.375683 39.931593,116.375747 39.93026,116.375806
39.926574)
```

It is important to note that coordinate reference systems can not be defined using CSV format and no information about the data structure can be included.

Due to its simplicity CSV formatted data can be efficiently transferred through the web, easily parsed and evaluated. It is well suited for bulk loading of geo-data into databases.

4 CONCLUSIONS

In this chapter, the major standards used for the interoperable sharing of geodata in web applications have been presented and their most important properties have been defined. Basic applications of each standard have been illustrated by examples of code written in XML and other grammars.

GML, in particular, provides almost limitless capability to define geographic objects. However, it also has some drawbacks. It is not easy to learn quickly—the documentation of GML 3.2.1 is more than 420 pages long. In addition, it can be extended according to the user's needs. The size of files can become very large, especially when extensive XML tags surround the actual data. This can make the transfer of and access to GML instances through the Internet difficult. Finally, GML format is inappropriate for presenting images.

KML, on the other hand, can be very useful to represent geographic features in 3D. The downside of this format is, however, that it can not support descriptions of complex features, a task easily handled in GML. KML is supported by commercial mapping APIs such as Google Maps or BING, and also open source mapping APIs such as OpenLayers.

GeoRSS, in turn is the simplest of the three XML-based formats. It is not difficult to learn, can be quickly implemented, and files can be easily transferred through the Internet. Yet, GeoRSS is more limited than the other standards. In contrast to GML, it cannot display most forms of information about geographic objects. Therefore, as in the case of KML, commercial and open source mapping APIs support GeoRSS, primarily as an import data format. GeoRSS is used by most of the web-mapping applications, including OpenLayers, Yahoo! Maps, Google Maps, Virtual Earth and others.

In general, XML-based formats differ mostly in capability and complexity, the usual trade-off when choosing between different markup languages.

WKT and GeoJSON provide other possibilities to present geographic information on the Internet. They generally follow the simple feature model of OGC, which makes the task of system coordination easy. Both are "lean" formats, without the baggage of many tags wrapped around the essential data. Compared to fully-fledged GML, they are also well suited to decrease network load. Both are supported by many applications. WKT, in particular, can be understood intuitively and is easy to read. It excels in describing the geodetic datum, geoid, coordinate system and map projection of geospatial objects. For these reasons it is extensively used in many GIS programs. Also it is used by web-mapping applications, e.g. OpenLayers. GeoJSON, in turn, does not need a special parser, since JavaScript can be parsed by any browser directly. GeoJSON format is supported by some open source APIs, including Yahoo's FireEagle and OpenLayers. At present the commercial APIs such as Google Maps or Virtual Earth do not support this format. GeoJSON is also used by some desktop applications, such as FME.

Overall, WKT, GeoJSON and GeoRSS offer the best compromise between capability and complexity. They support almost every type of geometrical feature, and can be implemented quickly without much training or prior experience.

Actually, no vector data format for Internet based geoservices is perfect, and each has its advantages and disadvantages. That is why it is so important to be familiar with their different capabilities and limitations, in order to decide which is most appropriate for the task on hand.

In the future, coding in Web GIS standards will become even more widespread than today. Other non-geographic formats are also likely to increasingly incorporate information about geographic locations.

REFERENCES

Bray, T., Paoli, J., Sperberg-McQueen, C.M., Maler, E. & Yergeau, F. (2008) Extensible Markup Language (XML) 1.0 (Fifth Edition). [Online] Available from: http://www.w3.org/TR/2008/REC-xml-20081126/, [Last accessed 1st June 2010].

Bray, T., Paoli, J., Sperberg-McQueen, C.M., Maler, E., Yergeau, F. & Cowan, J. (2006) Extensible Markup Language (XML) 1.1 (Second Edition). [Online] Available from: http://www.w3.org/TR/2006/REC-xml11-20060816/, [Last accessed 1 st June 2010].

Graf, G. (2004) *Vektordatenausgabe im Format SVG am Beispiel der Ausgabe von Thematischen Kartenfachdaten (TKFD)*. [Online] Available from: http://www.lv-bw.de/LVShop2/produktinfo/wir-ueber-uns/links/svg/VektordatenAusgabeImFormatSVG_110105.pdf, [Last accessed 2nd June 2010]

Kettemann, R. (2005) GIS im Intra-/Internet und Web-Dienste für Geoinformationssysteme. In: Kettemann, Rainer, Coors. & Volker, (2005) *Aktuelle Entwicklungen in der Geoinformatik*. Tagungsband 5. Vermessungsingenieurtag, Hochschule für Technik, Stuttgart.

Lake, R., Burggraf, D.S. & Trninic, Rae, L. (2004) *GML, Geography Mark-Up Language: Foundation for Geo-Web*. Wiley.

OGP, (2010) *OGP Geomatics Committee*. [Online] Available from: http://www.epsg.org, [Last accessed 16th December 2010].

Open Geospatial Consortium. (2001) *OpenGIS® Implementation Specification: Coordinate Transformation Services*. [Online] Available from: http://portal.opengeospatial.org/files/?artifact_id=999, [Last accessed 22nd May 2009].

Open Geospatial Consortium (2002) *Web Feature Service Implementation Specification*. de la Vretanos, P.A., (eds.), [Online] Available from: http://portal.opengeospatial.org/files/?artifact_id=8339, [Last accessed 10th June 2009]

Open Geospatial Consortium. (2004) *OGC Web Map Service Interface*. de la Beaujardiere, J., (eds.), [Online] Available from: http://portal.opengeospatial.org/files/?artifact_id=4756, [Last accessed 10th June 2009

Open Geospatial Consortium. (2006) *An Introduction to GeoRSS: A Standards Based Approach for Geo-enabling RSS feeds*. Reed, C., (eds.), [Online] Available from: http://www.opengeospatial.org/pt/06-050r3, [Last accessed 2nd June 2010].

Open Geospatial Consortium. (2007) *OpenGIS® Geography Markup Language (GML) Encoding Standard*. Portele, C., (eds.), [Online] Available from: http://portal.opengeospatial.org/files/?artifact_id=20509, [Last accessed 1st June 2010].

Open Geospatial Consortium. (2008) *OGC® KML*. Wilson, T., (eds.), [Online] Available from: http://portal.opengeospatial.org/files/?artifact_id=27810, [Last accessed 1st June 2010].

Open Geospatial Consortium. (2010) *OpenGIS® Implementation Standard for Geographic information - Simple feature access - Part 1: Common architecture*. [Online] Available from: http://portal.opengeospatial.org/files/?artifact_id=25355, [Last accessed 7th September 2010]

Ordnance Survey. (2010) What is the default OS MasterMap® style? [Online] Available from: http://www.ordnancesurvey.co.uk/oswebsite/products/osmastermap/faqs/data009.html, [Last accessed 1st June 2010]

Schaub, T. Doyle, A., Daly, M., Gillies, S. & Turner, A. (2008) GeoJSON draft version 6. [Online] Available from: http://wiki.geojson.org/GeoJSON_draft_version_6, [Last accessed 10th June 2009]

Seifert, M. (2005) Das AFIS-ALKIS-ATKIS-Anwendungsschema als Komponente einer Geodateninfrastruktur. *Zeitschrift für Vermessungswesen*, 2/2005.

Ueberschär, N. & Winter, A.M. (2006) *Visualisieren von Geodaten mit SVG im Internet*. Wichmann, p. 256.

Watt, A. & Lilley, C. (2001) *SVG Unleashed*. SAMS, ISBN 0-672-32429-6. p. 1117.

Worldkit. (2006) WorldKit easy web mapping user manual. http://worldkit.org/doc/rss.php, [Last accessed 2nd June 2010].

World Wide Web Consortium. (2005) *Document Object Model (DOM)*. http://www.w3.org/DOM, [Last accessed 10th June 2009].

World Wide Web Consortium. (2007) *XML Path Language (XPath) 2.0*. [Online] Available from: http://www.w3.org/TR/xpath20, [Last accessed 27th August 2010].

World Wide Web Consortium. (2009) *Scalable Vector Graphics (SVG) 1.1 Specification*. Ferraiolo, J., Fujisawa, J. & Jackson, D., (eds.), [Online] Available from: http://www.w3.org/TR/2003/REC-SVG11-20030114, [Last accessed 2nd June 2010].

Zerndl, M. (2009) Personal Communications, January 2009.

*Advances in Web-based GIS, Mapping Services
and Applications – Li, Dragićević & Veenendaal (eds)*
© 2011 Taylor & Francis Group, London, ISBN 978-0-415-80483-7

Geospatial catalogue web/grid service

Aijun Chen & Liping Di
Center for Spatial Information Science and Systems, George Mason University, Fairfax, USA

ABSTRACT: Catalogue service provides an efficient solution to the challenge of huge volumes of geospatial resources faced by the scientific researchers. It can help manage, discover, retrieve, and share distributed geospatial data, services, applications, and their replicas over the Web. Open Geospatial Consortium, Inc. (OGC) proposed catalogue service specification and a series of profiles for facilitating the interoperability of geospatial resources. In this chapter, we follow an e-business XML Registry Information Model (ebRIM)-based catalogue service for web profile to implement a catalogue web service. Geospatial metadata standards are used to extend the ebRIM to provide detailed descriptions for geospatial data and services. Any OGC catalogue service compatible client can access it to retrieve and acquire registered data and services. Geospatial metadata standards include datasets metadata, which describes geospatial datasets in the format of vector and raster, and service metadata which describes online available geospatial web services. Grid technology provides a ubiquitous computing infrastructure on the Web. A new catalogue service information model is also proposed for registering, retrieving and sharing geospatial data and services in grid environment. The catalogue service is implemented first as a web service. Then, it is wrapped into a grid service by following grid service specifications. The implemented catalogue web service has been used at GMU/CSISS to manage more than 11 terabytes geospatial data and geospatial web services from OGC. At the same time, the catalogue grid service works as a prototype system in the grid environment to manage the same data and geospatial grid services wrapped from geospatial web services. This work promotes the sharing and interoperability of geospatial resources in web/grid environment, and extends grid technology into the geoscience community.

Keywords: Geography, GIS, metadata, standards, web-based, services, distributed, grid

1 INTRODUCTION

Geospatial resources, such as geospatial data, information, knowledge and related storage, computing and transferring instruments, have been increasingly accumulated in front of professionals. It is a critical issue how to effectively manage and share those resources to make them easily accessible and usable. Catalogue service has always been playing an increasingly important role in resolving this issue. Thus, a web-based geospatial catalogue service is an efficient solution to managing, sharing and finding online geospatial resources. It can be implemented based on geospatial catalogue service standards from Open Geospatial Consortium, Inc. (OGC). Any OGC catalogue service compatible client can access it to acquire desired data and services. Geospatial metadata standards include geospatial datasets metadata and service metadata. Geospatial datasets metadata can be used to describe and share most geospatial datasets, in vector or raster format. Geospatial service metadata is used for describing online available geospatial web services.

OGC has been devoting its effort to promote the interoperability of geospatial data and services in the distributed computing environment through the development of consensus-based implementation specifications since it was established. Today, OGC specifications are being widely used by geoscience communities for many geospatial resource-involved activities. And some specifications have become International Standard Organization (ISO) standards. OGC Web Services (OWS) is one of the initiatives proposed by OGC for addressing the above issues. This initiative has resulted in a set of web-based data, services and systems interoperability specifications, such as Web Map Service (WMS), Web Coverage Service (WCS), Web Feature Service (WFS), Catalogue Service, and Geo Processing Workflow (GPW) specifications. Catalogue service is based on e-business Registry Information Model (ebRIM) and aims at providing an object-oriented registry system for registry, management and retrieval of geospatial resources. Each of those services is a standard web service that are orchestrated together to fulfill complicated geospatial applications (OGC 2008). WCS, WMS, and WFS define interfaces for access to geospatial resources. Catalogue Services for Web (CSW) specifies interfaces, HTTP protocol bindings, and a framework for defining application profiles used for publishing and accessing to catalogues of metadata of geospatial resources (OGC 2007a).

Grid computing, as a web-based information infrastructure, brings together geographically and organizationally dispersed computational resources, such as CPUs, storage systems, communication systems, data and software sources, instruments, and human collaborators to securely provide advanced distributed high-performance computing to users in a Virtual Organization (VO) that uses the authority and authentication security policy (Foster 2001, 2002). It aims at providing a global computing space with global resources while addressing the complete integration of heterogeneous computing systems and data resources. Supported by the Globus Project, Globus Toolkit represents the most popular and widely used grid software. The Globus 4.0 is based on Web Service Resource Framework (WSRF) and is fully grid service-oriented. Globus provides many functional modules both as grid services and non-grid services. We used some of these services in our research, such as the monitoring and discovery service (MDS2 & MDS4), the grid security infrastructure (GSI), the GridFTP and the Replica Location Service (RLS) (Globus 2009).

This chapter focuses on: 1) introducing recognized and popular geospatial standards from OGC, International Standards Organization (ISO), and US Federal Geospatial Data Committee (FGDC), such as catalogue service standard, metadata standards; 2) depicting how to extend ebRIM-based catalogue service's information model to accommodate ISO, FGDC and NASA metadata standards for describing geospatial resources; 3) describing how to utilize the extended information model to implement a geospatial catalogue web service, which will be a standard web service and accessible via web pages and OGC web service-compatible clients; 4) depicting how to closely combine the extended information model with grid computing technology to apply the implemented catalogue web service to grid environment; and 5) demonstrating the use of the implemented catalogue web/grid service in managing and providing access to more than 11 terabytes of geospatial satellite data.

2 RELATED WORK

The information model of the catalogue service plays a key role in describing and organizing metadata and information that describes geospatial datasets and services. Currently, there are mainly two information models used within the web service realm. They are ebXML model and the Universe Description, Discovery and Integration (UDDI) model (OASIS 2003). Both models provide multiple query patterns for catalogue querying, for example, browse and drill-down, or filtered queries against specified registry objects (OGC 2004b). The UDDI model focuses more on business entities and their associated service descriptions. However, the ebRIM model is object-oriented and more general and extensible. The OGC Catalogue Service uses ebXML information model to manage metadata based on the ISO standards. It extends the capabilities of ebXML model to address the catalogue service for the geospatial community (OGC 2005). The profiles of the catalogue service define how to incorporate FGDC Content Standards of Digital Geospatial Metadata (CSDGM) and ISO19115/ISO19119 datasets and service metadata with the ebRIM model for registry, management, discovery and retrieval of geospatial resources (OGC 2006, 2007b).

Wei et al. (2007) mapped the NASA Earth Observation System (EOS) Core System (ECS) metadata items to ISO 19115 metadata items for implementing a catalogue service for registering NASA EOS data. The metadata from geospatial data can be converted into ISO metadata items and expressed in XML file. Finally, the ISO metadata items can be automatically registered into the catalogue. Their work is upgraded in this chapter. Studies have also been reported on applying grid technology to geospatial discipline to enhance the retrieval of geospatial resources via geospatial catalogue service. Deelman et al. (2004) proposed a data model for capturing the complexity of the data publication and discovery process. Based on the model, a set of interfaces and operations were provided to support metadata management. A grid metadata service was then implemented. The proposed metadata management service was general although the paper mentioned both Dublin core metadata and FGDC metadata standards. O'Neill et al. (2004) introduced a metadata model for the NERC (Natural Environment Research Council) DataGrid for registering and discovering geospatial data. The model defines five distinct metadata objects for covering the registry and discovery of data. The model supports NASA Global Change Master Directory (GCMD) Directory Interchange Format (DIF), the Dublin Core, the GEO profile of Z39.50 and the Catalogue Interoperability Protocols (CIP). However, they did not follow the ISO 191xx series of standards and OGC catalogue service specifications although they are implemented, being compliant with ISO and OGC standards. Bai et al. (2007) proposed a geospatial catalogue service federation that federates three catalogues to provide a uniform access interface to different geospatial data.

3 GEOSPATIAL CATALOGUE SERVICE FOR WEB (CSW)

The OGC Catalogue Service (OGC 2004a) is defined as a standard web service. Its profile of geospatial Catalogue Service for Web (CSW) defines detailed specification for implementing a web-based catalogue service. The CSW has to be implemented based on the Service-Oriented Architecture (SOA) for supporting fundamental interactions: publishing resource descriptions so that they are accessible to prospective users (publish); discovering resources of interest according to some set of search criteria (discover); and then interacting with the resource provider to access the desired resources (bind). Using the SOA, the catalogue service provides publication and search functionality, enabling a requester to discover a suitable resource without requiring the requester to have advanced knowledge about the resources provider. So, the goal of the CSW is to enable users to find, download, and make use of geospatial resources in web/grid environments by providing functionalities of retrieving, storing, and managing geospatial resources.

The e-Business Registry Information Model (ebRIM) (OASIS 2005) is the core information model of the catalogue service in OGC CSW specification. It provides abstract descriptions of fundamental objects and possible logic relationships between two objects. We introduced ISO 19115/19119 (ISO 2003, 2005), NASA ECS (NASA, 1994) and FGDC metadata information model (FGDC, 2002) to extend the ebRIM model for constructing the geospatial metadata repository to describe geospatial resources. The extended information model makes it easy and normative to publish, manage and retrieve geospatial resources, especially for NASA HDF EOS data. The catalogue service is implemented based on this extended information model and provides OGC Web Service-compliant interfaces (details are available in Section 3.3). We also provided web-based functions for automatically generating and absorbing metadata in the format of XML. The extended CSW information model also plays a key role in the catalogue grid service that supports grid computing.

3.1 *OGC catalogue web service*

The ebRIM model, which is from Organization for the Advancement of Structured Information Standards (OASIS) version 3.0 (OASIS, 2005), defines how catalogue contents are structured and interrelated, and specifies formally how objects are organized, constrained and interpreted based on the conceptual structure of the domain. Figure 1 shows the ebRIM model and its extension to CSW.

The RegistryObject class is an abstract base class used by most classes in the model. It provides minimal metadata for registry objects, such as name, object type, identifier and so on. The Association class is inherited from the RegistryObject class that is used to define many-to-many associations between objects in the information model. It uses an "associationType" attribute to identify the relationship between a source "RegistryObject" and a target "RegistryObject". The "ClassificationScheme" class defines a tree structure made up of "ClassificationNode" to describe a structured way for classifying or categorizing "RegistryObject". An "ExtrinsicObject" provides required metadata about the content being submitted to the registry, thus allowing any type of object to be cataloged. And the "CSWExtrinsicObject" class adds the "repositoryItem" attribute in order to refer to the content stored in remote repositories outside of the registry. A dataset service can be tightly-coupled with a dataset by specifying the value "operatesOn" to the "associationType" attribute. The "Slot" instances provide a dynamic way to add arbitrary attributes to a registry object (Chen et al. 2005).

The "CSWExtrinsicObject" class adds the optional repositoryItem attribute in order to specify the network location of a resource located in a repository that may not be intrinsic to the catalogue service. The getRepositoryItem() operation returns the content as the entity body within an HTTP response message. The Geometry class may be used to indicate the geometric characteristics of registry objects. It extends CSWExtrinsicObject and adds a few attributes based on the simple geometry model. We use the repositoryItem attribute of CSWExtrinsicObject to link the geospatial metadata information model.

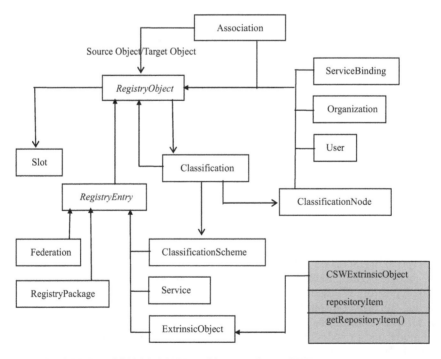

Figure 1.   The ebRIM model (OASIS 2005) and its extension to CSW.

The OGC CSW defines web-based interfaces: CSW Discovery and CSW Transaction for finding, binding and publishing registered objects in CSW. These interfaces provide a series of operations to operate the catalogue service: adding, deleting, updating and querying objects; modifying classification scheme, changing registry object classification; etc.

3.2 *Extending CSW with geospatial metadata standards*

There are usually three geospatial metadata standards being used to describe, manage, publish, and retrieve geospatial resources. These three metadata standards are ISO 19115 dataset metadata and 19119 service metadata, FGDC dataset metadata standards and its extension for remote sensing, and NASA ECS metadata for HDF EOS data. OGC has proposed "OpenGIS Catalogue Service Specifications 2.0—ISO19115/ISO19119 Application Profile for CSW2.0". This CSW-ISO profile information model is based on the international standard for metadata description ISO 19115:2003. And, the catalogue uses a service metadata description of ISO 19119:2003 standard to facilitate the management of geospatial services. However, in order to make the information model of the catalogue compatible to ISO19115/19119, FGDC, and NASA ECS metadata standards, we extended the OGC ebRIM-based CSW for serving the geospatial resources via the OGC standard interfaces. We did not use the CSW-ISO information models for describing the geospatial resources because it only complies with ISO19115/19119 standards. Another reason is the complexity of the CSW-ISO information models because of the big amount of ISO19115/19119 metadata entries. Only a little more than 60 entries are needed as core queryable entries in the catalogue. Lower complexity means higher efficiency and more convenient maintenance and usage for the catalogue service. Therefore, we simplified and synthesized the above mentioned three metadata standards on the efforts of complying with the ISO standards.

There are three kinds of metadata in ISO19115/19119 standards: mandatory, conditional, and optional. We selected all mandatory and part of the recommended conditional entries from ISO19115 and most of the entries of ISO19119. In order to describe NASA HDF EOS dataset, items from NASA ECS metadata are mapped to existing ones from ISO and new ones are added. Some additional entries for describing raster data are added from FGDC standard extension for remote sensing. So, this information model conforms to three geospatial metadata standards and provides the support for publishing, managing and querying of NASA HDF EOS data. Figure 2 shows the new dataset metadata information model that is integrated into ebRIM-derived CSW models. Only the main entries are shown here; others are omitted. We also omit service metadata information model here.

There are two ways for extending geospatial metadata informational model to the ebRIM-derived CSW model. The first is to derive new metadata classes from existing ebRIM classes by importing new classes into the ebRIM class tree structure. For example, class CSWExtrinsicObject is derived from an existing class in the ebRIM—ExtrinsicObject—to represent all the metadata objects describing objects that may not be intrinsic to the catalogue. The new attribute in class CSWExtrinsicObject, repositoryItem, is used to specify the location of the object described by a CSWExtrinsicObject object. Class MD_DatasetMetadata is derived from CSWExtrinsicObject in order to describe geospatial datasets. All metadata entries from the above dataset metadata are added to MD_DatasetMetadata class. The inherited repositoryItem attribute may be used to specify the location of the described geospatial dataset. Two other classes, SV_OperationMetadata and SV_Parameter, are derived from class CSWExtrinsicObject. These two new classes come from ISO 19119standards to provide more detailed description of geospatial service interfaces.

The second way to extend ebRIM is to use class Slots to extend an existing class. Every class extended from class RegistryObject has the capability to add Slots into itself. The class Service included in the ebRIM can be used to describe geospatial service but the available attributes in the class Service are not sufficient. Thus, new attributes derived from ISO 19119 are added to the class Service through Slots (Wei et al. 2005). We used the first way to extend

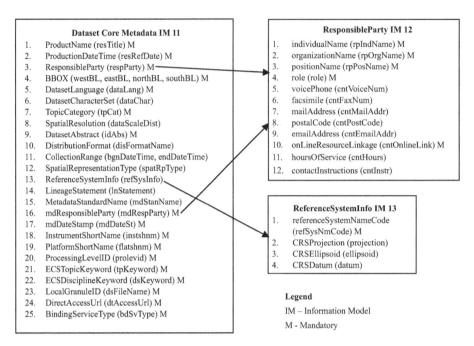

Dataset Core Metadata IM 11

1. ProductName (resTitle) M
2. ProductionDateTime (resRefDate) M
3. ResponsibleParty (respParty) M
4. BBOX (westBL, eastBL, northBL, southBL) M
5. DatasetLanguage (dataLang) M
6. DatasetCharacterSet (dataChar)
7. TopicCategory (tpCat) M
8. SpatialResolution (dataScaleDist)
9. DatasetAbstract (idAbs) M
10. DistributionFormat (disFormatName)
11. CollectionRange (bgnDateTime, endDateTime)
12. SpatialRepresentationType (spatRpType)
13. ReferenceSystemInfo (refSysInfo)
14. LineageStatement (lnStatement)
15. MetadataStandardName (mdStanName)
16. mdResponsibleParty (mdRespParty) M
17. mdDateStamp (mdDateSt) M
18. InstrumentShortName (instshnm) M
19. PlatformShortName (flatshnm) M
20. ProcessingLevelID (prolevid) M
21. ECSTopicKeyword (tpKeyword) M
22. ECSDisciplineKeyword (dsKeyword) M
23. LocalGranuleID (dsFileName) M
24. DirectAccessUrl (dtAccessUrl) M
25. BindingServiceType (bdSvType) M

ResponsibleParty IM 12

1. individualName (rpIndName) M
2. organizationName (rpOrgName) M
3. positionName (rpPosName) M
4. role (role) M
5. voicePhone (cntVoiceNum)
6. facsimile (cntFaxNum)
7. mailAddress (cntMailAddr)
8. postalCode (cntPostCode)
9. emailAddress (cntEmailAddr)
10. onLineResourceLinkage (cntOnlineLink) M
11. hoursOfService (cntHours)
12. contactInstructions (cntInstr)

ReferenceSystemInfo IM 13

1. referenceSystemNameCode (refSysNmCode) M
2. CRSProjection (projection)
3. CRSEllipsoid (ellipsoid)
4. CRSDatum (datum)

Legend

IM – Information Model

M - Mandatory

Figure 2.   Dataset metadata IM from ISO19115, NASA ECS and FGDC for remote sensing.

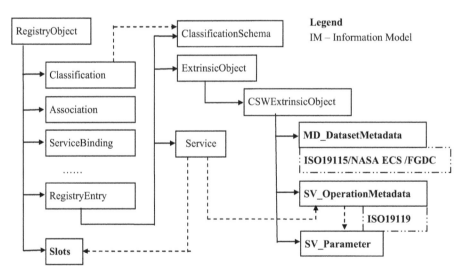

Figure 3.   Extension of ebRIM-derived CSW IM for serving geospatial resources.

the catalogue information model for accommodating datasets metadata and the second way to extend the information model for describing service metadata. A new metadata information model and its extension to ebRIM-derived CSW are illustrated in Figure 3.

3.3   *Geospatial catalogue web service interfaces*

The OGC catalogue service specification defines the interfaces of the catalogue service. The interfaces specify how clients interact with the catalogue service, for example, register, update, manage, query, etc. We followed the OGC catalogue service's profile—the Catalogue Service for Web (CSW) version 2.0.0 specification to implement the interfaces. All interfaces are based on the formal HTTP protocol binding and provide operations that a web-based catalogue service should comply. Three interfaces, OGC_Service, Discovery, and Manager, are implemented. Each interface has its own operations whose request, response and exception messages are defined. Interface OGC_Service has one operation—'getCapabilities' that is used to query catalogue service metadata. The metadata provides an OGC compatible XML document describing the capabilities and related information of the catalogue service. There are four operations related to the Discovery interface. These four operations allow a client to query the catalogue to get back metadata information about geospatial resources, including information, such as uniform resource identifier (URI), which is directly used to access geospatial resources. Interface Manager allows a client to update catalogue content using either a 'push' (operation transaction) or a 'pull' (operation 'harvestRecords') style of publication (Wei et al. 2007, OGC 2004a). We provide two kinds of HTTP protocol bindings: Keyword-Value Pair (KVP) encoding and XML encoding. The KVP encoding is for HTTP GET binding, and the XML encoding is for HTTP POST binding.

4   GEOSPATIAL CATALOGUE GRID SERVICE

Based on the implemented geospatial catalogue web service and grid technology, a geospatial catalogue grid service is designed and implemented in grid environment. The grid technology is closely integrated into the catalogue web service at the level of information model. It means that the information model of catalogue web service is extended to be grid-supported. In fact, this extension makes best use of the grid technology for facilitating the publishing, management and retrieval of geospatial resources by using grid computing resources.

Globus Toolkit is selected as grid computing platform to be integrated with catalogue web service. Globus toolkit provides lots of components covering security, data management, information service, and job management. Considering the important role of the catalogue service for managing data and providing information, one of Globus data management components, the replica location service (RLS), and one of the information service components, web service-compatible monitoring and discovery service (WS MDS), are selected to work with catalogue web service.

Globus' Replica Location Service (RLS) utilizes a pyramidal structure to maintain and provide access to distributed mapping information from logical names to physical names of data. The RLS consists of Local Replica Catalogs (LRC) and Replica Location Index (RLI). The RLC saves mapping between Logical File Names (LFN) of data and Physical File Names (PFN) associated with those LFNs at its local storages system. And the RLI contains mappings from LFNs to RLCs. The Globus WS MDS, named MDS4, consists of an Index Service and a Trigger Service. The Index Service collects data from various sources and provides an interface for query or subscription to that data. And, the Trigger Service collects data from various sources and can be configured to take action based on that data. The Index Service is a registry, providing the ability of discovering the properties of such resources as machines, operating systems, file systems, computing power and network that are used in a grid environment. Using the MDS4, the optimal resources in a grid environment can be found, selected and used for authorized user/client request (Globus 2009).

In order to integrate the catalogue information model with the RLS, a mapping between the Universal Unique ID (UUID) of geospatial metadata object, which is derived from MD_DatasetMetadata class for storing geospatial metadata, and the Logical File Name (LFN) of the RLS entries, is established. This LFN is mapped to a PFN in a RLC or a URL address of RLC in RLI. If the LFN is mapped to the URL address of RLC, two times requests to RLS are needed for getting the PFN which is the actual accessible address of user required data or services that is capable of providing the user required data. In order to take advantage of optimal resources in the grid environment, the PFNs of the RLS entries are combined with the network interface information of MDS. Based on this combination, any request with PFNs can be executed through the Index Service to know about what, which and where computational resources in the grid environment is the most optimal one and can be most efficiently and securely used. Therefore, by integrating Globus' RLS and MDS4, users can transparently utilize the optimal resources to access to datasets, their replicas and services in a grid environment.

The catalogue web service is wrapped as a catalogue grid service by strictly following the grid service specifications. In fact, a grid service is a specific case of a web service. The interfaces and behaviors of the catalogue grid service are complied with the specification of web service resource framework (WSRF). The implemented catalogue grid service can be run and called by other standard grid services in grid environment. The WSDL (web service description language) file of both catalogue web/grid services are provided for public access to them.

5 A PROTOTYPE SYSTEM

Prototypes of both catalogue web/grid services are implemented. The catalogue service works together with other web services, e.g. WCS and WMS, to share and provide access to large volumes of geospatial resources. The running grid environment is a Virtual Organization (VO) consisting of 7 grid nodes from 3 Certificates Authorities (CAs), which are George Mason University/Center for Spatial Information Science and Systems (GMU/CSISS), US National Aeronautics and Space Administration/Ames Research Center (NASA/ARC), and Lawrence Livermore National Laboratory/Earth System Grid (LLNL/ESG). GMU/CSISS provides one Sun Solaris server, one Apple G5 server with 5 cluster nodes and 5 RAIDs of total 10 terabytes storage capacity, and one Linux machine. They are connected through a local area network with 100 M cable. NASA/ARC provides two Linux machines, and LLNL ESG also provides two Linux machines.

A template XML file conforming to ISO 19139 is provided first as a sample XML file in which the values of all attributes and elements are empty. ISO 19139 is a Geographic Information—Metadata—XML Schema Specification (ISO 2007) that provides an XML schema that describes how ISO 19115 metadata should be stored in XML format. A transformation program is implemented to harvest metadata from geospatial data (e.g. NASA data in the format of Hierarchical Data Format—Earth Observing System (HDF-EOS)) into the XML template, and then automatically register to the catalogue database via harvest interface. The XML file is parsed to get back the values of every attributes and elements, and then register the values into the table MD_DatasetMetadata of the catalogue. The WSDL file of catalogue service and other services are parsed to get back values of attributes and elements and register to catalogue database. Client can search both dataset and service objects by providing OGC catalogue service compatible request messages via the discovery interfaces. The response messages contain a list of qualified Dataset objects in the format of XML (Wei et al. 2007). A friendly web interface is implemented for querying the catalogue service and get back datasets. Figure 4 is a screen copy of the web interface.

In order to make the catalogue grid service accessible to any OGC-compatible client, a catalogue service portal is implemented to provide OGC standards-compatible access interfaces, so that the merit of grid technology can be easily utilized by geoscience community. OGC client can use this portal to transparently access the catalogue grid service and grid resources behind it. The client submits standard request in XML format and get back standard response in XML format. Finally, the catalogue service can be accessed via three interfaces:

1. OGC standard interfaces without grid technology support;
2. Grid WSRF-compatible interfaces in grid environment; and
3. OGC standard interfaces with grid technology support (the portal).

We did the performance evaluation from a client's point of view when accessing the catalogue service via the above three interfaces. Access via the interfaces 1 is actually the same

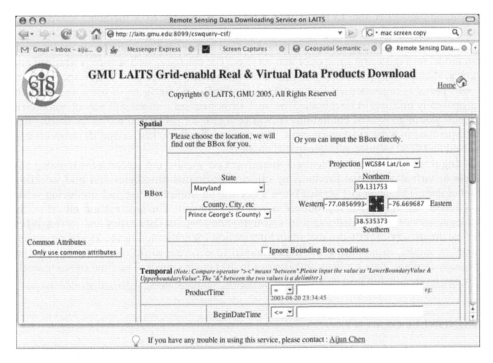

Figure 4. Discovery interface of geospatial catalogue web/grid service.

as access directly to a web service (Catalogue service) and access via the interfaces 2 and 3 is the same as access to a grid service. Taking a same size of payload, the evaluation shows that access to the catalogue service via the interfaces 2 and 3 only takes 600 milliseconds more processing time than access via the interface 1. Also, with the increasing of the size of the payload, the difference is converged to a small number comparing with the overall processing time. There is only few milliseconds difference of processing time between the access via the interface 2 and interface 3 (Chen et al. 2009). The benefits of using grid computing include not only the security of authentication and authority of access to resources in a grid VO, but the selection of optimal data and services in the grid environment. However, grid software has to be installed and grid services deployed at the system host. Grid certificates have to be issued from the host, and end users must install the certificate client package and the system host will keep records of the certificates and user names of the end users.

6 CONCLUSION

Integrating Grid technology with geospatial data and service standards technology is promising for solving data- and computing-intensive complicated geospatial issues. This chapter provides a pioneer research to take good advantages of both technologies. An OGC standards-compatible catalogue service is first implemented in web environment. Then, it is integrated with grid services to provide more powerful catalogue service in grid environment. The implementation of the geospatial web/grid catalogue services is based on extended catalogue information models. First, the catalogue information model is extended to accommodate three geospatial metadata standards for describing geospatial datasets and services. The model is then extended to be closely integrated with the data management component (RLS) and information service components (MDS4) of Globus toolkit for making best use of the merit of the grid technology in geoscience community. The extended catalogue information model is able to ingest ISO, FGDC, and NASA ECS metadata standards to facilitate the publishing, managing and querying of the NASA HDF EOS data in web/grid environment. ISO 19119 service metadata standard is extended into the catalogue information model for registering geospatial web services. The catalogue web service is wrapped and made grid-enabled by providing grid interfaces that is comply with grid WSRF specification. A web portal is implemented for providing standard web interfaces to users and let them transparently utilize grid resources. This work demonstrates a successful way of using Grid technologies in geospatial community. It will be useful for geospatial resources sharing and interoperating in web/grid environment. This research finally makes geospatial resources grid enabled and Grid technologies geospatial enabled.

Using this implemented catalogue web/grid service, we also did research for building and access to virtual geospatial products by utilizing geospatial ontologies and grid workflow technologies. The catalogue service plays a key role in implementing and producing workflow-based virtual geospatial products in grid environment. The information model of the catalogue service is extended again to accommodate abstract data types and service types. The catalogue service provides metadata information for constructing the logical processing workflow model, instantiating the logical model to concrete workflow model by selecting optimally available data replicas and services in grid environment. The final concrete workflow can be executed in grid environment to return results to users (Chen et al. 2009).

ACKNOWLEDGEMENT

This project was supported by grants from the NASA Earth Science Data and Information System Project (ESDISP) and the NASA Earth Science Technology Office (ESTO) (PI: Liping Di). Additional funding was provided by the Open Geospatial Consortium (OGC) for the development of Web Coverage Server, Web Map Server and Catalogue Service - Web as a part of OGC WMT II and OWS-II (PI: Liping Di).

REFERENCES

Bai, Y., Di, L., Chen, A., Liu, Y. & Wei, Y. (2007) Towards a Geospatial Catalogue Federation Service. *Photogrammetric Engineering & Remote Sensing*, 73 (6), 699–708.

Chen, A., Di, L., Wei, Y., Bai, Y., Wei, Y. & Liu, Y. (2009) Modeling Virtual Geospatial Products based on Grid Computing. *International Journal of Geographic Information Science*, 23 (5), 581–604.

Chen, A., Di, L., Wei, Y., Liu, Y., Bai, Y., Hu, C. & Mehrotra, P. (2005) Grid computing enabled geospatial catalogue web service. *Proc. of American Society for Photogrammetry and Remote Sensing* 7–11 th March, Baltimore, USA.

Deelman, E., Singh, G., Atkinson, M.P., Chervenak, A., Hong, N.P.C., Kesselman, C., Patil, S., Pearlman, L. & Su, M. (2004) Grid-Based Metadata Services. *Proc. of 16th international conference on scientific and statistical database management (SSDBM 04)*, 21–23th June, Santorini Island, Greece.

FGDC (2002). Content Standard for Digital Geospatial Metadata: Extensions for Remote Sensing Metadata. [Online] Available from: http://www.fgdc.gov/standards/status/ csdgm_rs_ex.html, Dec. 2002.

Foster, I., Kesselman, C. & Tuecke S. (2001) The Anatomy of the Grid – Enabling Scalable Virtual Organizations. *Intl. J. of High Performance Computing Applications*, 15 (3), 200–222.

Foster, I., Kesselman, C., Nick J.M. & Tuecke S. (2002) The Physiology of the Grid: *An open Grid services architecture for distributed systems integration*. 17th Feb., Technical report, Argonne National Laboratory, Chicago, USA.

Globus, (2004) The Globus Alliance. RLS Documentation. *The Globus Alliance*. [Online] Available from: http://www-unix.globus.org/toolkit/docs 3.2/rls/index.html [Accessed in May 2009].

Globus, (2009) The Globus Toolkit 4.0 Release Manuals. *The Globus Alliance*. [Online] Available from: http://www.globus.org/toolkit/docs/4.0/ [Accessed in May 2009].

ISO/TS211, (2007) Geographic information -- Metadata -- XML schema implementation, 19139:2007. *International Standards Organization (ISO) Technical Committee 211*, [Online] Available from: http://www.iso.org/iso/iso_catalogue/catalogue_tc/catalogue_tc_browse.htm?commid=54904 [Accessed in May 2009].

ISO/TC211, (2005) Geographic Information – Services, 19119:2005. *International Standards Organization (ISO) Technical Committee 211*, [Online] Available from: http://www.iso.org/iso/iso_catalogue/catalogue_tc/catalogue_tc_browse.htm? commid=54904 [Accessed in June 2009].

ISO/TC211, (2003) Geographic Information – Metadata, 19115:2003. *International Standards Organization (ISO) Technical Committee 211*, [Online] Available from: http://www.iso.org/iso/iso_catalogue/catalogue_tc/catalogue_tc_browse.htm?commid=54904 [Accessed in June 2009].

NASA, (1994) Proposed ECS Core Metadata Standard Release 2.0. [Online] Available from: http://edhs1.gsfc.nasa.gov/waisdata/docsw/pdf/tp4200105.pdf *U.S. National Aeronautics and Space Administration*. [Accessed in May 2009].

OASIS, (2003) UDDI Version 3.0.1. UDDI Spec Technical Committee Specification. OASIS *UDDI Specifications TC – Committee*. [Online] Available from: http://uddi.org/pubs/uddi-v3.0.1-20031014.pdf [Accessed in June 2009].

OASIS, (2005) EbXML Registry Information Model Version 3.0, OASIS Standard. [Online] Available from: http://www.oasis-open.org/committees/download.php/13591/docs.oasis-open.orgregrepv3.0specsregrep-rim-3.0-os.pdf [Accessed in June 2009].

OGC, (2004a) OGC® catalogue services specification, (OGC® 04-021r3), version 2.0.0 with corrigendum. *Open Geospatial Consortium, Inc.*

OGC, (2004b) Filter Encoding Implementation Specification, (OGC® 04-095), version 1.1.0. *Open Geospatial Consortium, Inc.* [Online] Available from: http://portal.opengeospatial.org/files/? artifact_id=8340 [Accessed in June 2009].

OGC, (2005) OpenGIS® Catalogue Services — ebRIM (ISO/TS 15000-3) profile of CSW. (OGC 05-025r3), version 1.0.0. Open Geospatial Consortium, Inc.

OGC, (2006) OpenGIS® catalogue service implementation specification 2.0.1 -- FGDC CSDGM Application Profile for CSW 2.0. (OGC® 06-129r1), version 0.0.12. Open Geospatial Consortium, Inc.

OGC, (2007a) OpenGIS® catalogue services specification, (OGC® 07-006r1), version 2.0.2. Open Geospatial Consortium, Inc.

OGC, (2007b) OGC Catalogue Services Specification 2.0.1 (with Corrigendum) - ISO19115/ISO19119 Application Profile for CSW 2.0, (OGC® 07-045), version 1.0. Open Geospatial Consortium, Inc. [Online] Available from: https://portal.opengeospatial.org/files/?artifact id=20727. [Accessed in June 2009].

OGC, (2008) OGC Web Services, Phase 6. [Online] Available from: http://www.opengeospatial.org/standards/requests/50. Open Geospatial Consortium, Inc.

O'Neill, K., Cramer, R., Gutierrez, M., Kleese van Dam, K., Kondapalli, S., Latham, S., Lawrence, B., Lowry, R. & Woolf A. (2004) A specialized metadata approach to discovery and use of data in the NERC DataGrid. *Proc. of the U.K. e-science All Hands Meeting*. Nottingham, UK. [Online] Available from: http://ndg.nerc.ac.uk/public_docs/AHM-NDGDMand MDMSplitFinal.pdf [Accessed in May 2009].

Wei, Y., Di, L., Zhao, B., Liao G. & Chen, A. (2007) Transformation of HDF-EOS metadata from ECS model to ISO 19115-based XML. *Computers & Geosciences*, 33 (2007), 238–247.

Wei, Y., Di, L., Zhao, B., Liao G., Chen, A. Bai, Y. & Liu, Y. (2005) The Design and Implementation of a Grid-enabled Catalogue Service. *Proc. of IGARSS 2005*. 25–29th July, Seoul, South Korea. p. 4.

Advances in Web-based GIS, Mapping Services and Applications – Li, Dragićević & Veenendaal (eds)
© 2011 Taylor & Francis Group, London, ISBN 978-0-415-80483-7

Author index

Anton, F. 311

Barker, T. 109
Behr, F.-J. 349
Bertolotto, M. 139
Bishop, I.D. 293
Bisier, J. 71
Boley, H. 311

Chen, A. 369
Chen, H. 293
Chen, J. 85
Chow, T.E. 15

Daniel, S. 185
Delfos, J. 171
Di, L. 369
Dragićević, S. 3
Dürrfeld, J. 71

Gao, S. 311
Gartner, G. 153, 207
Gong, J. 85
Guan, W. 255

Hall, G.B. 229
Harrap, R.M. 185

Holschuh, K. 349
Huang, H. 153
Huang, Q. 121

Kim, I.-H. 55
Kumari, J. 277

Leahy, M.G. 229
Lewis, B. 255
Li, J. 121
Li, S. 3
Li, W. 121
Li, Y. 153
Li, Z. 121
Liang, S.H.L. 37
Liu, Y. 85

Mavedati, S. 277
McArdle, G. 139
Miao, L. 121
Mioc, D. 311

Nino-Ruiz, M. 293

Paelke, V. 207
Pebesma, E. 71
Pullar, D. 109

Reed, C. 327
Rinner, C. 277

Schall, G. 207
Schöning, J. 207
Stock, C. 293
Sun, M. 121

Tan, T. 171
Torpie, D. 109
Tsou, M.H. 55

Veenendaal, B. 3, 171

Wagner, D. 349
Wang, J. 37
Wang, P. 293
Wang, X. 37
Wu, H. 121

Xiang, L. 85

Yang, C. 121
Yi, X. 311
Yue, P. 85

Zlotnikova, R. 349

*Advances in Web-based GIS, Mapping Services
and Applications – Li, Dragićević & Veenendaal (eds)*
© 2011 Taylor & Francis Group, London, ISBN 978-0-415-80483-7

Keyword index

A
Advances 3
AfricaMap 255
Argumentation mapping
    277
Augmented reality 185, 207

C
Cartography 139
Cloud computing 121
Clustering 37
Cyberinfrastructure 55

D
Data reduction 139
Data structures 349
Decision making 293
Decision support 277
Distributed 369
Distributed geographic
    information
        processing 121

E
Education 185

F
Format 349
Future 3

G
GeoGlobe 85
Geography 369
Geography 2.0 15
Geolocating 171
GeoOD-P2P 85
Geospatial clustering 37
Geospatial collaboration
    255
Geospatial
    cyberinfrastructure
    121
Geospatial standards 327
Geospatial web 229, 277

G
GIS 3, 139, 185, 207, 229,
    255, 369
GIServices 15, 55
Grid 369
Grid computing 55

H
Health applications 311

I
Information systems 109
Internet 349
Internet GIS 15
Interoperability 109, 327,
    349
IP positioning 171
Issues 3

L
Load-balancing 85, 153
Local knowledge 277
Location based services 171

M
Map-chatting 229
Mapping 3, 109, 139, 207
Mapping technologies 255
Mash-up 15
Metadata 369
Mobile 185
Mobile interaction 207
Multi-resolution pyramid
    85
Multisensor 207

N
Network communication
    cost 153

O
Open source 229, 255
Open source software 311
Open standards 327
Organic farming 277

P
Performance 121
Positioning 171
Positioning techniques
    171
Privacy 311

R
Real-time 139
Representation 349

S
Services 369
SIEVE 293
Simulation models 55
Spatial analysis 153
Spatial data infrastructure
    327
Spatial data mining 37
Spatial data modeling
    153
Spatial infrastructure 109,
    349
Spatial location 171
Spatial web portal 121
Spatial web services 109
Standards 349, 369
Standards development
    327
SVG-based spatial extended
    SQL 153

T
TeraGrid 55
Three-dimensional 207
Tracking 207

U
Urban 185

V
Visualization 139, 185, 207
Virtual environments 293
Virtual globe 85

W
Web 2.0 15
Web collaboration 229
Web based 139, 185
Web-based 3
Web-based collaboration
    293

Web-based GIS 311, 369
Web GIS 37, 121, 171, 349
WebGIS 153
Web service 37, 327

X
XML/GML/SVG 153

ISPRS Book Series

1. Advances in Spatial Analysis and Decision Making (2004)
   Edited by Z. Li, Q. Zhou & W. Kainz
   ISBN: 978-90-5809-652-4 (HB)

2. Post-Launch Calibration of Satellite Sensors (2004)
   Stanley A. Morain & Amelia M. Budge
   ISBN: 978-90-5809-693-7 (HB)

3. Next Generation Geospatial Information: From Digital Image Analysis to Spatiotemporal
   Databases (2005)
   Peggy Agouris & Arie Croituru
   ISBN: 978-0-415-38049-2 (HB)

4. Advances in Mobile Mapping Technology (2007)
   Edited by C. Vincent Tao & Jonathan Li
   ISBN: 978-0-415-42723-4 (HB)
   ISBN: 978-0-203-96187-2 (E-book)

5. Advances in Spatio-Temporal Analysis (2007)
   Edited by Xinming Tang, Yaolin Liu, Jixian Zhang & Wolfgang Kainz
   ISBN: 978-0-415-40630-7 (HB)
   ISBN: 978-0-203-93755-6 (E-book)

6. Geospatial Information Technology for Emergency Response (2008)
   Edited by Sisi Zlatanova & Jonathan Li
   ISBN: 978-0-415-42247-5 (HB)
   ISBN: 978-0-203-92881-3 (E-book)

7. Advances in Photogrammetry, Remote Sensing and Spatial Information Science. Congress
   Book of the XXI Congress of the International Society for Photogrammetry and Remote
   Sensing, Beijing, China, 3–11 July 2008 (2008)
   Edited by Zhilin Li, Jun Chen & Manos Baltsavias
   ISBN: 978-0-415-47805-2 (HB)
   ISBN: 978-0-203-88844-5 (E-book)

8. Recent Advances in Remote Sensing and Geoinformation Processing for Land Degradation
   Assessment (2009)
   Edited by Achim Röder & Joachim Hill
   ISBN: 978-0-415-39769-8 (HB)
   ISBN: 978-0-203-87544-5 (E-book)

9. Advances in Web-based GIS, Mapping Services and Applications (2011)
   Edited by Songnian Li, Suzana Dragićević & Bert Veenendaal
   ISBN: 978-0-415-80483-7 (HB)
   ISBN: 978-0-203-80566-4 (E-book)